W9-CSP-093

Chemical Engineering Fluid Mechanics

Chemical Engineering Fluid Mechanics

Ron Darby
Department of Chemical Engineering
Texas A&M University
College Station, Texas

Marcel Dekker, Inc.   New York • Basel • Hong Kong

Library of Congress Cataloging-in-Publication Data

Darby, Ron
   Chemical engineering fluid mechanics / Ron Darby.
      p.   cm.
   Includes bibliographical references and index.
   ISBN 0-8247-9628-4   (hardcover : alk. paper)
   1. Chemical processes. 2. Fluid dynamics. I. Title.
TP155.5.D28 1996
660' .284--dc20

                                                              95-49403
                                                                  CIP

The publisher offers discounts on this book when ordered in bulk quantities. For more information, write to Special Sales/Professional Marketing at the address below.

This book is printed on acid-free paper.

Marcel Dekker, Inc.
270 Madison Avenue, New York, New York 10016

Current printing (last digit):
10 9 8 7 6 5 4 3 2 1

PRINTED IN THE UNITED STATES OF AMERICA

Preface

The objectives of this book are twofold: (1) for the student, to show how the fundamental principles underlying the behavior of fluids (with emphasis on one-dimensional macroscopic balances) can be applied in an organized and systematic manner to the solution of practical engineering problems, and (2) for the practicing engineer, to provide a ready reference of current information and basic methods for the analysis of a variety of problems encountered in practical engineering situations.

The scope of coverage includes internal flows of Newtonian and non-Newtonian incompressible fluids, adiabatic and isothermal compressible flows (up to sonic or choking conditions), external flows (e.g. drag), and flow in porous media. Applications include dimensional analysis and scale-up, piping systems with fittings for Newtonian and non-Newtonian fluids (for unknown driving force, unknown flow rate, unknown diameter, or most economical diameter), compressible flows in piping systems up to choked flow, flow measurement and control, pumps, compressors, fluid-particle separation methods (e.g. centrifugal, sedimentation, filtration), packed columns, and fluidized beds. The treatment is from the viewpoint of the process engineer, who is concerned with equipment operation, performance, and sizing and selection, as opposed to the details of the mechanical design or flow patterns in such situations.

For the student, this is a basic text for a first-level course in process engineering fluid mechanics, which emphasizes the systemic application of fundamental principles (e.g. macroscopic mass, energy, and momentum balances and economics) to the analysis of a variety of fluid problems of a practical nature. Methods of analysis of many of these operations have been taken from the recent technical literature and have not previously been available in textbooks. A large number of problems that illustrate these applications have been included.

For the practicing engineer, the book serves as a useful reference for the working equations that govern many applications of practical interest, as well as a source of basic principles needed to analyze other fluid systems not covered explicitly in the book. The objective here is not to provide a "mindless set of recipes" for rote application, but to demonstrate an organized approach to problem analysis beginning with basic principles and ending with results of very practical applicability.

The book is based on notes that I have compiled and continuously revised while teaching the junior-level fluid mechanics course for chemical engineering students at Texas A&M University over the last 30 years. It has been my experience that, when being introduced to a new subject, students learn best by starting with simple, special cases that they can easily relate to and then progressing to more generalized formulations and more complex problems. That is the philosophy adopted in this book. This will certainly be criticized by some, since it is contrary to the usual procedure followed by most textbooks, in which the basic principles are presented first in the most general form and mathematical notation (the divergence theorem, Reynolds transport theorem, etc.), and the special cases are then derived from these. Esoterically, it is very appealing to progress from the general to the specific rather than vice versa. However, having taught from both perspectives, it is my observation that most beginning students do not gain an appreciation or understanding from the very general, mathematically complex theoretical vector expressions until they have gained a certain physical "feel" for how fluids behave, and the laws governing their behavior, in special situations to which they can easily relate. They also understand and appreciate the principles much better if they see how they can be applied to the analysis of practical and useful situations, with results that actually "work" in practice. That is why the multidimensional vector generalizations of the basic conservation laws have been eschewed in favor of the simpler component and one-dimensional form of these laws.

It is also important to maintain a balanced perspective between fundamental "theoretical" and empirical information, for the practicing engineer must utilize both to be effective. It has been said that all of the tools of mathematics and physics in the world are not sufficient to enable one to calculate how much water will flow in a given time from a kitchen tap when it is opened. However, by proper formulation and utilization of certain experimental observations, this is a routine problem for the engineer. The engineer must be able to solve certain problems by direct application of theoretical principles only (e.g. laminar flow in uniform conduits), other problems by utilizing hypothetical ''models'' that account for a limited understanding of the basic flow phenomena by incorporation of empirical parameters (e.g. turbulent flow in conduits and fittings), and still other problems in which important information is purely empirical (pump efficiencies, two-phase flow in packed columns, etc.). In many of these problems (of all types), application of dimensional analysis (or the principle of ''conservation of dimensions'') for generalizing the results of specific analysis, guiding experimental design, and scaling up both theoretical and experimental results can be a very powerful tool.

It is the aim of this book to provide a first step in this direction. If it can provide the reader a useful introduction to the simplified form of the basic governing equations, and an illustration of a consistent method of applying these to the analysis of a variety of practical flow problems, it will have served its purpose. Hopefully, the serious student will use this as a starting point to delve deeper into the limitless expanse of the world of fluid mechanics, and the practicing engineer will find it a useful introduction to a variety of practical applications.

Ron Darby

Contents

Unit Conversion Factors

Dimension	Equivalent Units
Mass	1 kg = 1000 g = 0.001 metric ton (tonne) = 2.20461 lb_m = 35.27392 oz
	1 lb_m = 453.593 g = 0.453593 kg = 5×10^{-4} ton = 16 oz
Force	1 N = 1 kg m/s^2 = 10^5 dyn = 10^5 g cm/s^2 = 0.22418 lb_f
	1 lb_f = 32.174 lb_m ft/s^2 = 4.4482 N = 4.4482×10^5 dyn
Length	1 m = 100 cm = 10^6 μm = 10^{10} Å = 39.37 in. = 3.2808 ft = 1.0936 yd = 0.006214 mi
	1 ft = 12 in. = 1/3 yd = 0.3048 m = 30.48 cm
Volume	1 m^3 = 1000 liters = 10^6 cm^3 = 35.3145 ft^3 = 264.17 gal
	1 ft^3 = 1728 in.^{3} = 7.4805 gal = 0.028317 m^3 = 28.317 liters = 28,317 cm^3
Pressure	1 atm = 1.01325×10^5 N/m^2 (Pa) = 1.01325 bar = 1.01325×10^6 dyn/cm^2 = 760 mm Hg @ 0°C (torr) = 10.333 m H$_2$O @ 4°C = 33.9 ft H$_2$O @ 4°C = 29.921 in. Hg @ 0°C = 14.696 lb_f/in.^{2} (psi)
Energy	1 J = 1 N m = 10^7 erg = 10^7 dyn cm = 2.667×10^{-7} kWh = 0.23901 cal = 0.7376 ft lb_f = 9.486×10^{-4} Btu [550 ft lb_f/(hp s)]
Power	1 W = 1 J/s = 0.23901 cal/s = 0.7376 ft lb_f/s = 9.486×10^{-4} Btu/s = 1×10^{-3} kW = 1.341×10^{-3} hp
Flow Rate	1 m^3/s = 35.3145 ft^3/s = 264.17 gal/s = 1.585×10^4 gal/min (gpm) = 10^6 cm^3/s
	1 gpm = 6.309×10^{-5} m^3/s = 2.228×10^{-3} ft^3/s = 63.09 cm^3/s

Example: The factor to convert Pa to psi is 14.696 psi/(1.01325×10^5 Pa)
Some values of the gas constant: R = 8.314×10^3 kg m^2/(s^2 kg mol K)
$$= 8.314 \times 10^7 \text{ g cm}^2/(\text{s}^2 \text{ g mol K})$$
$$= 82.05 \text{ cm}^3 \text{ atm/(g mol K)}$$
$$= 1.987 \text{ cal/(g mol K) or Btu/(lb mol °R)}$$
$$= 1545 \text{ ft } lb_f/(\text{lb mol °R})$$
$$= 10.73 \text{ ft}^3 \text{ psi/(lb mol °R)}$$
$$= 0.730 \text{ ft}^3 \text{ atm/(lb mol °R)}$$

Chemical Engineering Fluid Mechanics

1

Basic Concepts

I. FUNDAMENTALS

A. BASIC LAWS

The fundamental principles that apply to the analysis of fluid flows are few and may be represented by the "conservation laws", i.e.

1. Conservation of mass
2. Conservation of energy (first law of thermodynamics)
3. Conservation of momentum (Newton's second law)
4. Second law of thermodynamics

To these may also be added:

5. Conservation of dimensions ("fruit salad law")
6. Conservation of dollars (economics)

These conservation laws are basic and are the starting point for the solution of every problem.

The second law of thermodynamics states that a process can occur spontaneously only if it goes from a state of higher energy to one of lower energy. In practical terms, this means that energy is dissipated (i.e. transferred from useful mechanical form to low level thermal energy) by any system that is in a dynamic (nonequilibrium) state. In other words, the energy associated with "friction loss" is always positive. In more mundane terms, this law tells us that, for example, water will run downhill spontaneously but cannot run uphill unless it is "pushed" (i.e. unless energy is supplied from an external source).

B. EXPERIENCE

The conservation of dimensions (or "fruit salad") law enables us to design experiments, to acquire and organize data (i.e. "experience") obtained in a test or "model" system in the most efficient and general form, and to apply it to the solution of problems in similar systems that may involve different properties on a different scale. Since the vast majority of problems in fluid mechanics cannot be solved without resort to experience (i.e. empirical knowledge), this is a very important principle, and it will be used extensively.

II. OBJECTIVE

It is the intent of this book to show how these basic laws can be applied, along with pertinent knowledge of system properties, operating conditions, etc., to the analysis of a wide variety of practical problems involving the flow of fluids. It is the author's belief that the engineer is much more versatile, valuable, and capable if he or she approaches the problem-solving process from a basic perspective, starting from first principles to develop a solution rather than looking for a "similar" problem as an example to follow (that may or may not be applicable). It is this philosophy, along with the objective of arriving at workable solutions to practical problems, upon which this work is based.

III. PHENOMENOLOGICAL RATE OR TRANSPORT LAWS

In addition to the conservation laws for mass, energy, momentum, etc., there are additional laws that govern the *rate* at which these quantities are transported from one region to another in a continuous medium. These are called *phenomenological laws* because they describe certain phenomena but cannot be derived from more fundamental principles. These laws may be written for all conserved quantities (e.g. mass, energy, momentum, and electric charge) and may be expressed in the general form as

$$\text{Rate of transport} = \frac{\text{Driving force}}{\text{Resistance}}$$
$$= \text{Conductance} \times \text{Driving force} \qquad (1\text{-}1)$$

This expression applies to the transport of any conserved quantity, e.g. mass, energy, momentum, or charge, which we will refer to as Q. The rate of transport of any quantity Q, per unit area normal to the direction of transport, is called the *flux* of Q. The "driving force" for the transport, on the microscopic or molecular level, is the negative of the gradient of the concentration of Q (with respect to the direction of transport). That is, Q flows "downhill", from a region of high concentration to a region of low concentration. This is expressible in the form

$$\text{Flux of Q in the y direction} = K_T \left[-\frac{d(\text{conc. of Q})}{dy} \right] \tag{1-2}$$

where K_T is the *transport coefficient* for the quantity Q and, for microscopic (molecular) transport, is a property only of the medium. For macroscopic systems, involving complex flows (e.g. turbulence), the driving force is a representative difference in the concentration of Q, and the transport coefficient is a function of flow conditions as well as the properties of the medium.

Example 1-1: What are the dimensions of the transport coefficient, K_T? If we denote the dimensions of a quantity by brackets, e.g. the dimensions of Q are [Q], a dimensional equation corresponding to Eqn (1-2) can be written as follows:

$$[\text{Flux of Q}] = \frac{[K_T][Q]}{[\text{Volume}][y]}$$

Since $[\text{Flux of Q}] = [Q]/(L^2 t)$, $[\text{Volume}] = L^3$, and $[y] = L$, we have

$$[K_T] = \frac{L^2}{t}$$

because [Q] cancels out from the equation. That is, the dimensions of the transport coefficient are independent of the specific quantity that is being transported.

A. FOURIER'S LAW OF HEAT CONDUCTION

As an example, Fig. 1-1 illustrates two horizontal parallel plates with a "medium" (either solid or fluid) between them. If the top plate is kept at a temperature T_1 which is higher than the temperature (T_0) of the bottom plate, there will be a transport of thermal energy ("heat") from the upper plate to the lower plate through the medium, in the $-y$ direction. If the flux of heat in the y direction is denoted by q_y, then our transport law can be written

$$q_y = -\alpha_T \frac{d(\rho c_v T)}{dy} \tag{1-3}$$

where α_T is called the *thermal diffusion coefficient*, and $(\rho c_v T)$ is the concentration of "heat". If neither the density (ρ) nor the heat capacity (c_v) varies with

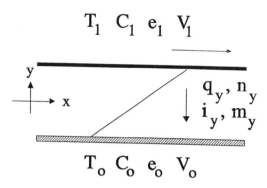

FIGURE 1-1 Transport of energy, mass, charge, and momentum from upper surface to lower surface.

position (which is normally the case), this equation can be written in the simpler form

$$q_y = -k \frac{dT}{dy}$$
(1-4)

where $k = \alpha_T \rho c_v$ is the "thermal conductivity" of the medium. This was formalized by Fourier in 1822 and is known as *Fourier's law* of heat conduction.

B. FICK'S LAW OF DIFFUSION

An analogous situation can be envisioned if the medium is a stationary fluid and the temperature difference $(T_1 - T_0)$ is replaced by a concentration difference $(C_1 - C_0)$ for some species which is soluble in the fluid (e.g. a top plate of pure salt in contact with water). If the diffusing species (e.g. the salt) is A, and the flux of A in the y direction is denoted by n_{Ay}, then the transport law is given by

$$n_{Ay} = -D_{AB} \frac{dC_A}{dy}$$
(1-5)

where D_{AB} is the molecular diffusivity of species A in the medium B. Here n_{Ay} is negative, because species A is diffusing in the $-y$ direction. Equation (1-5) is known as *Fick's law of diffusion* (even though it is the same as Fourier's law, with the symbols changed) and was formulated in 1855.

C. OHM'S LAW OF ELECTRICAL CONDUCTIVITY

The same transport law can be written for electric charge (which is another conserved quantity). In this case, the top plate is at a potential e_1 and the bottom plate is at potential e_0. The resulting "charge flux" (i.e. current density) from the top plate to the bottom is i_y (which is negative). The transport law for this situation is known as *Ohm's law* (1827) and is given by

$$i_y = -k_e \frac{de}{dy} \tag{1-6}$$

where k_e is the "electrical conductivity" of the medium between the plates. Note that the potential or voltage (e) is equivalent to a "concentration of charge".

D. NEWTON'S LAW OF VISCOSITY

Another quantity which is conserved is momentum, and we can write an equivalent expression for the transport of momentum. We must be careful here, however, because velocity and momentum are vectors, unlike mass, energy, and charge, which are scalars. Hence, even though we may draw some analogies between the one-dimensional transport of these quantities, this analogy is not likely to hold in multidimensional systems with complex geometry. In this case, we consider the top plate to be subject to a force in the x direction that causes it to move with a velocity v_x, and the lower plate is stationary. Since "x-momentum" at any point is mv_x, the concentration of momentum must be ρv_x. If we denote the flux of x-momentum in the y direction by $(\tau_{yx})_{mf}$, the transport equation is

$$(\tau_{yx})_{mf} = -\nu \frac{d(\rho v_x)}{dy} \tag{1-7}$$

where ν is called the *kinematic viscosity*. It should be evident that $(\tau_{yx})_{mf}$ is negative, because the faster fluid (at the top) drags the slower fluid (below) along with it, so that x-momentum is being transported in the $-y$ direction by virtue of this "drag". If the density does not vary with position, this can also be written

$$(\tau_{yx})_{mf} = -\mu \frac{dv_x}{dy} \tag{1-8}$$

where $\mu = \rho \nu$ is the fluid viscosity. Equation (1-8) is known as *Newton's law of viscosity* and was formulated in 1687! It applies directly to a class of (common) fluids called *Newtonian fluids* which we shall discuss in detail subsequently.

1.   Momentum Flux and Shear Stress

Newton's law of viscosity and the conservation of momentum are also related to Newton's second law of physics, which is commonly written $F_x = ma_x = $

$d(mv_x)/dt$. For a steady-state system, this is equivalent to $F_x = \dot{m}v_x$, where $\dot{m} = dm/dt$ is the mass flow rate. If F_x is the force acting in the x direction on the top plate in Fig. 1 to make it move, it must also be the "driving force" for the transport of x-momentum, which flows from the faster to the slower fluid (in the $-y$ direction). Thus the force F_x acting on a unit area of surface A_y is equivalent to a "flux of x-momentum" in the $-y$ direction [e.g. $-(\tau_{yx})_{mf}$]. (Note that A_y is the area of the surface bounding the fluid volume of interest, for which the *outward* normal is in the y direction.) F_x/A_y is also the "shear stress", τ_{yx}, which acts on the fluid, that is, the force $+F_x$ (in the $+x$ direction), that acts on the area A_y of the $+y$ surface. It follows that a *positive* shear stress is equivalent to a *negative* momentum flux, since $\tau_{yx} = -(\tau_{yx})_{mf}$. [In the following chapter, we will define the rheological (mechanical) properties of materials in terms which are common to the field of mechanics, i.e. by relations between the *stresses* which act upon the material and the resulting material *deformation*.] It follows that an equivalent form of Newton's law of viscosity can be written

$$\tau_{yx} = \mu \frac{dv_x}{dy} \tag{1-9}$$

It is important to distinguish between momentum flux and shear stress because of the difference in sign. Some references define viscosity (i.e. Newton's law of viscosity) by Eqn (1-8), whereas others use Eqn (1-9) (which we will follow). It should be evident that these definitions are equivalent because of the fact that $\tau_{yx} = -(\tau_{yx})_{mf}$.

2. Vectors vs Dyads

All of the above transport laws are described by the same equation (in one dimension), with different symbols (i.e. the same game, with different jerseys on the players). However, there are some unique features to Newton's law of viscosity that distinguish it from the other laws and are very important when applying it. First of all, as pointed out above, momentum is fundamentally different from the other conserved quantities. This is because mass, energy, and electric charge are all *scalar* quantities with no directional properties, whereas momentum is a *vector* with directional character. Since the gradient (i.e. the "directional derivative" dq/dy or, more generally, ∇q) is a vector, it follows that the gradient of a scalar (e.g. concentration of heat, mass, charge) is a vector. Likewise, the flux of mass, energy, and charge are vectors. However, Newton's law of viscosity involves the gradient of a vector (e.g. velocity or momentum), which implies *two* directions: the direction of the vector quantity (momentum or velocity) and the direction in which it varies (the "gradient" direction). Such quantities are referred to as "dyads" or "second-order tensors". Hence, momentum flux is a dyad, with the direction of the momentum (e.g. x), as well as the direction in which this momentum is transported (e.g. $-y$). It is also evident that the equivalent shear stress (τ_{yx}) has equivalent directions corresponding to the direction in which the force acts (x) and the direction (i.e. "orientation") of the surface on which it acts (y). [Note that all surfaces are vectors because of their orientation, their directions being defined by the (outward) vector which is normal to the surface which bounds the fluid volume of interest.] This is very significant when it comes to generalizing these one-dimensional laws to two or three dimensions, in which case much of the analogy between Newton's law of viscosity and the other transport laws is lost.

3. Newtonian vs Non-Newtonian Fluids

It is also evident that this "phenomenological" approach to transport processes leads to the conclusion that fluids should behave in the fashion that we have called Newtonian, which does not account for the occurrence of so-called non-Newtonian behavior, which is fairly common. This is because the phenomenological laws inherently assume that the molecular "transport coefficients" depend at most upon the thermodynamic state of the material (i.e. temperature, pressure, and density), but not upon its "dynamic state", i.e. the state of stress, strain, or strain rate. This assumption is false for fluids with complex structure, e.g. non-Newtonian fluids, as we shall illustrate in subsequent chapters.

A completely general description of the flow and deformation (e.g. rheological) properties of both Newtonian and non-Newtonian fluids is beyond the scope of this book, and the reader is referred to the more advanced literature for details. However, quite a bit can be learned, and many problems of a practical nature solved, by considering relatively simple models for the fluid viscosity, even for fluids with complex properties, provided the complexities of elastic behavior can be avoided. These properties can be measured in the laboratory, with proper attention to the data interpretation, and represented by any of several relatively simple mathematical expressions. We will not attempt to delve in detail into the molecular or structural origins of complex fluid properties, but will make use of information that can be readily obtained through routine measurements and simple modeling. Hence, we will consider non-Newtonian fluids along with, and in parallel with, Newtonian fluids in many of the flow situations that we will be analyzing.

PROBLEMS

1. A slider bearing consists of a sleeve surrounding a cylindrical shaft, which is free to move axially within the sleeve. A lubricant (e.g. grease) is in the gap between the sleeve and the shaft, to isolate the metal surfaces and support the axial stress resulting from the shaft motion. The diameter of the shaft is 1 in., and the sleeve has a diameter of 1.02 in. and a length of 2 in.

(a) If you want to limit the total force on the sleeve to less than 0.5 lb_f when the shaft is moving at a velocity of 20 ft/s, what should the viscosity of the grease be? What is the magnitude of the flux of momentum in the gap, and which direction is the momentum being transported?

(b) If the lubricant is a grease with a viscosity of 400 cP (centipoise), what is the force exerted on the sleeve when the shaft is moving at 20 ft/s?

(c) The sleeve is cooled to a temperature of 150°F, and it is desired to keep the shaft temperature below 200°F. What is the cooling rate (i.e. the rate that heat must be removed by the coolant), in Btu/hr, to achieve this?

Properties of the grease may be assumed to be: Specific heat = 0.5 BTU/(lb_m °F); SG (specific gravity) = 0.85; thermal conductivity = 0.06 BTU/(hr ft °F).

(d) If the grease becomes contaminated, it could be corrosive to the shaft metal. Assume that this occurs, and the surface of the shaft starts to corrode away at a rate of 0.1 μm/yr. Determine the maximum concentration of metal ions in the grease when this begins. Properties of the shaft metal may be assumed to be: MW = 65; SG = 8.5; Diffusivity of metal ions in grease = 8.5 x 10^{-5} cm^2/s.

2. By making use of the analogies between the molecular transport of the various conserved quantities, describe how you would set up an experiment to solve each of the following problems by making electrical measurements (e.g. describe the design of the experiment, how and where you would measure voltage and current, and how the measured quantities are related to the desired quantities):

(a) Determine the rate of heat transfer from a cylinder to a fluid in which it is immersed, and which is flowing slowly normal to the cylinder axis, if the surface of the cylinder is at temperature T_0 and the fluid far away from the cylinder is at temperature T_1. Also determine the temperature distribution within the fluid.

(b) Determine the rate at which a (spherical) mothball evaporates when it is immersed in a slowly flowing stream of air, and also the concentration distribution of the evaporating compound in the air stream.

(c) Determine the local stress as a function of position on the surface of a wedge-shaped body immersed in a fluid stream which is flowing slowly parallel to the surface. Also, determine the local velocity distribution in the fluid as a function of position in the fluid.

2

Dimensional Analysis and Scale-up

I. INTRODUCTION

This chapter considers the concepts of dimensions, units, and the various systems in use for describing these quantities. In particular, the distinction between scientific and engineering systems of dimensions is explained, and the various metric and English units used in each system are discussed. It is important that the engineer be familiar with all of these systems, as they are all in common use in various fields of engineering and will be for the indefinite future. It is common to encounter a variety of units, in different systems, when analyzing a given problem, and the engineer must be adept at dealing with all of these.

These concepts will also be extended to the dimensional analysis of engineering systems, by application of the law of conservation of dimensions and the application of the results to the design and interpretation of laboratory "model" measurements for simulation of larger scale ("field") systems (this is also known as *similitude*). These concepts are presented early on, because we shall make frequent use of them in describing the results of both theoretical and experimental analyses of various systems in a form that is the most concise, general, and useful. This will also provide guidance to the best approach to take in the solution of many complex problems.

II. DIMENSIONS AND UNITS

A. DIMENSIONS

The dimensions of a quantity identify the physical character of that quantity, e.g. force (F), mass (M), length (L), time (t), temperature (T), electrical charge (e), etc. On the other hand, ''units'' identify the reference scale by which the magnitude of the respective physical quantity is measured. Many different reference scales (units) can be defined for a given dimension. For example, the dimension of length can be measured in units of miles, centimeters, inches, meters, yards, angstroms, furlongs, light years, kilometers, etc.

Dimensions can be classified as either *fundamental* or *derived*. *Fundamental* dimensions cannot be expressed in terms of any other dimensions. They include length (L), time (t), temperature (T), mass (M), and/or force (F) (depending upon the *system* of dimensions used). *Derived* dimensions can be expressed in terms of other fundamental dimensions, for example area ($[A] = L^2$), volume ($[V] = L^3$), energy ($[E] = FL = ML^2/t^2$), power ($[P] = FL/t = ML^2/t^3$), viscosity ($[\mu] = Ft/L^2 = M/Lt$), etc.[1]

There are two systems of fundamental dimensions in use (with their associated units), which may be referred to as *scientific* and *engineering* systems. These two systems differ basically in the manner in which the dimension of force is defined. In both systems, mass, length, and time are fundamental dimensions. Furthermore, Newton's second law provides a relation between the dimensions of force, mass, length, and time:

Force = Mass × Acceleration

i.e.

$$F = ma \tag{2-1}$$

or

$$[F] = [ma] = ML/t^2$$

In scientific systems, this is accepted as the definition of force; that is, force is a derived dimension, being identical to ML/t^2.

In engineering systems, however, force is considered in a more practical or ''pragmatic'' context as well. This is because the mass of a body is not usually measured directly, but is instead determined by its ''weight'' (W), i.e. the gravitational force which is exerted by mutual attraction between two bodies of mass m_1 and m_2:

$$W = G\,(m_1 m_2/r^2) \tag{2-2}$$

---

[1] The notation [ ] means ''the dimensions of'' whatever is in the [ ].

Here, G is a constant having a value of 6.67×10^{-11} N m^2/kg^2, and r is the distance between the centers of m_1 and m_2. If m_2 is the mass of the earth and r is its radius at a certain location on earth, then W is the "weight" of mass m_1 at that location:

$$W = m_1 g \tag{2-3}$$

The quantity g is called the *acceleration due to gravity*, and is equal to $m_2 G/r^2$. At sea level and 45° latitude on the earth, the value of g is 32.17 ft/s^2 or 9.81 m/s^2 ("standard gravity", g_{std}). The value of g is obviously different on the moon (different r and m_2) and varies slightly over the surface of the earth as well (since the radius of the earth varies with both elevation and latitude).

Since the mass of a body is determined indirectly by its weight (i.e. the gravitational force acting on the mass) under specified gravitational conditions, engineers decided that it would be more practical and convenient if a system of dimensions were defined in which "what you see is what you get", i.e. the numerical magnitudes of mass and weight are equal under these conditions. This must not violate Newton's law, however, so both Eqns (2-1) and (2-3) are valid. Now since the value of g is not unity when expressed in common units of length and time, the only way that the numerical values of weight and mass can be the same under any conditions is by introduction of a "conversion factor" which forces this equivalence. This factor is designated g_c and is incorporated into Newton's second law for engineering systems (sometimes referred to as "gravitational systems") as follows:

$$F = \frac{ma}{g_c}, \qquad W = \frac{mg}{g_c} \tag{2-4}$$

This additional definition of force is equivalent to defining F as a fundamental dimension, the redundancy being accounted for by the conversion factor g_c. Thus, if a unit for the weight of mass m is to be defined so that the numerical values of F and m are identical under standard gravity conditions (i.e. $a = g_{std}$), it follows that the numerical magnitude of g_c must be identical to that of g_{std}. However, it is important to distinguish between g and g_c, because they are fundamentally different quantities. As explained above, g is not a constant; it is a variable, that depends on both m_2 and r (Eqn 2-2). However, g_c is a constant, since it is merely a *conversion factor* which is defined by standard gravity. Note that these two quantities are also physically different, since they have different dimensions, i.e.

$$[g] = \frac{L}{t^2}, \qquad [g_c] = \frac{ML}{Ft^2} \tag{2-5}$$

The factor g_c can be considered a conversion factor which relates equivalent "force and mass (ML/t^2) units" in engineering systems. In such systems, both force and mass are fundamental dimensions, since they are related by two separate (but compatible) definitions: Newton's second law and the engineering definition of weight. The conversion factor g_c thus accounts for the redundancy in these two definitions.

B. UNITS

There are several sets of units which are used in both scientific and engineering systems of dimensions. These can be classified as either metric (SI and cgs) or English (fps). Although the internationally accepted standard is the SI scientific system, English engineering units are still very common and probably will remain so for the foreseeable future. Therefore, the reader should, at least, master these two systems and become adept in converting between them. These systems are illustrated in Table 2-1. Note that there are two different English scientific systems, one in which M, L, and t are fundamental and F is derived, and another in which F, L, and t are fundamental and M is derived. In one English scientific system, the "slug" mass unit is fundamental, and in the other the "poundal" force unit is fundamental. However, these systems are archaic and rarely used in practice, and the metric engineering systems involving the kg_f and g_f have been generally replaced by the SI system. Thus, the most common systems in general use are the scientific metric (e.g. SI) and the English engineering systems.

Since Newton's second law is satisfied identically in scientific units with no conversion factor (i.e. $g_c = 1$), the following identities hold for scientific units:

$$g_c = 1 \ \frac{kg \ m}{N \ s^2} = 1 \ \frac{g \ cm}{dyn \ s^2} = 1 \ \frac{slug \ ft}{lb_f \ s^2} = 1 \ \frac{lb_m ft}{poundal \ s^2}$$

In engineering systems, both F and M may be considered to be fundamental because of the engineering definition of weight in addition to Newton's second law. However, this results in a redundancy that requires the conversion factor g_c to rectify. The value of this conversion factor in the various engineering units provides the following identities:

$$g_c = 9.81 \ \frac{kg_m \ m}{kg_f \ s^2} = 981 \ \frac{g_m \ cm}{g_f \ s^2} = 32.17 \ \frac{lb_m \ ft}{lb_f \ s^2}$$

C. CONVERSION FACTORS

Conversion factors, which relate the magnitudes of different units with common dimensions, are actually identities, i.e. 1 foot is identical to 12 inches, 1 BTU

TABLE 2-1   Systems of Dimensions/Units

	Scientific				Engineering			
	L	M	F	g_c	L	M	F	g_c
English	ft	lb_m	poundal	1	ft	lb_m	lb_f	32.2
	ft	slug	lb_f	1				
Metric (SI)	m	kg	N	1	m	kg_m	kg_f	9.8
(cgs)	cm	g	dyn	1	cm	g_m	g_f	980

Conversion factors: g_c [ML/Ft^2] (F = ma/g_c)
g_c = 32.17 lb_m ft/(lb_f s^2) = 9.8 kg_m m/(kg_f s^2) = 980 g_m cm/(g_f s^2) = 1 kg m/(N s^2) = 1 g cm/
(dyn s^2) = 1 slug ft/(lb_f s^2) = 1 lb_m ft/(poundal s^2) = 12 in./ft = 60 s/min = 30.48 cm/ft = 778
ft lb_f/Btu = $\cdots$ = 1 [0]

is identical to 778 ft lb_f, etc. Since any identity can be expressed as a ratio with a magnitude but no dimensions, the same holds for conversion factors, i.e.:

$$12 \frac{in.}{ft} = 778 \frac{ft\ lb_f}{BTU} = 30.48 \frac{cm}{ft} = 14.7 \frac{psi}{atm} = 10^5 \frac{dyn}{N} = 1\ (0),\ etc.$$

A table of commonly encountered conversion factors is provided in the front of the book. The value of any quantity expressed in a given set of units can be converted to any other equivalent set of units by multiplying or dividing by the appropriate conversion factor to cancel the unwanted units.

Example 2-1:   To convert a quantity X measured in feet to the equivalent in miles:

$$\frac{X\ feet}{5280\ feet/mile} = \frac{X}{5280}\ miles$$

Note that the conversion factor relating mass units in scientific systems to those in engineering systems can be obtained by equating the appropriate g_c values from the two systems, e.g.:

$$g_c = 1 \frac{slug\ ft}{lb_f\ s^2} = 32.2 \frac{lb_m\ ft}{lb_f\ s^2}$$

Thus, the conversion factor relating slugs to lb_m is seen to be 32.2 lb_m/slug.

III. CONSERVATION OF DIMENSIONS

Physical laws, theories, empirical relations, etc. are normally expressed by equations which relate the significant variables and parameters in the system. These equations usually contain a number of terms. For example, the relation between the vertical elevation (z) and the horizontal distance (x) at any time for a projectile fired from a gun may be expressed in the form

$$z = ax + bx^2 \tag{2-6}$$

This equation can be derived from the laws of physics, in which case the parameters a and b depend upon such factors as the muzzle velocity, angle of inclination of the gun, and wind resistance. The equation may be also considered to be empirical if measured values of z versus x can be represented by an equation of this form, with no reference to the laws of physics.

For any equation to be valid, every term in the equation must have the same physical character, i.e. the same net dimensions (and consequently the same units, in a consistent system of units). This is known as the *law of conservation of dimensions* (otherwise known as the ''fruit salad law''; i.e. you can't add apples and oranges unless you are making fruit salad). Let us look further at Eqn (2-6). Since both z and x have dimensions of length, e.g. $[x] = L$, $[z] = L$, it follows from the fruit salad law that the dimensions of a and b must be

$$[a] = 0, \; [b] = 1/L$$

(i.e. a has no dimensions—it is dimensionless, and the dimensions of b are 1/length, or length^{-1}). For the sake of argument, let us assume that x and z are measured in feet and that the values of a and b in the equation are 5 and 10 ft^{-1} respectively. Thus if x = 1 ft,

$$z = (5)(1 \text{ ft}) + (10 \text{ ft}^{-1})(1 \text{ ft})^2 = 15 \text{ ft}$$

On the other hand, if we choose to measure x and z in inches, the value of z for x = 1 in. is

$$z = (5)(1 \text{ in.}) + (10 \text{ ft}^{-1}) \left(\frac{1}{12 \text{ in./ft}} \right) (1 \text{ in.})^2 = 5.83 \text{ in.}$$

This is still in the form of Eqn (2-6), i.e.

$$z = ax + bx^2$$

but now a = 5 and b = (10/12) in.$^{-1}$. Thus the magnitude of a has not changed, but the magnitude of b *has* changed. This simple example illustrates two important principles:

(1) *Conservation of dimensions* (''fruit salad law'')—All terms in a given equation must have the same net dimensions for the equation to be valid.

(2) *Scaling*—The fact that the value of the dimensionless parameter a has the same magnitude regardless of the units (e.g. scale) used in the problem illustrates the universal nature of dimensionless quantities. That is, the magnitude of any dimensionless quantity will always be independent of the scale of the problem or the system of (consistent) units used. This principle is the basis for the application of dimensional analysis, which permits information and relationships determined in one size system (e.g. a model) to be applied directly to a similar system of a different size if the system variables are expressed in dimensionless form. This process is known as "scale-up".

The "universality" of certain dimensionless quantities is often taken for granted. For example, the exponent 2 in the last term of Eqn (2-6) has no dimensions and hence has the same magnitude regardless of the scale or units used for measurement. In a like manner, the kinetic energy per unit mass of a body moving with a velocity v is given by

$$\text{ke} = \frac{1}{2} v^2$$

Both of the numerical quantities in this equation, e.g. 1/2 and 2, are dimensionless, so they always have the same magnitude regardless of the units used to measure v.

A. NUMERICAL VALUES

Ordinarily, numerical quantities that appear in equations which have a theoretical basis, such as that above, are dimensionless and hence "universal". However, many valuable engineering relations have an empirical rather than a theoretical basis, in which case this conclusion does not always hold. For example, a very useful expression for the (dimensionless) friction loss coefficient (K_f) for valves and fittings is given by

$$K_f = \frac{K_1}{N_{Re}} + K_\infty \left(1 + \frac{1}{\text{I.D.}} \right)$$

Here, N_{Re} is the Reynolds number, which is dimensionless, as are the constants K_1 and K_∞. However, the term I.D. is the internal diameter of the fitting, with dimensions of length. By the "fruit salad" law, the constant 1 in the term 1/I.D. must also have dimensions of length, and so is not independent of scale, i.e. it has a magnitude of 1 only in specific units. In fact, it has a value of 1 in., so I.D. must also be measured in inches for this form of the equation to be valid. If I.D. were to be measured in centimeters, for example, this term would be 2.54/I.D., since (1 in.)(2.54 cm/in.) = 2.54 cm.

B. CONSISTENT UNITS

The conclusion that dimensionless numerical values are universal is valid only if a consistent system of units is used for all quantities in a given equation. If such is not the case, then the numerical quantities may include conversion factors relating the different units. For example, the velocity (V) of a fluid flowing in a pipe can be related to the volumetric flow rate (Q) and the internal pipe diameter (D) by any of the following equations:

$$V = 183.3 \ Q/D^2 \qquad (2\text{-}7)$$

$$V = 0.408 \ Q/D^2 \qquad (2\text{-}8)$$

$$V = 0.286 \ Q/D^2 \qquad (2\text{-}9)$$

$$V = 4A\pi D^2 \qquad (2\text{-}10)$$

Although the dimensions of V (L/t) are the same as those of Q/D^2 (L^3/tL^2 = L/t), it is evident that the numerical coefficient is not universal despite the fact that it is dimensionless. This is because a consistent system of units is not used in each case. In each of these equations, the units of V are ft/s and D is in inches. However, in Eqn (2-7) Q is in ft^3/s, while in Eqn (2-8) Q is in gallons/min (gpm) and in Eqn (2-9) it is in barrels/hr (bbl/hr). Thus, although the dimensions are consistent in each term of all of these equations, the units are not, so the numerical coefficients include unit conversion factors. Only in Eqn (2-10) are the units assumed to be all from the same consistent system, so that the factor $4/\pi$ is both dimensionless and unitless and is hence universal. It is always advisable to write equations in a form which is universally valid to avoid confusion; i.e. all quantities should be expressed in consistent units.

IV. DIMENSIONAL ANALYSIS

The law of conservation of dimensions may be applied to group the variables and parameters which are important in a given system into a set of *dimensionless groups* of these variables. The original (dimensional) variables may then be replaced by the corresponding set of dimensionless groups, and any valid relationship (theoretical or empirical) between the original variables can be expressed in terms of these dimensionless groups. This has two important advantages, namely:

(1) Dimensionless quantities are universal, as we have seen, so that any relationship involving dimensionless variables is independent of the size or scale of the system. Consequently, information obtained from a model (e.g. small) system which is represented in dimensionless

form can be applied directly to geometrically and dynamically similar systems of any size or scale. This allows us to translate information directly from laboratory models to large-scale equipment or plant operations (scale up). Geometric similarity requires that the two systems have the same shape (geometry), and dynamical similarity requires them to be operating in the same hydrodynamic mode (i.e. both either laminar or turbulent). This will be discussed in more detail later.

(2) The manner in which a given system depends upon the separate variables that influence the system can be presented in much simpler and more general form in terms of the dimensionless variables than in terms of the original variables. This is because the number of dimensionless groups is invariably less than the number of original variables involved in the problem. Thus, the amount of effort required to represent a relationship between the groups is much less than that required to relate each of the variables independently, and the resulting relation will be simpler in form. For example, a relation between two variables requires two dimensions (x vs y) to represent, whereas a relation between three variables requires three dimensions (x vs y vs z) or a family of two dimensional "curves" (a set of x vs y values, each for a given z). The difference is the same as the difference between one page and a book (of many pages). Relating four variables would obviously require many books or volumes. Thus, reducing the number of variables from, say, four to two would greatly simplify any problem.

It is important to realize that the process of dimensional analysis only replaces the set of original (dimensional) variables by an equivalent (smaller) set of dimensionless variables (i.e. the groups). It does not tell how these variables are related—the relation must be determined either theoretically, by application of basic scientific laws, or empirically, by measurements and data analysis. However, dimensional analysis is a very powerful tool, in that it can provide a direct guide for experimental design and scale-up and for expressing operating relationships in the most general and useful form.

There are a wide variety of approaches to dimensional analysis. The classical method is the "Buckingham Π theorem", so called because Buckingham used the symbol Π (capital Greek pi) to represent the dimensionless groups[1]. Another classical approach, which involves a more direct application of the law

---

[1] We shall use the recommended notation of the AIChE for dimensionless groups that are named after their originator, i.e. a capital N with a subscript identifying the group. However, there are a variety of other dimensionless quantities that are identified by other symbols.

of conservation of dimensions, is attributed to Lord Rayleigh. Numerous variations on these methods have also been presented in the literature. The one thing all of these methods have in common is that they require a knowledge of the variables which are important to the problem as a starting point. This can be determined through common sense, intuition, experience, physical reasoning, or by asking someone who is more experienced or knowledgeable. They can also be determined from a knowledge of the physical principles which govern the system (e.g. the conservation of mass, energy, momentum, etc., as written for the specific system to be analyzed) along with the fundamental equations which describe these principles. These equations may be macroscopic or microscopic (e.g. coupled sets of partial differential equations, along with their boundary conditions). However, this knowledge often requires as much (or more) insight, intuition, and/or experience as is required to compose the list of variables from logical deduction. The analysis of any engineering problem requires key assumptions which distinguish those factors that are important in the problem from those that are insignificant. [This is sometimes referred to as the "bathwater" rule—it is necessary to separate the "baby" from the "bathwater" in any problem and to retain the significant elements (the "baby") and discard the insignificant ones (the "bathwater") and not vice versa!] The talent required to do this depends much more upon sound understanding of fundamentals and the exercise of good judgment than upon mathematical ability, and the best engineer is most often the one who is able to make the most appropriate assumptions to simplify a problem (i.e. discard the "bathwater" and save the "baby"). Many problem statements, as well as solutions, involve implicit assumptions, i.e. assumptions that are implied but not stated. One should always be on the lookout for such implicit assumptions and try to identify them wherever possible, since they set corresponding limits on the applicability of the results.

The method we will use to illustrate the dimensional analysis process is one which involves a minimum of manipulations. It does require an initial knowledge of the variables which are important in the system and the dimensions of these variables. The objective of the process is to determine an appropriate set of dimensionless groups of variables, which can then be used in place of the original individual variables for the purpose of describing the behavior of the system. The process will be explained by means of an example, and the results will be used to illustrate the application of dimensional analysis to experimental design and scale-up.

A. PIPELINE ANALYSIS

The procedure for performing a dimensional analysis will be illustrated by means of an example concerning the flow of fluid through a circular pipe. In this example, we will determine an appropriate set of dimensionless groups which may be used to represent the relationship between the flow rate of an incompressible fluid in a pipeline, the properties of the fluid, the dimensions of the pipeline, and the driving force for moving the fluid, as illustrated in Figure 2-1. The procedure is as follows:

Step 1—Identify the important variables in the system. The flow rate may be represented by the total volumetric flow rate (Q) or the average velocity in the pipeline (V). We shall choose the latter. The driving force can be represented by the difference in the pressure between the upstream (P_1) and downstream (P_2) ends of the pipe ($\Delta P = P_1 - P_2$). The pipe dimensions are the diameter (D) and length (L), and the fluid properties are the density (ρ) and viscosity (μ). It is also possible that the "texture" of the pipe wall (i.e. the surface roughness, ε) may be important. This is the most important step in the process. The appropriate variables can be identified by various means, including experience, judgment, brainstorming, intuition, or examination of the basic equations that describe the fundamental physical principles that govern the system along with appropriate boundary conditions. It is also important to include only those variables which are fundamental, as opposed to being derivable from others. For example, we might choose to include the fluid velocity (V), the pipe diameter (D), and the volumetric flow rate (Q) among our variables. However, since by definition $Q = \pi D^2 V/4$, these three variables are not all independent, any one of them being derivable from the other two by this identity, and only two of them would be permitted.

Step 2—List all of the problem variables, along with their dimensions. The procedure is the simplest if the most fundamental dimensions (in a scientific

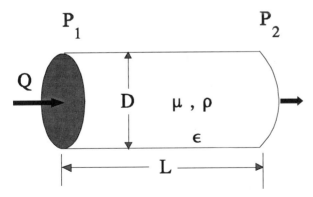

FIGURE 2-1 Flow in a pipeline.

system, i.e. M, L, t) are used (e.g. energy should be converted to $FL = ML^2/t^2$, etc.):

Variable	Dimensions
V	L/t
ΔP	$F/L^2 = M/(Lt^2)$
D	L
L	L
ε	L
ρ	M/L^3
μ	M/(Lt)
7	3

The number of dimensionless groups that will be obtained is equal to the number of variables less the minimum number of fundamental dimensions involved in these variables (7 − 3 = 4 groups).

Step 3—Choose a set of reference variables. The actual variables chosen are arbitrary, except that the following criteria must be satisfied:

(1) The number of reference variables must be equal to the minimum number of fundamental dimensions in the problem (in this case, three).

(2) No two reference variables should have exactly the same dimensions.

(3) All of the dimensions that appear in the problem variables must also appear collectively in the dimensions of the reference variables.

In general, the procedure is simplest if the reference variables chosen have the simplest possible dimensions consistent with the above criteria. In this problem, we have three dimensions (M, L, t), so we need three reference variables. The variables D, ε, and L all have the dimension of length, so let us choose D (arbitrarily) as one reference variable:

$$[D] = L$$

The dimension t appears in V, ΔP, and μ, but V has the simplest dimensions, so we choose it as our second reference variable:

$$[V] = L/t$$

We also need a reference variable containing the dimension M, which could be either ρ or μ. Since ρ has the simplest dimensions, we choose it for the third reference variable:

$$[ρ] = M/L^3$$

Our three reference variables are now D, V, and ρ.

Step 4—Solve the above reference variable "dimensional equations" for the dimensions (L, t, M) in terms of the variables (D, V, ρ), i.e.

$$L = [D], \qquad t = [D/V], \qquad M = [\rho D^3]$$

Step 5—Write the dimensional equations for each of the remaining variables, and substitute the results of step 4 for the dimensions in terms of the reference variables:

$$[\varepsilon] = L = [D]$$

$$[L] = L = [D]$$

$$[\mu] = \frac{M}{Lt} = \left[\frac{\rho D^3}{(D)(D/V)} \right] = [\rho V D]$$

$$[\Delta P] = \frac{M}{Lt^2} = \left[\frac{\rho D^3}{(D)(D/V)^2} \right] = [\rho V^2]$$

Step 6—Since these equations are each a dimensional identity, dividing one side by the other results in one dimensionless group from each equation:

$$N_1 = \frac{\varepsilon}{D} \qquad \text{or} \qquad \frac{D}{\varepsilon}$$

$$N_2 = \frac{L}{D} \qquad \text{or} \qquad \frac{D}{L}$$

$$N_3 = \frac{\mu}{DV\rho} \qquad \text{or} \qquad \frac{DV\rho}{\mu}$$

$$N_4 = \frac{\Delta P}{\rho V^2} \qquad \text{or} \qquad \frac{\rho V^2}{\Delta P}$$

B. UNIQUENESS

Note that these results are not unique, since the reciprocal of each group is just as valid as the initial group. In fact, any combination of these groups will be dimensionless and will be just as valid as any other combination as long as all of the variables are represented among the groups. Thus these four groups can be replaced by any combination of these groups, and, indeed, a different combination would have resulted if we had used a different set of reference variables. However, any valid set of groups can be derived by a suitable combination of any other valid set. As we shall see, the most appropriate set of groups will depend on the particular problem to be solved, i.e. which of the variables are known and which are unknown. Specifically, it is most appropriate to arrange the groups so that the unknown variables each appear in only one group, if possible. It should be noted that the variables that were *not* chosen as the reference variables each appeared in only one group.

C. DIMENSIONLESS VARIABLES

All of the variables in this problem have now been replaced by an equivalent (smaller) set of dimensionless groups of variables. For example, if it is desired to determine the driving force required to transport a given fluid at a given rate through a given pipe, the solution could be represented as

$$\Delta P = fn(V, D, L, \varepsilon, \rho, \mu)$$

or, in terms of the equivalent dimensionless groups,

$$N_4 = fn(N_1, N_2, N_3)$$

Note that the number of variables has been reduced from the original seven to four (groups). Furthermore, the relation between these dimensionless variables (groups) is independent of scale. That is, any two *similar* systems will be exactly equivalent, regardless of size or scale, if the values of all dimensionless variables (groups) are the same in each. By ''similar'' is meant that both systems must have the same geometry or shape (which is actually another dimensionless ''variable''), and both must be operating under comparable dynamic conditions (e.g. either laminar or turbulent—this will be elaborated upon later). Also, the fluids must be rheologically similar (e.g. Newtonian). [The difference between Newtonian and non-Newtonian fluids will be explained subsequently. For the present, a Newtonian fluid is one that requires only one rheological property, the viscosity (μ), to determine flow behavior, whereas a non-Newtonian fluid requires a rheological ''function'', which contains two or more parameters. Each of these parameters is a rheological property, so that in place of the viscosity for a Newtonian fluid, the non-Newtonian fluid would require two or more ''rheological properties'', depending upon the specific model that describes the fluid, with a corresponding increase in the number of dimensionless groups.]

D. PROBLEM SOLUTION

It should be emphasized that the specific relationship between the variables (groups) that is implied above is not determined by dimensional analysis. It must be determined from theoretical or experimental analysis. Dimensional analysis gives only an appropriate set of dimensionless groups that can be used as generalized variables in these relationships. However, because of the universal generality of the dimensionless groups, any functional relationship between them that is valid in any system must also be valid in any other *similar* system.

E. ALTERNATIVE GROUPS

The above set of dimensionless groups is convenient for representing the behavior of a pipeline if it is desired to determine the driving force (ΔP) required to move a given fluid at a given rate through a given pipeline, since the unknown quantity (ΔP) appears in only one group (N_4), which can be considered the "dependent" group. However, the same variables apply to the case where the driving force is known and it is desired to determine the flow rate (V) that would result for a given fluid through a given pipe. In this case, V is the dependent (unknown) variable, but it appears in more than one group (N_3 and N_4). Therefore, there is no single obvious dependent group. However, since this set of groups is not unique, we can rearrange them into another equivalent set in which the unknown velocity appears in only one group. This can easily be done, for example, by combining groups N_3 and N_4 to form a group that does not contain v:

$$N_5 = (N_3)^2 N_4 = \left(\frac{DV\rho}{\mu}\right)^2 \left(\frac{\Delta P}{\rho V^2}\right) = \frac{\Delta P D^2 \rho}{\mu^2}$$

This new group can then be used in place of either N_3 or N_4, along with N_1 and N_2, to complete the required set of four groups for which the unknown V appears in only one. If we replace N_4 by N_5, the implied relation is

$$N_3 = fn(N_1, N_2, N_5) \qquad \text{or} \qquad \frac{DV\rho}{\mu} = fn\left(\frac{\varepsilon}{D}, \frac{L}{D}, \frac{\Delta P D^2 \rho}{\mu^2}\right)$$

in which the unknown (V) appears only in the group on the left.

Let us reexamine our original problem for a moment. If the pipeline is relatively long, operating at steady state, and the fluid is incompressible, then the conditions over any given length of the pipe will be the same as along any other segment of the same length, except for the regions very near the entrance and exit. If these regions are small relative to the rest of the pipe (e.g. L >> D) they can be neglected, and the pressure drop per unit length of pipe should be the same over any given segment of the pipe. Thus, the only significance of the pipe length is to spread the total pressure drop over the entire length, and hence the two variables ΔP and L are not independent and can therefore be combined into one, the pressure *gradient*, $\Delta P/L$. This reduces the total number of variables from seven to six, and the number of groups from four to three. These three groups can be derived by following the original procedure. However, since ΔP and L each appear in only one of the original groups (N_2 and N_4), dividing one of these by the other will automatically produce a group with the desired variable in the resulting group, which will replace N_2 and N_4:

$$N_6 = \frac{N_4}{N_2} = \frac{D\Delta P/L}{\rho V^2}$$

The three groups are now N_1, N_3, and N_6:

$$N_1 = \frac{\varepsilon}{D}, \qquad N_3 = \frac{DV\rho}{\mu}, \qquad N_6 = \frac{D\Delta P/L}{\rho V^2}$$

Group N_6 is also known as a *friction factor* (f), since the driving force (ΔP) is required to overcome "friction" (i.e. the energy dissipated) in the pipeline, and N_3 is known as the *Reynolds number* (N_{Re}). There are various definitions of the friction factor, each of which are some multiple of N_6; e.g. the Fanning friction factor is $N_6/2$, and the Darcy friction factor is $2N_6$. Group N_4 is also known as the *Euler number*.

V. SCALE-UP

We have previously stated that dimensional analysis results in an appropriate set of groups that may be used to describe the behavior of a system, but it does not tell how these groups are related. In fact, dimensional analysis does not determine any numbers related to the groups (except for exponents on the variables). The relationship between the groups, which represents the system behavior, must be determined by either theoretical analysis or experimentation. Even when theoretical results are possible, however, it is often necessary to obtain data to evaluate or confirm the adequacy of the theory. Because dimensionless groups are independent of scale, they also provide a guide for the proper design of an experiment which is intended to simulate another (larger scale) similar system, and for scaling up the results of model measurements to the full-scale system. For example, the behavior of our pipeline can be described by a functional relation of the form

$$N_6 = fn(N_1, N_3)$$

or

$$\frac{D\Delta P/L}{\rho V^2} = fn\left(\frac{\varepsilon}{D}, \frac{DV\rho}{\mu}\right)$$

This is valid for any Newtonian fluid in any (circular) pipe of any size (scale) under given dynamical conditions (e.g. laminar or turbulent). Thus, if the values of N_3 (or N_{Re}, the Reynolds number) and N_1 (ε/D) for an experimental model are identical to those for a full-scale system, it follows that the value of N_6 (or f, the friction factor) must also be the same in the two systems. In such a case, the model is said to be dynamically similar to the full-scale (field) system, and measurements of the variables in N_6 can be translated (scaled) directly from the model to the field system. In other words, the equality between the groups N_3 (N_{Re}) and N_1 (ε/D) in the model and in the field is a necessary condition for the dynamic similarity between the two systems.

Example 2-2: **Laminar Flow of a Newtonian Fluid in a Pipe**—It turns out (for reasons that we will explain later) that if the Reynolds number in pipe flow has a value less than about 2000, the flow follows a smooth, straight pattern—

called a *laminar* flow. In this case, the "friction loss" (i.e. the pressure drop) does not depend upon the pipe wall roughness (ε) or the density (ρ) (the reason for this will become clear when we examine the mechanism of pipe flow in more detail). With two fewer variables, we would have two fewer groups, so the system can be described completely by only one group (which does not contain either ε or ρ). The form of this group could be determined by repeating the dimensional analysis procedure or simply by eliminating these two variables from the three groups that we already have. This is easily done by multiplying the friction factor (f) and the Reynolds number (N_{Re}) to get the required group, which then becomes

$$N_7 = fN_{Re} = \frac{\Delta PD^2}{L\mu V} = \text{const.}$$

Since this is the only "variable" that is needed to describe this system, it follows that the value of this group must be a constant and therefore has the same value for the *laminar* flow of *any* Newtonian fluid at *any* flow rate in *any* pipe. This is in contrast to *turbulent pipe flow* ($N_{Re} > 4000$) in long pipes, which can only be described completely by three groups (e.g. f, N_{Re}, and ε/D). Thus, the flow in two different pipes must satisfy the same functional relation between these three groups, even though the actual values of the individual groups may be quite different. However, for laminar pipe flow, since only one group (fN_{Re}) is required, the value of that group must be the same in *all* laminar pipe flows, regardless of the values of the individual variables.

Example 2-3: **Scale-Up of Pipe Flow**—As an example of the application of dimensional analysis to experimental design and scale-up, consider the following. We would like to know the total pressure driving force (ΔP) required to pump oil ($\mu = 30$ cP, $\rho = 0.85$ g/cm^3) through a horizontal pipeline with a diameter (D) of 48 in. and a length (L) of 700 miles, at a flow rate of 1 million barrels per day. The pipe is to be commercial steel, which has an equivalent roughness (ε) of 0.0018 in. In order to get this information, we want to design a laboratory experiment in which the laboratory model and the full-scale field pipeline are operating under dynamically similar conditions, so that measurements of ΔP in the model can be scaled up directly to find ΔP in the field. The necessary condition for dynamical similarity for this system is

$$(N_1)_m = (N_1)_f \quad \text{or} \quad \left(\frac{\varepsilon}{D}\right)_m = \left(\frac{\varepsilon}{D}\right)_f$$

$$(N_3)_m = (N_3)_f \quad \text{or} \quad \left(\frac{DV\rho}{\mu}\right)_m = \left(\frac{DV\rho}{\mu}\right)_f$$

from which it follows that

$$(N_6)_m = (N_6)_f \quad \text{or} \quad \left(\frac{\Delta PD}{L\rho V^2}\right)_m = \left(\frac{\Delta PD}{L\rho V^2}\right)_f$$

where the subscript m represents the experimental model and f represents the full-scale field system. Since the volumetric flow rate (Q) is specified instead of the velocity (V), we can make the substitution $V = 4Q/(\pi D^2)$ to get the following equivalent groups:

$$\left(\frac{\varepsilon}{D}\right)_m = \left(\frac{\varepsilon}{D}\right)_f \tag{2-11}$$

$$\left(\frac{4Q\rho}{\pi D\mu}\right)_m = \left(\frac{4Q\rho}{\pi D\mu}\right)_f \tag{2-12}$$

$$\left(\frac{\pi^2 \Delta PD^5}{16L\rho Q^2}\right)_m = \left(\frac{\rho^2 \Delta PD^5}{16L\rho Q^2}\right)_f \tag{2-13}$$

Note that all the numerical coefficients in these equations cancel. By substituting the known values for the pipeline variables into Eqn (2-12), we find that the value of the Reynolds number for this flow is 2.6×10^6, which is highly turbulent. Thus, all of these groups are important.

Let us now identify the knowns and unknowns in the problem. The knowns obviously include all of the field variables except $(\Delta P)_f$. In addition, we will measure the pressure drop in the lab model $(\Delta P)_m$ after specifying the lab test conditions that simulate the field conditions. This will then be known and will be scaled up to find the unknown pressure drop in the field:

Knowns: (D, L, ε, Q, μ, ρ)$_f$, $(\Delta P)_m = 7$

Unknowns: (D, L, ε, Q, μ, ρ)$_m$, $(\Delta P)_f = 7$

There are seven unknowns but only three equations that relate these quantities. Therefore, four of the unknowns may be chosen "arbitrarily". This process is not really arbitrary, however, because we are constrained by certain practical considerations, such as a lab model that must be smaller than the field pipeline and the need to choose test materials that are convenient, inexpensive, and readily available. For example, the diameter of the pipe to be used in the model could, in principle, be chosen arbitrarily. However, it is related to the field pipe diameter by Eqn (2-11):

$$D_m = D_f\left(\frac{\varepsilon_m}{\varepsilon_f}\right)$$

Thus, if we were to use the same pipe material (commercial steel) for the model as in the field, we would also have to use the same diameter (48 in.). This is obviously not practical, but a smaller diameter for the model would require a much smoother material in the lab (since $D_m \ll D_f$ requires $\varepsilon_m \ll \varepsilon_f$). The smoothest material we can find would be glass or plastic or smooth drawn tubing such as copper or stainless steel, all of which have equivalent roughness values of the order of 0.00006 in. (see Table 6-1, p. 151). If we choose one of these (e.g. plastic), then the required lab diameter is set by Eqn (2-11):

$$D_m = D_f\left(\frac{\varepsilon_m}{\varepsilon_f}\right) = (48 \text{ in.})\left(\frac{0.00006}{0.0018}\right) = 1.6 \text{ in.}$$

Since the roughness values are only approximate, so is this value of D_m. Thus we could choose a convenient size pipe for the model with a diameter of the order of 1.6 in. (for example, from Appendix F, we see that a Schedule 40 $1\frac{1}{2}$ in. pipe has a diameter of 1.61 in.).

We still have five remaining unknowns, $(Q, \rho, \mu, L)_m$ and $(\Delta P)_f$, and only two remaining equations, so we still have three "arbitrary" choices. Of course, we will choose a pipe length for the model which is much less than the 700 miles in the field, but it only has to be much longer than its diameter to avoid end effects. Thus we can choose a convenient length that will fit into the lab (say 50 ft), which still leaves two "arbitrary" unknowns to specify. Since there are two fluid properties to specify (μ and ρ), this means that we can choose (arbitrarily) a different fluid for the lab test than the oil used in the field. Water is the most convenient, available, and inexpensive fluid, so if we specify water ($\mu = 1$ cP, $\rho = 1$ g/cm^3) we have used up all our "arbitrary" choices. The remaining two unknowns, Q_m and $(\Delta P)_f$, are determined by the two remaining equations. From Eqn (2-12):

$$Q_m = Q_f\left(\frac{\rho_f}{\rho_m}\right)\left(\frac{D_m}{D_f}\right)\left(\frac{\mu_m}{\mu_f}\right) = 10^6 \frac{\text{bbl}}{\text{day}}\left(\frac{0.85}{1.0}\right)\left(\frac{1.6}{48}\right)\left(\frac{1.0}{30}\right) = 944 \text{ bbl/day}$$

or

$$Q_m = \left(\frac{944 \text{ bbl}}{\text{day}}\right)\left(\frac{42 \text{ gal}}{\text{bbl}}\right)\left(\frac{1}{1440 \text{ min/day}}\right) = 27.5 \text{ gal/min}$$

Note that if the same units are used for the variables in both the model and the field, no conversion factors are needed because only ratios are involved.

Now our experiment has been designed: We use plastic pipe with an inside diameter of 1.6 in., 50 ft long, and pump water through it at a rate of 27.5 gpm. Then we measure the pressure drop through this pipe and use the final equation to scale up to find the field pressure drop. Let us say that the measured pressure drop with this system in the lab is 1.2 psi, then the pressure drop in the field pipeline, from Eqn (2-13), would be

$$(\Delta P)_f = (\Delta P)_m \left(\frac{D_m}{D_f}\right)^5 \left(\frac{L_m}{L_f}\right) \left(\frac{\rho_f}{\rho_m}\right) \left(\frac{Q_f}{Q_m}\right)^2$$

$$= (1.2 \text{ psi}) \left(\frac{1.6}{48}\right)^5 \left(\frac{700 \times 5280}{50}\right) \left(\frac{0.85}{1.0}\right) \left(\frac{10^6}{944}\right)^2 = 3480 \text{ psi}$$

This total pressure driving force would probably not be produced by a single pump, but would be apportioned among several pumps spaced along the pipeline.

This example illustrates the power of dimensional analysis as an aid in experimental design and the scale-up of resulting measurements. We have actually determined the pumping requirements for a large pipeline by applying the results of dimensional analysis, judicially selecting unspecified laboratory conditions to establish the design of a laboratory test model which simulates the field pipeline, making one measurement in the lab, and scaling up the result. We have not used any basic principles or engineering correlations other than the conservation of dimensions and the exercise of logical judgment. However, we shall see later that information is available to us, based upon similar experiments that have been conducted by others (and presented in dimensionless form), that we can use to solve problems like this without conducting any additional experiments.

VI. DIMENSIONLESS GROUPS IN FLUID MECHANICS

Table 2-2 lists a number of dimensionless groups that are commonly encountered in fluid mechanics problems. The name of the group, its symbol, definition, significance, and most common area of application are listed in the table. Wherever it is feasible, it is desirable to express basic relations (either theoretical or empirical) in dimensionless form, with the variables being dimensionless groups, since this represents the most general way of presenting results, which is independent of scale or specific system properties. We shall follow this guideline so far as is practical in this book.

VII. ACCURACY AND PRECISION

At this point, we digress slightly to make some observations about the accuracy and precision of experimental data. Since we, as engineers, continuously make use of data which represent measurements of various quantities, it is important that we understand and appreciate which of the numbers we deal with are useful and which are not.

First of all, we should make a clear distinction between accuracy and precision. *Accuracy* is a measure of how close a given value is to the "true" value, whereas *precision* is a measure of the uncertainty in the value or how "reproducible" the value is. For example, if we were to measure the width of a piece of paper using a ruler, we might find that it is 21.5 cm, give or take 0.1 cm. The "give or take" (i.e. the uncertainty) value of 0.1 cm is the precision of the measurement, which is determined by how close we are able to reproduce measurement with the ruler. However, it is possible that the ruler, when compared with a "standard" unit of measure, might be in error by, say, 0.2 cm. Thus the "accuracy" of the ruler is limited, which contributes to the uncertainty of our measurement, although we may not know what this limitation is unless we can compare our "instrument" to one which we know to be true.

Thus, the accuracy of a given value may be difficult to determine, although the precision of a measurement can be determined by a direct evaluation of reproducibility if multiple repetitions of the measurement are made. Unfortunately, when using values or data provided by others from handbooks, textbooks, journals, etc., we do not usually have access to either the "true" value or information on the reproduciblity of the measured values. However, we can make use of both common sense (i.e. reasonable judgment) and convention to estimate the implied precision of a given value. The number of decimal places (or the number of digits, when represented in scientific notation) should be indicative of the precision of a given number. For example, if the mileage from Dallas to Houston is stated as being 250 miles, and we drive at 60 mi/hr, should we say that it would take us 4.166667 (= 250/60) hours for the trip? This number implies that we can determine the answer to a precision of 0.0000005 hr, which is one part in 10^7, or less than 2 milliseconds! This is obviously ludicrous, since the mileage value is nowhere near that precise (is it ± 1 mile, ± 5 miles?— *exactly* where did we start and end?), nor can we expect to drive at a speed having this degree of precision (e.g. 0.000005 mph, or about 20 μm/s!).

It is conventional to assume that the precision of a given number is comparable to the magnitude of the smallest digit in that number. That is, we assume that the value of 250 miles implies 250 $\pm$ 1 mile (or perhaps $\pm$ 0.5 mph). However, unless the values are always given in scientific notation, so that the smallest digit can be associated with a specific decimal place, there will be considerable uncertainty, in which case common sense (judgment) should prevail. For example, if the diameter of a tank is specified to be 10.32 ft, we should assume that that value has a precision (or uncertainty) of about 0.005 ft (or 0.06 in., or 1.5 mm). However, if the diameter is said to be 10 ft, the number of digits cannot provide a guide to the precision of the number. It is unlikely that a tank of that size would be constructed to the precision of 1.5 mm, so we would probably assume the uncertainty to be about 0.5 in., or "roughly 10.0 ft". On the other hand, if I say that I have 5 fingers on my hand, this means

TABLE 2-2 Dimensionless Groups in Fluid Mechanics

Name	Symbol	Formula	Notation	Significance	Application
Archimedes number	N_{Ar}	$N_{Ar} = \dfrac{\rho_f g \Delta \rho d^3}{\mu^2}$	ρ_f = fluid density	(Bouyant $\times$ Inertial)/ (Viscous) forces	Settling particles, Fluidization
Bingham number	N_{Bm}	$N_{Bm} = \dfrac{\tau_0 D}{\mu_\infty V}$	τ_0 = yield stress μ_∞ = limiting viscosity	(Yield/Viscous) stresses	Flow of Bingham plastics
Bond number	N_{Bo}	$N_{Bo} = \dfrac{\Delta \rho d^2 g}{\sigma}$	σ = surface tension	(Gravity/Surface tension) forces	Rise or fall of drops or bubbles
Cauchy number	N_C	$N_C = \dfrac{\rho V^2}{K}$	K = bulk modulus	(Inertial/Compressibility) forces	Compressible flow
Euler number	N_{Eu}	$N_{Eu} = \dfrac{\Delta P}{\rho V^2}$	ΔP = pressure drop in pipe	(Pressure energy/Kinetic energy)	Flow in closed conduits

Name	Symbol	Formula	Definitions	Physical meaning	Application
Drag coefficient	C_D	$C_D = \dfrac{F_d}{\frac{1}{2}\rho V^2 A}$	F_d = drag force A = cross sectional area normal to flow	(Drag stress)/(1/2 Momentum flux)	External flows
Fanning (Darcy) friction factor	$f(f_D)$	$f = \dfrac{e_f D}{2V^2 L}$ $f_D = 4f$ or $f = \tau_w/(\rho V^2/2)$	e_f = friction loss (energy/mass) τ_w = wall stress	(Energy dissipated)/(KE of flow $\times$ 4L/D) or (Wall Stress/Momentum Flux)	Flow in pipes, channels, fittings, etc.
Froude number	N_{Fr}	$N_{Fr} = \dfrac{V^2}{gL}$	L = characteristic length	(Inertial/Gravity) forces	Free surface flows
Hedstrom number	N_{He}	$N_{He} = \dfrac{\tau_0 D^2 \rho}{\mu_\infty^2}$	τ_0 = yield stress μ_∞ = limiting viscosity	(Yield $\times$ Inertia/Viscous) stresses	Flow of Bingham plastics
Mach number	N_{Ma}	$N_{Ma} = \dfrac{V}{c}$	c = speed of sound	(Gas velocity/Speed of sound)	High-speed compressible flow

exactly 5, and no more or less (i.e. an "infinite" number of "significant digits"). In general, the number of decimal digits which are included in reported data, or the precision to which values can be read from graphs or plots, should serve as a guide to the actual precision of the data. Therefore, answers calculated from data with limited precision will likewise be limited in precision (the computer people have an acronym for this—"GIGO", which stands for "Garbage In, Garbage Out"). When the actual precision of data or given values is uncertain, a general rule of thumb is to report the results of calculations to no more than three digits, which corresponds to an uncertainty of about 0.1 to 1% (this is actually better precision than can be justified by most engineering measurements). Inclusion of more than three digits in your answer implies a greater precision than this and must be justified. Those who report values with a large number of digits which cannot be justified are usually making the implied statement "I just wrote down the numbers—I really didn't think about it." This is most unfortunate, because if these people don't think about the numbers they write down, can we be assured that they are thinking about other critical aspects of the problem?

Example 2-4: Our vacation time accrues by the hour, a certain number of hours of vacation time being credited per month worked. When we request leave or vacation, we are likewise expected to report it in increments of one hour. We received a statement from the accountants that we have accrued "128.00 hours of vacation time". What is the precision of this number?

The precision is implied by the smallest number to the right of the decimal point, i.e. 0.005 hr, or approximately 20 s. Does this imply that we must report leave taken to the closest 20 s? (We think not. It takes at least a minute to fill out the leave request form—would this time be charged against our accrued leave? The accountant just "wasn't thinking" when the numbers were reported.)

When combining values, each of which has a finite precision or uncertainty, it is important to be able to estimate the corresponding uncertainty of the result. Although there are various "rigorous" ways of doing this, a very simple method which gives good results as long as the relative uncertainty is a small fraction of the value is to use the approximation (which is really just the first term of a Taylor series expansion):

$$A(1 \pm a)^x \cong A(1 \pm xa + \cdots)$$

which is valid for any value of x, provided $a \leq 0.1$ (about). This assumes that the relative uncertainty of each quantity is expressed as a fraction of the given value, e.g. the fractional uncertainty in the value A is a or, equivalently, the percentage error in A is 100 a.

Example 2-5: Suppose we wish to calculate the shear stress on the bob surface in a cup-and-bob viscometer, from a measured value of the torque or moment on the bob. The equation for this is

$$\tau_{r\theta} = \frac{T}{2\pi R_i^2 L}$$

If the torque (T) can be measured to ±5%, the bob radius (R_i) is known to ± 1%, and the length (L) is known to 3%, the corresponding uncertainty in the shear stress is determined as follows:

$$\tau_{r\theta} = \frac{T(1 \pm 0.05)}{2\pi R_i^2 (1 \pm 0.01)^2 L(1 \pm 0.03)}$$

$$= \frac{T}{2\pi R_i^2 L} [1 \pm (0.05) \pm (2)(0.01) \pm (0.03)]$$

$$= \frac{T}{2\pi R_i^2 L} (1 \pm 0.1)$$

That is, there would be a 10% error, or uncertainty, in the answer. Note that even though terms in the denominator have a negative exponent, the maximum error due to these terms is still cumulative, since a given error may be either positive or negative, i.e. errors may either accumulate (giving rise to the maximum possible error) or cancel out (we should be so lucky!).

PROBLEMS

UNITS AND DIMENSIONS

1. The following formula for the pressure drop through a valve was found in a design manual:

$$h_L = \frac{522 \, K \, q^2}{d^4}$$

where h_L = the "head loss" in feet of fluid flowing through the valve
   K = dimensionless resistance coefficient for the valve
   q = flow rate through the valve, in ft^3/s
   d = diameter of the valve, in inches
   (a) Can this equation be used without changing anything if SI units are used for the variables? Explain.
   (b) What are the dimensions of "522" in this equation? What are its units?

(c) Determine the pressure drop through a 2-in. valve with a K of 4 for water at 20°C flowing at a rate of 50 gpm (gallons/min), in units of: (1) feet of water; (2) psi; (3) atm; (4) Pa; (5) dyn/cm^2; and (6) inches of mercury.

2. The energy balance on the fluid in a stream tube is known as Bernoulli's equation and can be written

$$\frac{\Delta P}{\rho} + g\Delta z + \frac{1}{2}\Delta(V^2) + e_f + w = 0$$

where $-w$ is the work done on a unit mass of fluid and e_f is the energy dissipated by friction in the fluid per unit mass, including all thermal energy effects due to heat transfer or internal generation. If $P_1 = 25$ psig, $P_2 = 10$ psig, $z_1 = 5$ m, $z_2 = 8$ m, $V_1 = 20$ ft/s, $V_2 = 5$ ft/s, $\rho = 62.4$ lb_m/ft^3, and $w = 0$, calculate the value of e_f in terms of each of the following systems of units:
   (a) SI
   (b) mks engineering (metric engineering)
   (c) English engineering
   (d) English scientific (with M as a fundamental dimension)
   (e) English thermal units (e.g. BTU)
   (f) Metric thermal units (e.g. calories)

3. Determine the weight of a 1 g mass at sea level in units of:
   (a) dynes
   (b) lb_f
   (c) g_f
   (d) poundals

4. One cubic foot of water weighs 62.4 lb_f under standard gravity.
   (a) What is its weight in dynes, poundals, and g_f?
   (b) What is its density in lb_m/ft^3 and $slugs/ft^3$?
   (c) What is its weight on the moon (g = 6 ft/s²) in lb_f?
   (d) What is its density on the moon?

5. You have purchased a 5 oz bar of gold (100% pure), at a cost of $400/oz. Because the bar was weighed in air, you conclude that you got a bargain, since its true mass is greater than 5 oz due to the buoyancy of air. If the true density of the gold is 1.9000 g/cm^3, what is the actual value of the bar based upon its true mass?

6. You purchased 5 oz of gold in La Paz, Bolivia ($g = 977.110$ cm/s^2) for \$400/oz. You then took the gold and the same spring scale on which you weighed it in La Paz to Reykjavik, Iceland ($g = 983.06$ cm/s^2), where you weighed it again, and sold it for \$400/oz. Did you make money, lose money, or break even? If you made or lost money, how much?

7. You weigh a body with a mass M on an electronic scale, which is calibrated with a known mass.
   (a) What does the scale actually measure, and what are its dimensions?
   (b) If the scale is calibrated in the appropriate system of units, what would the scale reading be if the mass of M is: (1) 1 slug; (2) 1 lb$_m$ (in scientific units); (3) 1 lb$_m$ (in enginering units); (3) 1 g$_m$ (in scientific units); (4) 1 g$_m$ (in engineering units)?

8. The acceleration due to gravity on the moon is about 5.4 ft/s^2. If your weight is 150 pounds on earth:
   (a) What is your mass, in slugs, on the moon?
   (b) What is your weight on the moon, in SI units?
   (c) What is your weight on earth, in poundals?

CALCULATIONS, CONVERSION FACTORS

9. Determine the value of the Reynolds number for SAE 10 lube oil flowing at 100°F at 2000 gpm through a 10 in. Schedule 40 pipe. The oil SG is 0.92, and its viscosity can be found in Appendix A. If the pipe is made of commercial steel ($\varepsilon = 0.0018$ in.), use the Moody diagram (Fig. 6-4, p. 149) to determine the friction factor, f, for this system. Estimate the precision of your answer, based upon the information and procedure you used to determine it (i.e. give what you determine should be reasonable upper and lower bounds, or the corresponding percentage variation, in the value of f based on the information you used).

10. The Peclet number (N_{Pe}) is defined as

$$N_{Pe} = N_{Re} N_{Pr} = \left(\frac{DV\rho}{\mu}\right)\left(\frac{c_p\mu}{k}\right) = \frac{DGc_p}{k}$$

where D = pipe diameter, G = mass velocity = ρV, c_p = specific heat, k = thermal conductivity, μ = viscosity. Calculate the value of N_{Pe}

for water at 60°F flowing through a 1 cm diameter tube at a rate of 100 lb_m/hr. (Use the most accurate data you can find, and state your answer in the appropriate number of digits.)

11. The heat transfer coefficient (h) for a vapor bubble rising through a boiling liquid is given by

$$h = A\left[\frac{kV\rho c_p}{d}\right]^{1/2} \quad \text{where } V = \left[\frac{\Delta\rho g\sigma}{\rho_v^2}\right]^{1/4}$$

where h = heat transfer coefficient [e.g. BTU/(hr °F ft^2)]
  c_p = liquid heat capacity [e.g. cal/(g °C)]
  k = liquid thermal conductivity [e.g. J/(s °R m)]
  σ = liquid-vapor surface tension [e.g. dyn/cm]
  $\Delta\rho = \rho_{liquid} - \rho_{vapor} = \rho_l - \rho_v$
  d = bubble diameter
  g = acceleration due to gravity
  (a) What are the fundamental dimensions of V and A?
  (b) If the value of h is 1000 BTU/(hr ft^2 °F) for a 5 mm diameter steam bubble rising in boiling water at atmospheric pressure, determine the values of V and A in SI units. You must look up values for the other quantities you need, and be sure to cite the sources you use for these data.

12. Determine the value of the gas constant, R, in units of ft^3 atm/(lb mol °R), starting with the value of the standard molar volume of a perfect gas.

13. The conditions at two different positions along a pipeline (at points 1 and 2) are related by the Bernoulli equation

$$\frac{P_2 - P_1}{\rho} + g(Z_2 - Z_1) + \frac{\alpha}{2}(V_2^2 - V_1^2) + e_f + w = 0$$

where P is the pressure, Z is the elevation, V is the velocity at the reference points, $-w$ is the work done on the fluid, and e_f is the energy lost by friction in the fluid between the two reference points. For flow in a pipe,

$$e_f = \left(\frac{4fL}{D}\right)\left(\frac{V^2}{2}\right)$$

where D is the pipe diameter and L is the pipe length between points 1 and 2. If the flow is laminar ($N_{Re} < 2000$) the value of α is 2 and

$f = 16/N_{Re}$, but for turbulent flow in a smooth pipe α is 1 and $f = 0.0791/N_{Re}^{1/4}$. The work done by a pump on the fluid is related to the power delivered to the fluid (H_P) and the mass flow rate of the fluid ($\dot{m}$) by $H_P = -w\dot{m}$.

Consider water ($\rho = 1$ g/cm^3, $\mu = 1$ cP) being pumped at a rate of 150 gpm through a 2000 ft long, 3 in. diameter pipe. The water is transported from a reservoir ($Z = 0$) at atmospheric pressure to a condenser at the top of a column which is at an elevation of 30 ft and a pressure of 5 psig. Determine:

(a) The value of the Reynolds number in the pipe
(b) The value of the friction factor in the pipe (assuming that it is smooth)
(c) The power that the pump must deliver to the water, in hp (horsepower)

14. Calculate the value of the Reynolds number for sodium flowing at a rate of 50 gpm through a 1/2 in. I.D. tube, at 400°F.

DIMENSIONAL ANALYSIS AND SCALE-UP

15. A simple pendulum consists of a small, heavy ball of mass M on the end of a long string of length L. The period of the pendulum should depend upon these factors, as well as gravity, since that is the driving force for making it move. What information can you get about the relationship between these variables from a consideration of their dimensions? Suppose you measured the period, T_1, of a pendulum with mass M_1 and length L_1. Could you use this to determine the period of a different mass and length? What would be the ratio of the pendulum period on the moon to that on the earth? How could you use a pendulum to determine the variation of g on the earth's surface?

16. In the steady flow of a Newtonian fluid through a long uniform circular tube, if $N_{Re} < 2000$ the flow is laminar and the fluid elements move in smooth, straight, parallel lines. Under these conditions, it is known that the relation between the flow rate and the pressure drop in the pipe does not depend upon the fluid density or the pipe wall material.

(a) Perform a dimensional analysis of this system to determine the dimensionless groups that apply. Express your result in a form in which the Reynolds number can be identified.

(b) If water is flowing at a rate of 0.5 gpm through a pipe with an I.D. of 1 in., what is the value of the Reynolds number? If the diameter is doubled at the same flow rate, what will be the effect on the Reynolds number and on the pressure drop?

17. It is known that the power required to drive a fan depends upon the impeller diameter D; the impeller rotational speed ω; the fluid density ρ; and the volume flow rate Q. (Note that the fluid viscosity is not important for gases under normal conditions.)
    (a) What is the minimum number of fundamental dimensions required to define all of these variables?
    (b) How many dimensionless groups are required to determine a relation between the power and all the other variables? Find these groups by dimensional analysis, and arrange the results so that the power and the flow rate each appear in only one group.

18. A gas bubble with diameter D rises with a velocity V in a liquid of density ρ and viscosity μ.
    (a) Determine the dimensionless groups that include the effects of all of the significant variables, in such a form that the liquid viscosity appears in only one group. Note that the driving force for the bubble motion is buoyancy, which is equal to the weight of the displaced fluid.
    (b) You want to know how fast a 5 mm diameter air bubble will rise in a liquid with a viscosity of 20 cP and a density of 0.85 g/cm^3. You want to simulate this system in the laboratory using water ($\mu = 1$ cP, $\rho = 1$ g/cm^3) and air bubbles. What size air bubble should you use?
    (c) You perform the experiment, and measure the velocity of the air bubble in water (V_m). What is the ratio of the velocity of the 5 mm bubble in the field liquid (V_f) to that in the lab (V_m)?

19. Perform a dimensional analysis to determine the groups which relate the variables which are important in determining the settling rate of solid particles falling in a liquid. Note that the driving force for moving the particles is gravity, which determines the *net* weight of the particle. At very slow settling velocities, it is known that the velocity is independent of the fluid density. Show that this also requires that the velocity be inversely proportional to the fluid viscosity.

20. You need to predict the performance of a large industrial mixer under various operating conditions. To obtain the necessary data, you decide to run a laboratory test on a small-scale model of the unit. You have deduced that the power (P) required to operate the mixer depends upon the following variables:

    Tank diameter D   Impeller diameter d
    Impeller rotational speed N   Fluid density ρ
    Fluid viscosity μ

    (a) Determine the minimum number of fundamental dimensions required by these variables, and the number of dimensionless groups that can be defined by them.
    (b) Find an appropriate set of dimensionless groups, such that D and N each appear in only one group. If possible, identify one or more of the groups with groups commonly encountered in other systems as well.
    (c) You need to know how much power would be required to run a mixer in a large tank, 6 ft in diameter, using an impeller with a diameter of 3 ft operating at a speed of 10 rpm, containing a fluid with a viscosity of 25 cP and a specific gravity of 0.85. To do this, you run a lab test on a model of the system, using a scale model of the impeller which is 10 in. in diameter. The only appropriate fluid that you have in the lab has a viscosity of 15 cP and a specific gravity of 0.75. Can this fluid be used for the test? Explain.
    (d) If the above lab fluid is used, what size tank should be used in the lab, and how fast should the lab impeller be rotated?
    (e) With the lab test properly designed and the proper operating conditions chosen, you run the test and find that it takes 150 W to operate the lab test model. How much power would be required to operate the larger field mixer under the plant operating conditions?

21. When an open tank with a free surface is stirred with an impeller, a vortex will form around the shaft. It is important to prevent this vortex from reaching the impeller, since entrainment of air in the liquid tends to cause foaming. The shape of the free surface depends upon (among other things) the fluid properties, the speed and size of the impeller, the size of the tank, and the depth of the impeller below the free surface.

    (a) Perform a dimensional analysis of this system to determine an appropriate set of dimensionless groups which can be used to describe the system performance. Arrange the groups so that the impeller speed appears in only one group.

(b) In your plant you have a 10 ft diameter tank containing a liquid which is 8 ft deep. The tank is stirred by an impeller which is 6 ft in diameter, located 1 ft from the tank bottom. The liquid is Newtonian, with a viscosity of 100 cP and a specific gravity of 1.5. You need to know the maximum speed at which the impeller can be rotated without entraining the vortex. To find this out, you design a laboratory test using a scale model of the impeller which is 8 in. in diameter. What, if any, limitations are there on your freedom to select a fluid for use in the lab test?

(c) Select an appropriate fluid for the lab test, and determine how large the tank used in the lab should be and the location of the impeller, using this fluid.

(d) The lab impeller is run at a speed where the vortex just reaches the impeller. What is the relation between this speed and that at which entrainment would occur in the tank in the plant?

22. The variables which are involved in the performance of a centrifugal pump include the fluid properties (μ and ρ), the impeller diameter (d), the casing diameter (D), the impeller rotational speed (N), the volumetric flow rate of the fluid (Q), the head (H) developed by the pump ($\Delta P = \rho g H$), and the power required to drive the pump (HP).

(a) Perform a dimensional analysis of this system to determine a set of dimensionless groups which would be appropriate to characterize the pump. Arrange the groups so that the fluid viscosity and the pump power each appear in only one group.

(b) You want to know what pressure a pump will develop with a liquid having a specific gravity of 1.4 and a viscosity of 10 cP, at a flow rate of 300 gpm. The pump has an impeller with a diameter of 12 in., which is driven by a motor running at 1100 rpm. (It is known that under these conditions the pump performance is independent of fluid viscosity unless the viscosity is greater than about 50 cP.) You want to run a lab test using a similarly scaled pump, with an impeller diameter of 6 in. and a 3600 rpm motor, which simulates the operation of the larger field pump. Should you use the same liquid in the lab as in the field, or can you use a different liquid? Why?

(c) If you use the same liquid, what flow rate should be used in the lab to simulate the operating conditions of the field pump?

(d) If the lab pump develops a pressure of 150 psi at the proper flow rate, what pressure will the field pump develop with the field fluid?

(e) What pressure would the field pump develop with water at a flow rate of 300 gpm?

23. A centrifugal pump with an 8 in. diameter impeller operating at a rotational speed of 1150 rpm requires 1.5 hp to deliver water at a rate of 100 gpm and a pressure of 15 psi. Another pump for water, which is geometrically similar but has an impeller diameter of 13 in., operates at a speed of 1750 rpm. Estimate the pump pressure, flow capacity, and power requirements of this second pump. Under these conditions, the performance of both pumps is independent of the fluid viscosity.

24. When a ship moves through the water, it causes waves. The energy and momentum in these waves must come from the ship and is manifested as a "wave drag" force on the ship. It is known that this drag force (F) depends upon the ship speed (V), the fluid properties (ρ, μ), the length of the waterline (L), and the beam width (W), as well as the shape of the hull, among other things. (There is at least one "other thing" that is important, which relates to the "wave drag", i.e. the energy which is required to create and sustain the waves from the bow and the wake. What is this additional variable?) Note that "shape" is a dimensionless parameter, which is implied by the requirement of geometric similarity. If two geometries have the same shape, the ratio of each corresponding dimension of the two will also be the same.

    (a) Perform a dimensional analysis of this system to determine a suitable set of dimensionless groups which could be used to describe the relationship between all of the variables. Arrange the groups such that viscous and gravitational parameters each appear in separate groups.

    (b) It is assumed that "wave drag" is independent of viscosity and that "hull drag" is independent of gravity. You wish to determine the drag on a ship having a 500 ft long waterline moving at 30 mph through seawater (SG = 1.1). You can make measurements on a scale model of the ship, 3 ft long, in a towing tank containing fresh water. What speed should be used for the model to simulate the wave drag and the hull drag?

25. You want to determine the thickness of the film when a Newtonian fluid flows uniformly down an inclined plane at an angle θ with the horizontal at a specified flow rate. To do this, you design a laboratory experiment from which you can scale up measurements to any other Newtonian fluid under corresponding conditions.
    (a) List all of the independent variables which are important in this problem, with their dimensions. If there are any variables which are not independent but act only in conjunction with one another, list only the net combination which is important.
    (b) Determine an appropriate set of dimensionless groups for this system, in such a way that the fluid viscosity and the plate inclination each appear in only one group.
    (c) Decide what variables you would choose for convenience, what variables would be otherwise specified, and what you would measure in the lab.

26. You would like to know the thickness of a syrup film as it drains at a rate of 1 gpm down a flat surface which is 6 in. wide and is inclined at an angle of 30° from the vertical. The syrup has a viscosity of 100 cP and a specific gravity of 0.9. In the laboratory, you have a fluid with a μ = 70 cP and SG = 1.0 and a 1 ft wide plane inclined at an angle of 45° from the vertical.
    (a) At what flow rate, in gpm, would the laboratory conditions simulate the specified conditions?
    (b) If the thickness of the film in the laboratory is 3 mm at the proper flow rate, what would the thickness of the film be for the 100 cP fluid at the specified conditions?

27. The size of liquid droplets produced by a spray nozzle depends upon the nozzle diameter, the fluid velocity, and the fluid properties (which may, under some circumstances, include the surface tension).
    (a) Determine an appropriate set of dimensionless groups for this system.
    (b) You want to know what size droplets will be generated by a fuel oil nozzle with a diameter of 0.5 mm at an oil velocity of 10 m/s. The oil has a viscosity of 10 cP, a specific gravity of 0.82, and a surface tension of 35 dyn/cm. You have a nozzle in the lab with a nozzle diameter of 0.2 mm which you want to use in a lab experiment to find the answer. Can you use the same fuel oil in the lab test as in the field? If not, why not?
    (c) If the only fluid is water, tell how you would design the lab experiment. Water has a viscosity of 1 cP and a specific gravity

of 1, but its surface tension can be varied by adding small amounts of additives which do not affect the viscosity or density.

(d) Determine what conditions you would use in the lab, what you would measure, and the relationship between this and the unknown droplet diameter.

28. An ethylene storage tank in your plant explodes. The distance that the blast wave travels from the blast site (R) depends upon the energy released in the blast (E), the density of the air (ρ), and time (t). Use dimensional analysis to determine:

    (a) The dimensionless group(s) that can be used to describe the relationship between the variables in the problem

    (b) The ratio of the velocity of the blast wave at a distance of 2000 ft from the blast site to the velocity at a distance of 500 ft from the site

    The pressure difference across the blast wave (ΔP) also depends upon the blast energy (E), the air density (ρ), and time (t). Use this information to determine:

    (c) The ratio of the blast pressure at a distance of 500 ft from the blast site to that at a distance of 2000 ft from the site.

29. You want to find the drag force due to the air on a new automobile design, at various speeds. To do this, you test a 1/30 scale model of the car in the lab. You must design an experiment whereby the drag force measured in the lab can be scaled up directly to find the force on the full-scale car at a given speed.

    (a) What is the minimum number of (dimensionless) variables which are required to completely define the relationship between all of the variables in the problem? Determine the appropriate variables (e.g. the dimensionless groups).

    (b) The only fluids that you have available in the lab are air and water. Could you use either one of these, if you wanted to? Why (or why not)?

    (c) Tell which of these fluids you would use in the lab, and then determine what the velocity of this fluid past the model car would have to be so that the experiment would simulate the drag on the full-scale car at 40 mph. If you decide that it is possible to use either one of the two fluids, determine the answer for both of them.

    (d) What is the relationship between the measured drag force on the model and the drag force on the full scale car? If the experiment is possible, determine this relationship for both air and water. Repeat for a speed of 70 mph.

(e) It turns out that for very high values of the Reynolds number, the drag force is independent of the fluid viscosity. Under these conditions, if the speed of the car doubles, by what factor does the power required to overcome wind drag change?

30. The power required to drive a centrifugal pump and the pressure that the pump will develop depend upon the size (diameter) and speed (angular velocity) of the impeller, the volumetric flow rate through the pump, and the fluid properties. However, if the fluid is not too viscous (e.g. over about 100 cP), the pump performance is essentially independent of the fluid viscosity.

    (a) Perform a dimensional analysis to determine the dimensionless groups which would be required to define the pump performance. Arrange the groups so that the power and pump pressure each appear in only one group.

    You have a pump with an 8 in. diameter impeller which develops a pressure of 15 psi and requires 1.5 hp to operate when running at 1150 rpm with water at a flow rate of 100 gpm. You also have a similar pump with a 13 in. diameter impeller, driven by a 1750 rpm motor.

    (b) If the second pump is to be operated under equivalent (similar) conditions to the first one, what should the flow rate be?

    (c) If this pump is operated at this flow rate, what pressure would it develop, and what power would be required to drive it when pumping water?

31. Determine the value of the Reynolds number for water flowing at a rate of 0.5 gpm through a 1 in. I.D. pipe. If the diameter of the pipe is doubled for the same flow rate, how much will each of the following change?

    (a) The Reynolds number
    (b) The pressure drop
    (c) The friction factor

32. The pressure drop for a fluid with a viscosity of 5 cP and a density of 0.8 g/cm^3, flowing at a rate of 30 g/s in a 50 ft long 1/4 in. diameter pipe is 10 psi. Use this information to determine the pressure drop for water at 60°F flowing at 0.5 gpm in a 2 in. diameter pipe. What is the value of the Reynolds number for each of these cases?

33. In a distillation column, vapor is bubbled through the liquid to provide good contact between the two phases. The bubbles are formed when the vapor passes through a hole (orifice) in a plate (tray) which is in contact with the liquid. The size of the bubbles depends upon the diameter of the orifice, the velocity of the vapor through the orifice, the viscosity and density of the liquid, and the surface tension between the vapor and the liquid.

    (a) Determine the dimensionless groups required to completely describe this system, in such a manner that the bubble diameter and the surface tension do not appear in the same group.

    You want to find out what size bubbles would be formed by a hydrocarbon vapor passing through a 1/4 in. orifice at a velocity of 2 ft/s, in contact with a liquid having a viscosity of 4 cP and a density of 0.95 g/cm³ (the surface tension is 30 dyn/cm). To do this, you run a lab experiment using air and water (surface tension 60 dyn/cm).

    (b) What size orifice should you use, and what should the air velocity through the orifice be?

    (c) If you design and run this experiment and find that the air bubbles are 0.1 in. diameter, what size will the vapor bubbles be in the organic fluid with the 1/4 in. orifice?

34. If the viscosity of the liquid is not too high (e.g. less than about 100 cP), the performance of many centrifugal pumps is not very sensitive to the fluid viscosity. You have a pump with an 8 in. diameter impeller, which develops a pressure of 15 psi and consumes 1.5 hp when running at 1150 rpm pumping water at a rate of 100 gpm. You also have a similar pump with a 13 in. diameter impeller, driven by a 1750 rpm motor, and you would like to know what pressure that pump would develop with water and how much power it would take to drive it.

    (a) If the second pump is to be operated under conditions similar to that of the first, what should the flow rate be?

    (b) When operated at this flow rate with water, what pressure should it develop, and what power would be required to drive it?

35. A concentrated slurry is prepared in an open 8 ft diameter mixing tank, using an impeller with a diameter of 6 ft located 3 ft below the surface. The slurry is non-Newtonian and can be described as a Bingham plastic with a yield stress of 50 dyn/cm², a limiting viscosity of 20 cP, and a density of 1.5 g/cm³. A vortex is formed above the impeller, and if the speed is too high the vortex can reach the

blades of the impeller, entraining air and causing problems. Since this condition is to be avoided, you need to know how fast the impeller can be rotated without entraining the vortex. To do this, you conduct a lab experiment using a scale model of the impeller which is 1 ft in diameter. You must design the experiment so that the critical impeller speed can be measured in the lab and scaled up to determine the critical speed in the larger mixer.

(a) List all of the variables that are important in this system, and determine an appropriate set of dimensionless groups.

(b) Determine the diameter of the tank that should be used in the lab and the depth below the surface that the impeller should be located.

(c) Should you use the same slurry in the lab model as in the field? If not, what properties should the lab slurry have?

(d) If the critical speed of the impeller in the lab system is ω (rpm), what is the critical speed of the impeller in the large tank?

36. You would like to know the thickness of a paint film as it drains at a rate of 1 gpm down a flat surface that is 6 in. wide and is inclined at an angle of 30° to the vertical. Assume that the paint is Newtonian, with a viscosity of 100 cP and a density of 0.9 g/cm³. You have data from the laboratory for the film thickness of a fluid having a viscosity of 70 cP and a density of 1 g/cm³ flowing down a plane 1 ft wide inclined at an angle of 45° to the vertical, at various flow rates.

(a) At what flow rate (in gpm) will the laboratory system correspond to the conditions of the other system?

(b) If the film thickness of the laboratory fluid is 3 mm at these conditions, what would the film thickness be for the other system?

3

Fluid Properties in Perspective

I. CLASSIFICATION OF MATERIALS AND FLUID PROPERTIES

What is a fluid? It isn't a solid—but, what is a solid? Perhaps it is easier to define these materials in terms of how they respond in a specific situation, such as *simple shear*, as illustrated in Fig. 3-1. We envision the material contained between two infinite parallel plates, the bottom one being fixed and the top one subject to an applied force parallel to the plate, which is free to move in its plane. The material is assumed to adhere to the plates, and its properties can be classified by the way the top plate responds when the force is applied.

The mechanical behavior of a material, and its corresponding mechanical or rheological[1] properties, may be defined in terms of the *shear stress* (τ_{yx}) (force per unit area) and the *shear strain* (γ_{yx}) (which is a relative displacement). These are related to the total force (F_x) on the plate and the displacement (U_x) of the plate as follows:

$$\tau_{yx} = F_x/A_y \tag{3-1}$$

$$\gamma_{yx} = U_x/h_y = \frac{du_x}{dy} \tag{3-2}$$

---

[1]Rheology is the study of the deformation and flow behavior of materials, both fluids and solids.

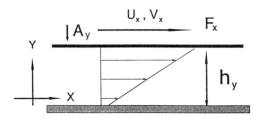

FIGURE 3-1   Simple shear.

The response of the material to this situation, in terms of how the shear stress and shear strain are related, defines the mechanical or *rheological* properties by which the material may be classified. It is noted that the shear stress has dimensions (and hence units) of force/area (Pa, dyn/cm², lb_f/ft^2, etc.) and that shear strain is dimensionless (with no units).

   If the material between the plates is, for example, a perfectly *rigid solid* (e.g. a brick), it will not move at all no matter how much force is applied (unless it breaks). Thus, the equation that defines this material is

$$\gamma_{yx} = 0 \tag{3-3}$$

   However, if the top plate moves a certain distance, which is proportional to the applied force, and then stops, the material is called a *linear elastic (Hookean) solid* (e.g. rubber). The equation which defines such a material is

$$\tau_{yx} = G\gamma_{yx} \tag{3-4}$$

where G is a constant called the *shear modulus.*

   On the other hand, if the displacement of the top plate is not directly proportional to the applied force (it may be either more or less than proportional to the force), the material is said to be a *nonlinear (non-Hookean) elastic solid.* It may be represented by an equation of the form

$$G = \tau_{yx}/\gamma_{yx} = fn(\tau \text{ or } \gamma) \tag{3-5}$$

where G is still the shear modulus but is no longer a constant. It is, instead, a *function* of how far the plate moves (γ_{yx}) or of the magnitude of the applied force (τ_{yx}), i.e. $G(\gamma)$ or $G(\tau)$. The particular form of the function depends upon the specific nature of the material.

   At the other extreme, if the molecules of the material are so far apart that they exert negligible attraction on each other (e.g. a low-pressure gas), the plate can be moved by the application of a negligible force, and the equation that describes this material is

$$\tau_{yx} = 0 \tag{3-6}$$

This material is called an *inviscid* (*Pascalian*) *fluid*. However, if the molecules do exhibit a significant attraction as they move past one another, such that the *relative rate* of movement (i.e. the velocity gradient) is proportional to the applied force (e.g. shear stress), the material is known as a *Newtonian fluid* and is described by

$$\tau_{yx} = \mu \dot{\gamma}_{yx} \tag{3-7}$$

where $\dot{\gamma}_{yx}$ is the *shear rate* or *rate of shear strain*:

$$\dot{\gamma}_{yx} = \frac{d\gamma_{yx}}{dt} = \frac{dv_x}{dy} = V_x/h_y \tag{3-8}$$

and μ is the fluid *viscosity*. Note that the viscosity is defined by Eqn (3-7), i.e. $\mu = \tau_{yx}/\dot{\gamma}_{yx}$, and has dimensions of Ft/L^2 [with corresponding units of Pa s, dyn $s/cm^2 = g/(cm\ s) = poise$, $lb_f\ s/ft^2$, etc.].

If the properties of the fluid are such that the shear stress and shear rate are not proportional but are related in some other fashion, the fluid is called *non-Newtonian*. In this case, the fluid viscosity may still be defined as $\tau_{yx}/\dot{\gamma}_{yx}$ but is not a constant. It is, instead, a function of either the shear rate or shear stress and is designated by η:

$$\eta = \frac{\tau_{yx}}{\dot{\gamma}_{yx}} = fn(\tau\ or\ \dot{\gamma}) \tag{3-9}$$

Most common fluids of simple structure are Newtonian (water, air, glycerin, oils, etc.). However, fluids with complex structure (polymer melts or solutions, suspensions, emulsions, foams, etc.) are generally non-Newtonian. Some very common examples are mud, paint, ink, mayonnaise, shaving cream, dough, mustard, and toothpaste.

Actually, there are fluids and solids which have both elastic (solid) properties and viscous (fluid) properties. These are called *viscoelastic* and are most notably materials composed of high polymers. The complete description of the rheological properties of these materials may involve a function relating stress and strain as well as their derivatives or integrals with respect to time. Since the elastic properties of these materials (fluids as well as solids) endow them with a tendency to return to a preferred configuration after being deformed, they are often termed "memory" materials and exhibit time-dependent properties.

This classification of material behavior is summarized in Table 3-1 (in which the subscripts have been omitted for simplicity). Since we are concerned with fluids, we will concentrate primarily on the flow behavior of Newtonian and non-Newtonian fluids. However, we will also illustrate some of the unique characteristics of viscoelastic fluids, such as the ability of solutions of certain high polymers to flow through pipes in turbulent flow with much less energy expenditure than the solvent alone.

TABLE 3-1 Classification of Materials

Purely elastic solids[a]			Viscoelastic fluids and solids (non-linear) $\tau = \text{fn}(\gamma, \dot{\gamma}, \ldots)$	Purely viscous fluids[b]		
Rigid solid (Euclidian)	Linear elastic solid (Hookean)	Nonlinear elastic solid (Non-Hookean)		Nonlinear viscous fluid (Non-Newtonian)	Linear viscous fluid (Newtonian)	Inviscid fluid (Pascalian)
$\gamma = 0$	$\tau = G\,\gamma$	$\tau = \text{fn}\,(\gamma)$		$\tau = \text{fn}\,(\dot{\gamma})$	$\tau = \mu\,\dot{\gamma}$	$\tau = 0$
	or	or		or	or	
	$G = \tau/\gamma$	$G = \tau/\gamma$		$\eta = \tau/\dot{\gamma}$	$\mu = \tau/\dot{\gamma}$	
	Shear modulus (constant)	Modulus (function of γ or τ)		Viscosity (function of $\dot{\gamma}$ or τ)	Viscosity (constant)	

[a] Elastic deformations store energy.
[b] Viscous deformations dissipate energy.

II. DETERMINATION OF FLUID VISCOUS (RHEOLOGICAL) PROPERTIES

As previously discussed, the flow behavior of fluids is determined by their rheological properties, which govern the relation between shear stress and shear rate. In principle, these properties could be determined by measurements in a "simple shear" situation, as illustrated in Fig. 3-1. One would put the "unknown" fluid in the gap between the plates, subject the upper plate to a specified velocity (V), and measure the required force (F) (or vice versa). The shear stress (τ) would be determined by F/A, the shear rate ($\dot{\gamma}$) is given by V/h, and the viscosity (η) is given by $\tau/\dot{\gamma}$. The experiment is repeated for different values of V (and the corresponding F is measured) to determine the viscosity at various shear rates (or shear stresses). However, this geometry is not convenient to work with, because it is hard to keep the fluid in the gap with no confining walls, and correction for the walls is not simple. There are more convenient geometries for measuring viscous properties, which we will describe. The working equations used to obtain viscosity from measured quantities will be given here, although the development of these equations will be delayed until after the appropriate fundamental principles have been discussed.

A. CUP-AND-BOB (COUETTE) VISCOMETER

As the name implies, this viscometer consists of two concentric cylinders, the outer "cup" and the inner "bob", with the test fluid in the annular gap (see Fig. 3-2). One cylinder (preferably the cup) is rotated at a fixed angular velocity (Ω). The force is transferred to the sample, causing it to deform, and is transferred by the fluid to the other cylinder (i.e. the bob). This force results in a torque (T) which can be measured by a torsion spring, for example. Thus, the known quantities are the radius of the inner bob (R_i), the radius of the outer cup (R_o), the length of surface in contact with the sample (L), and the measured annular velocity (Ω) and torque (T). From these quantities, we must determine the corresponding shear stress and shear rate to find the fluid viscosity. The shear stress is determined by a balance of moments on a cylindrical surface within the sample (at a distance r from the center) and the torsion spring, i.e.:

T = Force × Lever arm = Shear stress × Surface area × Radius

or

$$T = (\tau_{r\theta})(2\pi rL)(r)$$

Solving for the shear stress:

$$\tau_{r\theta} = \frac{T}{2\pi r^2 L} = \tau \tag{3-10}$$

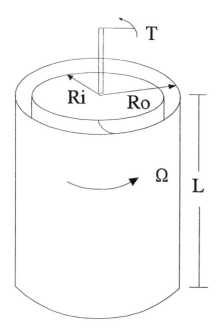

FIGURE 3-2 Cup-and-bob (Couette) viscometer.

Setting $r = R_i$ gives the stress on the bob surface, and $r = R_o$ gives the stress on the cup. If the gap is small [i.e. $(R_o - R_i)/R_i \leq 0.05$], the curvature can be neglected and the flow in the gap is equivalent to flow between parallel plates. In this case, the shear stress should be evaluated at the average r [i.e. $(R_i + R_o)/2$], and the average shear rate is given by

$$\dot{\gamma} = \frac{dV_\theta}{dr} \cong \frac{\Delta V}{\Delta r} = \frac{V_o - V_i}{R_o - R_i} = \frac{R_o \Omega}{R_o - R_i} = \frac{\Omega}{1 - R_i/R_o}$$

or

$$\dot{\gamma}_{r\theta} = \frac{dV_\theta}{dr} = \frac{\Omega}{1 - \beta} = \dot{\gamma} \tag{3-11}$$

where $\beta = R_i/R_o$. However, if the gap is not small, the shear rate must be corrected for the curvature in the velocity profile. This can readily be done by applying the following approximate expression for the shear rate at the bob (which is accurate to within 5% or so, e.g. Darby, 1985):

$$\dot{\gamma} = \frac{2\Omega}{n'(1 - \beta^{2/n'})} \tag{3-12}$$

Here,

$$n' = \frac{d(\log T)}{d(\log \Omega)} \tag{3-13}$$

is the point slope of the log-log plot of T versus Ω, at the value of Ω (or T) in Eqn (3-12). Thus a series of data points of T versus Ω must be obtained in order to determine the value of the slope (i.e. n') at each point, which is needed to determine the corresponding values of the shear rate. The viscosity at each shear rate (or shear stress) is then determined by dividing the shear stress at the bob (Eqn 3-10) by the shear rate at the bob (Eqn 3-12), for each data point.

Example 3-1: The following data were taken in a cup-and-bob viscometer with a bob radius of 2 cm, a cup radius of 2.05 cm, and a bob length of 15 cm. Determine the viscosity of the sample and the equation for the model which best represents this viscosity.

Torque (dyne cm)	Speed (rpm)
2,000	2
3,500	4
7,200	10
12,500	20
20,000	40

The viscosity is the shear stress at the bob, as given by Eqn (3-10), divided by the shear rate at the bob, as given by Eqn (3-12). The value of n' in Eqn (3-12) is determined from the *point slope* of the log T vs log rpm plot, at each data point. Such a plot is shown in Fig. 3-3. The line through the data is the best fit of all data points by linear least squares (this is easily done on a spreadsheet) and has a slope of 0.77 (with an R^2 of 0.999). In general, if the data do not fall on a straight line on this plot, the point slope (tangent) at each data point must be determined, giving a different value of n' for each data point.

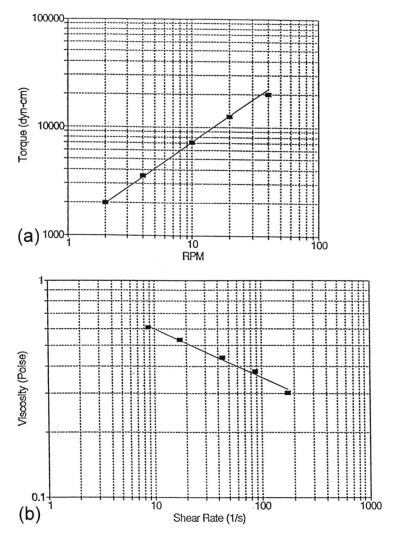

FIGURE 3-3 Examples of (a) cup-and-bob viscosity (torque vs. speed) and (b) viscosity vs. shear rate (lines are LMS best fit).

Using 0.77 for n' in Eqn (3-12) for the shear rate, and Eqn (3-10) for the shear stress, gives the following values:

Shear stress at bob (dyn/cm²)	Shear rate at bob (s⁻¹)	Viscosity (poise)
5.31	8.74	0.61
9.28	17.5	0.53
19.1	43.7	0.43
33.2	87.4	0.37
53.1	175	0.31

The plot of viscosity vs shear rate is shown, in which the line represents Eqn (3-24), with m = 1.01 "poise", and n = 0.77. In this case, the power law model represents the data quite well over the entire range of shear rate, so that n = n' is the same for each data point. If this were not the case, the *local* slope of log T vs log rpm would determine a different value of n' for each data point, and the power law model would not give the best fit over the entire range of shear rate. The shear rate and viscosity would still be determined as above (using the local value of n' for each data point), but the viscosity curve would be best fit by some other model, depending upon the trend of the data.

B. TUBE FLOW (POISEUILLE) VISCOMETER

Another common method of determining viscosity is by measuring the pressure drop ($\Delta\Phi = \Delta P + \rho g\,\Delta z$) and flow rate (Q) in steady laminar flow through a uniform circular tube of length L and diameter D (this is called *Poiseuille flow*). This can be done by using pressure taps through the wall to measure the pressure difference or by measuring the total pressure drop from a reservoir to the end of the tube, as illustrated in Fig. 3-4. The latter is more common, since very small diameter tubes are usually used, but this arrangement requires that correction factors be applied for the static head of the fluid in the reservoir and the pressure loss from the reservoir to the tube. As will be shown later, a momentum balance provides a relation between the shear stress at the tube wall (τ_w) and the measured pressure drop:

$$\tau_w = \frac{-\Delta\Phi}{4L/D} \tag{3-14}$$

The shear rate at the tube wall ($\dot{\gamma}_w$) is given by

$$\dot{\gamma}_w = \Gamma\left(\frac{3n' + 1}{4n'}\right) \tag{3-15}$$

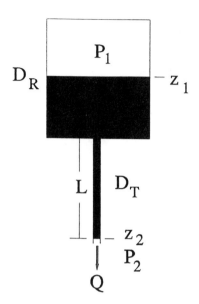

FIGURE 3-4 Tube flow (Poiseuille) viscometer.

where

$$\Gamma = \frac{32Q}{\pi D^3} = \frac{8V}{D} \qquad (3\text{-}16)$$

and

$$n' = \frac{d \log \tau_w}{d \log \Gamma} = \frac{d \log \Delta\Phi}{d \log Q} \qquad (3\text{-}17)$$

is the point slope of the log-log plot of $\Delta\Phi$ versus Q, evaluated at each data point. This n' is the same as that determined by the cup-and-bob viscometer, for a given fluid.

III. TYPES OF OBSERVED FLUID BEHAVIOR

When the measured values of shear stress or viscosity are plotted versus shear rate, various types of behavior may be observed, depending upon the fluid properties, as shown in Figs. 3-5 and 3-6. It should be noted that the shear stress and shear rate may both be either positive or negative, depending upon the direction of motion or the applied force, the reference frame, etc. (however, by our convention they are always the same sign). Since the viscosity must always be positive, the shear rate (or shear stress) argument in the viscosity function

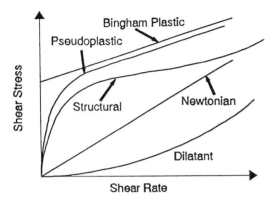

FIGURE 3-5 Shear stress vs shear rate for various fluids.

for a non-Newtonian fluid should be the absolute magnitude, regardless of the sign of the shear rate and shear stress.

A. NEWTONIAN FLUID

If the shear stress vs shear rate plot is a straight line through the origin (or a straight line with a slope of unity on a log-log plot), the fluid is Newtonian:

$$\text{Newtonian:} \quad \tau = \mu \dot{\gamma} \tag{3-18}$$

where μ is the viscosity.

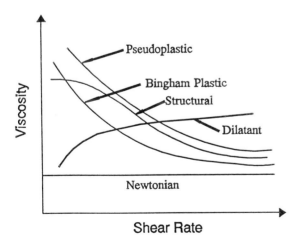

FIGURE 3-6 Viscosity vs shear rate for fluids in Fig. 3-5.

B. BINGHAM PLASTIC MODEL

If the data appear linear but do not extrapolate through the origin, intersecting instead at a shear stress value of τ_0, the material is called a Bingham plastic:

$$\text{Bingham plastic:} \quad \tau = \tau_0 + \mu_\infty \dot{\gamma} \qquad (3\text{-}19)$$

The *yield stress* τ_0, and the *high shear limiting (or plastic) viscosity* μ_∞ are the two rheological properties required to determine the flow behavior of a Bingham plastic. The viscosity function for the Bingham plastic is

$$\eta(\dot{\gamma}) = \frac{\tau_0}{\dot{\gamma}} + \mu_\infty \qquad (3\text{-}20)$$

or

$$\eta(\tau) = \frac{\mu_\infty}{1 - \tau_0/\tau} \qquad (3\text{-}21)$$

Since this material will not "flow" unless the shear stress exceeds the yield stress, these equations apply only when $|\tau| > \tau_0$. For smaller values of the shear stress, the material behaves as a rigid solid, i.e.,

$$\text{For } |\tau| < \tau_0: \quad \dot{\gamma} = 0 \qquad (3\text{-}22)$$

As is evident from Eqn (3-20) or (3-21), the Bingham plastic exhibits a *shear thinning* viscosity, i.e. the larger the shear stress or shear rate, the lower is its viscosity. This behavior is typical of many concentrated slurries and suspensions, such as muds, paints, foams, emulsions (e.g. mayonnaise), ketchup, blood, etc.

C. POWER LAW MODEL

If the data (either shear stress or viscosity) exhibit a straight line on a log-log plot, the fluid is said to follow the *power law* model, which can be represented as

$$\text{Power law:} \quad \tau = m\dot{\gamma}^n \qquad (3\text{-}23)$$

The two rheological viscous properties are m, the *consistency coefficient* and n, the *flow index*. The apparent viscosity function for the power law model in terms of shear rate is

$$\eta(\dot{\gamma}) = m\dot{\gamma}^{n-1} \qquad (3\text{-}24)$$

or, in terms of shear stress,

$$\eta(\tau) = m^{1/n}\tau^{(n-1)/n} \qquad (3\text{-}25)$$

It is evident that for n = 1, the power law fluid model reduces to a Newtonian fluid as a special case. If n < 1, the fluid is shear thinning (or *pseudoplastic*), and if n > 1 the model represents shear thickening (or *dilatant*) behavior. These characteristics are illustrated in Figs. 3-5 and 3-6. Most non-Newtonian fluids are shear thinning, whereas shear thickening behavior is relatively rare, being observed primarily for some concentrated suspensions of very small particles (e.g. starch suspensions) and some unusual polymeric fluids. The power law model is very popular for curve-fitting viscosity data for many fluids over one to three decades of shear rate. However, it is dangerous to extrapolate beyond the range of measurements using this model, since (for n < 1) it predicts an ever-increasing viscosity as the shear rate decreases, and an ever-decreasing viscosity as the shear rate increases, both of which are physically unrealistic.

D. STRUCTURAL VISCOSITY MODELS

A typical viscosity characteristic of many non-Newtonian fluids (e.g. polymeric fluids, flocculated suspensions, colloids, foams, gels) is illustrated by the curves labeled *structural viscosity* in Figs. 3-5 and 3-6. These fluids exhibit Newtonian behavior at very low and very high shear rates, with shear thinning or pseudo-plastic behavior at intermediate shear rates. In some materials this can be attributed to a reversible "structure" or network that forms in the "rest" or equilibrium state. When the material is sheared, the structure breaks down, resulting in a shear-dependent (shear thinning) behavior. Some actual examples of this type of behavior are shown in Fig. 3-7. As can be seen, structural viscosity behavior is exhibited by fluids as diverse as polymer solutions, blood, latex emulsions, and mud (sediment). Equations (i.e. models) that represent this type of behavior are discussed below.

1. Carreau Model

Structural viscosity behavior can be modeled by a variety of equations, including the Carreau model:

$$\text{Carreau:} \quad \eta(\dot{\gamma}) = \eta_\infty + \frac{\eta_0 - \eta_\infty}{[1 + (\lambda\dot{\gamma})^2]^p} \tag{3-26}$$

This model contains four rheological parameters: the low shear limiting viscosity, η_0; the high shear limiting viscosity, η_∞; a time constant, λ; and the shear thinning index, p. This is a very general viscosity model, and it can represent the viscosity function for a wide variety of materials. However, it may require data over a range of six to eight *decades* of shear rate to completely define the shape of the curve (and hence to determine all four parameters). As an example, Fig. 3-8 shows viscosity data for several polyacrylamide solutions over a range of about 10^6 in shear rate, with the lines through the data showing the Carreau model fit of the data. The corresponding values of the Carreau parameters for each of the curves are given in Table 3-2. In fact, over certain ranges of shear

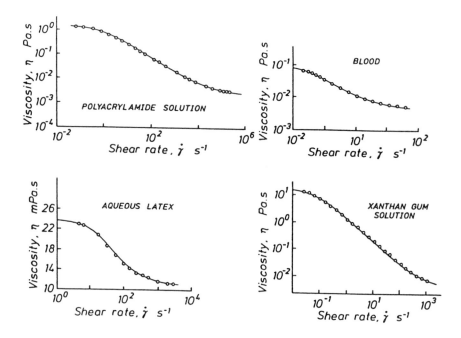

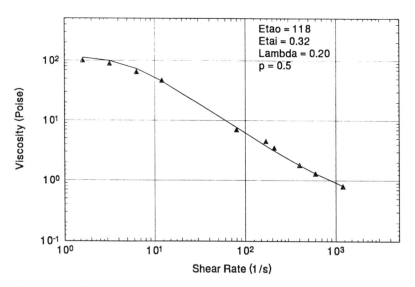

MISSISSIPPI MUD VISCOSITY
Mud Conc. 34.1%, SpGr = 1.42

Etao = 118
Etai = 0.32
Lambda = 0.20
p = 0.5

FIGURE 3-7   Some examples of structural viscosity behavior.

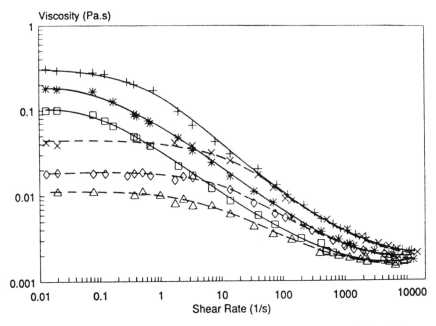

FIGURE 3-8 Viscosity data and Carreau model fit of polyacrylamide solutions (Darby and Pivsa-Art, 1991).

rate, the Carreau model reduces to various other popular models as special cases (including the Bingham plastic and power law models), as shown below.

Low to Intermediate Shear Rate Range—If $\eta_\infty \ll (\eta, \eta_0)$, the Carreau model reduces to a three-parameter model (η_0, λ, and p) which is equivalent to a power law model with a low shear limiting viscosity, which is called the *Ellis* model:

TABLE 3-2 Values of Carreau Parameters for Model Fit in Fig. 3-8

Solution	η_0 (Pa s × 10)	λ (s)	p	η_∞ (Pa s × 1000)
100 mg/kg (fresh)	1.113	11.89	0.266	1.30
250 mg/kg (fresh)	1.714	6.67	0.270	1.40
500 mg/kg (fresh)	3.017	3.53	0.300	1.70
100 mg/kg (sheared)	0.098	0.258	0.251	1.30
250 mg/kg (sheared)	0.169	0.167	0.270	1.40
500 mg/kg (sheared)	0.397	0.125	0.295	1.70

Source: Darby and Pivsa-Art (1991).

Ellis: $\qquad \eta(\dot\gamma) = \dfrac{\eta_0}{[1 + (\lambda\dot\gamma)^2]^p}$ (3-27)

Intermediate to High Shear Rate Range—If $\eta_0 \gg (\eta, \eta_\infty)$ and also $(\lambda\,\dot\gamma)^2 \gg 1$, the model reduces to the equivalent of a power law model with a high shear limiting viscosity, called the *Sisko* model:

Sisko: $\qquad \eta(\dot\gamma) = \eta_\infty + \dfrac{\eta_0}{(\lambda\dot\gamma)^{2p}}$ (3-28)

Although this appears to have four parameters, it is in reality a three-parameter model, since the combination $\eta_0/(\lambda)^{2p}$ is a single parameter, along with p and η_∞.
Intermediate Shear Rate Behavior—For $\eta_\infty \ll \eta \ll \eta_0$ and also $(\lambda\,\dot\gamma)^2 \gg 1$, the Carreau model reduces to the power law model:

Power law: $\qquad \eta(\dot\gamma) = \dfrac{\eta_0}{(\lambda\dot\gamma)^{2p}}$ (3-29)

where the power law parameters m and n are equivalent to the following combination of Carreau parameters:

$$m = \frac{\eta_0}{\lambda^{2p}}, \qquad n = 1 - 2p$$ (3-30)

Bingham Plastic Behavior—If the value of p is set equal to 1/2 in the Sisko model, the result is equivalent to the Bingham plastic model:

Bingham: $\qquad \eta(\dot\gamma) = \eta_\infty + \dfrac{\eta_0}{\lambda\,\dot\gamma}$ (3-31)

where the yield stress τ_0 is equivalent to η_0/λ, and η_∞ is the limiting (high shear) viscosity.

2. Meter Model

A stress-dependent viscosity model, which has the same general characteristics as the Carreau model, is the *Meter* model:

Meter: $\qquad \eta(\tau) = \eta_\infty + \dfrac{\eta_0 - \eta_\infty}{1 + (\tau/\sigma)^{2a}}$ (3-32)

where σ is a characteristic stress parameter, and a is the shear thinning index.

IV. TEMPERATURE DEPENDENCE OF VISCOSITY

All fluid properties are dependent upon temperature, and for most fluids the viscosity is the property that is most sensitive to temperature changes.

A. LIQUIDS

For liquids, as the temperature increases, the degree of molecular motion increases, reducing the short-range attractive forces between molecules and lowering the viscosity of the liquid. The viscosity of various liquids is shown as a function of temperature in Appendix A. For many liquids, this temperature dependence can be represented reasonably well by the Arrhenius equation:

$$\mu = A \exp(B/T) \tag{3-33}$$

where T is the absolute temperature. If the viscosity of a liquid is known at two different temperatures, this information can be used to evaluate the parameters A and B, which then permits the calculation of the viscosity at any other temperature. If the viscosity is known at only one temperature, this value can be used as a reference to establish the temperature scale on the graph in Appendix A-2, which can then be used to estimate the viscosity at any other temperature. Viscosity data for 355 liquids have been fit by Yaws et al. (1994) by the equation

$$\log_{10}\mu = A + B/T + CT + DT^2 \tag{3-34}$$

where T is in Kelvin, and the viscosity is in centipoise. The correlation parameters A, B, C, and D are given by Yaws et al. (1994) for each of these liquids.

For non-Newtonian fluids, any model parameter with the dimensions or physical significance of viscosity (the power law consistency m, the Carreau parameters η_∞ and η_0, etc.) will have a temperature dependence similar to that of a Newtonian fluid.

B. GASES

In contrast to the behavior of liquids, the viscosity of gases increases with temperature. This is because gas molecules are much farther apart so the short-range attractive forces are very small. However, as the temperature is increased, the molecular kinetic energy increases, resulting in a greater degree of interaction between the molecules and consequently a higher viscosity. The viscosity of gases is not as sensitive to temperature as that of liquids, however, and may often be represented by the equation:

$$\mu = aT^b \tag{3-35}$$

The parameters a and b may be evaluated from a knowledge of the viscosity at two different temperatures, and the equation can then be used to calculate the viscosity at any other temperature.

The value of the parameter b is often close to 1.5. In fact, if the viscosity (μ_1) of a gas is known at only one temperature (T_1), the following equation can be used to estimate the viscosity at any other temperature:

$$\mu = \mu_1 \left(\frac{T}{T_1}\right)^{3/2} \left(\frac{T_1 + 1.47\ T_B}{T + 1.47\ T_B}\right) \tag{3-36}$$

where the temperatures are in degrees Rankine and T_B is the boiling point of the gas.

V. DENSITY

In contrast to viscosity, the density of both liquids and gases decreases with increasing temperature, and the density of gases is much more sensitive to temperature than that of liquids. If the density of a liquid and its vapor are known at 60°F, the density at any other temperature can be estimated from the equation

$$\frac{(\rho - \rho_v)_T}{(\rho - \rho_v)_{60°F}} = \left(\frac{T_c - T}{T_c - 519.67}\right)^{1/N} \tag{3-37}$$

where the temperatures are in degrees Rankine and T_c is the critical temperature. The value of N is given in Table 3-3 for various gases.

The specific gravity of hydrocarbon liquids at 60°F is also often represented by the API gravity:

$$SG_{60°F} = \frac{141.5}{131.5 + °API} \tag{3-38}$$

For gases, if the temperature is well above the critical temperature and the pressure is below the critical pressure, the gas may usually be represented quite well by the ideal gas law:

$$\rho = \frac{PM}{RT} = \frac{M}{V_m}\left(\frac{T_{ref}}{T}\right)\left(\frac{P}{P_{ref}}\right) \tag{3-39}$$

where M is the gas molecular weight, temperatures and pressures are absolute, and V_m is the "standard molar volume" [22.4 m³/(kg mol) at 273 K and 1 atm,

TABLE 3-3 Parameter N to Be Used in Eqn (3-37)

Liquid	N
Water and alcohols	4
Hydrocarbons and ethers	3.45
Organics	3.23
Inorganics	3.03

359 ft^3/(lb mol) at 492°R and 1 atm, or 379.4 ft^3/(lb mol) at 520°R (60°F) and 1 atm]. The notation "SCF" (which stands for "standard cubic feet") is often used for hydrocarbon gases to represent the volume that would be occupied by the gas at 60°F and 1 atm pressure.

For other methods of predicting fluid properties and their temperature dependence, the reader is referred to the book by Reid et al. (1977).

REFERENCES

Barnes, H. A., J. F. Hutton, and K. Walters, *An Introduction to Rheology*, Elsevier (1989)

Darby, R., *J. Rheology*, **29**, 359 (1985)

Darby, R., and S. Pivsa-Art, *Canad. J. Chem. Eng.*, **69**, 1395 (1991)

Reid, R. C., J. M. Prausnitz, and T. K. Sherwood, *Properties of Gases and Liquids*, 3rd ed., McGraw-Hill (1977)

Yaws, C. L, X. Lin, and L. Bu, *Chem. Eng.*, p. 119, April (1994)

PROBLEMS

RHEOLOGICAL PROPERTIES

1. Using tabulated data for the viscosity of water and SAE 10 lube oil as a function of temperature, plot the data in a form which is consistent with each of the following equations:
    (a) $\mu = A \exp(B/T)$;
    (b) $\mu = aT^b$
    (c) Determine the values of A, B and a, b which give the best fit to the data for each fluid. (Note that T is absolute temperature.)

2. The viscosity of a fluid sample is measured in a cup-and-bob viscometer. The bob is 15 cm long with a diameter of 9.8 cm, and the cup has a diameter of 10 cm. The cup rotates, and the torque is measured on the bob. The following data were obtained:

Ω (rpm)	T (dyn cm)
2	3.6×10^5
4	3.8×10^5
10	4.4×10^5
20	5.4×10^5
40	7.4×10^5

    (a) Determine the viscosity of the sample.
    (b) What equation would be the most appropriate for describing the viscosity of this sample? Convert the above data to correspond-

ing values of viscosity versus shear rate, and plot it on appropriate axes consistent with the data and your equation. Use this plot to determine the values of each of the parameters which appear in your equation.

(c) What is the viscosity of this sample at a cup speed of 100 rpm in the viscometer?

3. A fluid sample is contained between two parallel plates, separated by a distance of 2 ± 0.1 mm. The area of the plates is 100 ± 0.01 cm^2. The bottom plate is stationary, and the top moves with a velocity of 1 cm/s when a force of 315 ± 25 dyn is applied to it, and at 5 cm/s with a force of 1650 ± 25 dyn.
   (a) Is the fluid Newtonian?
   (b) What is its viscosity?
   (c) What is the range of uncertainty for your answer to (b)?

4. The following materials exhibit flow properties which can be described by models which include a yield stress (e.g. Bingham plastic): (a) catsup; (b) toothpaste; (c) paint; (d) coal slurries; (e) printing ink. In terms of typical applications of these materials, describe how the yield stress influences their behavior, in contrast to how they would behave if they were Newtonian.

5. Starting with the equations for $\tau = \text{fn}(\dot{\gamma})$ that define the power law and Bingham plastic fluids, derive the equations for the viscosity functions for these models as a function of shear stress, i.e. $\eta = \text{fn}(\tau)$.

6. A paint sample is tested in a Couette (cup-and-bob) viscometer that has an outer radius of 5 cm, an inner radius of 4.9 cm, and a bob length of 10 cm. When the outer cylinder is rotated at a speed of 4 rpm, the torque on the bob is 0.0151 N m, and at a speed of 20 rpm it is 0.0226 N m.
   (a) What are the corresponding values of shear stress and shear rate for these two data points (in cgs units)?
   (b) What can you conclude about the viscous properties of the paint sample?
   (c) Which of the following models might be used to describe the paint: (1) Newtonian; (2) Bingham plastic; (3) power law ? Tell why.
   (d) Determine the values of the fluid properties (i.e. parameters) for any of the above models that could be used.

(e) What would the viscosity of the paint be at a shear rate of 500 s^{-1} (in poise)?

7. The quantities which are measured in a cup-and-bob viscometer are the rotation rate of the cup (rpm) and the corresponding torque transmitted to the bob. These quantities are then converted to corresponding values of shear rate ($\dot{\gamma}$) and shear stress (τ), which, in turn, can be converted to corresponding values of viscosity (η).
   (a) Show what a log-log plot of τ vs $\dot{\gamma}$ and η vs $\dot{\gamma}$ would look like for materials which follow the following models: (1) Newtonian; (2) power law (shear thinning); (3) power law (shear thickening); (4) Bingham plastic; (5) structural.
   (b) Show how the values of the parameters for each of these models can be evaluated from the respective plot of η vs $\dot{\gamma}$. That is, relate each of the model parameters to some characteristic or combination of characteristics of the plot such as the slope, specific values read from the plot, intersection of tangent lines, etc.

8. What is the difference between shear stress and momentum flux? How are they related? Illustrate each one in terms of the angular flow in the gap in a cup-and-bob viscometer, in which the outer cylinder (cup) is rotated, and the torque is measured at the stationary inner cylinder (bob).

9. A fluid is contained in the annulus in a cup-and-bob viscometer. The bob has a radius of 50 mm and a length of 10 cm and is made to rotate inside the cup by application of a torque on a shaft attached to the bob. If the cup inside radius is 52 mm and the applied torque is 0.03 ft lb_f, what is the shear stress in the fluid at the bob surface and at the cup surface? If the fluid is Newtonian, with a viscosity of 50 cP, how fast will the bob rotate (in rpm) with this applied torque?

10. A sample of a coal slurry is tested in a Couette (cup-and-bob) viscometer. The bob has a diameter of 10.0 cm and a length of 8.0 cm, and the cup had a diameter of 10.2 cm. When the cup is rotated at a rate of 2 rpm, the torque on the bob is 6.75×10^4 dyn cm, and at a rate of 50 rpm it is 2.44×10^6 dyn cm.
   (a) If the slurry follows the power law model, what are the values of the flow index and consistency coefficient?
   (b) If the slurry follows the Bingham plastic model, what are the values of the yield stress and the limiting viscosity?

(c) What would the viscosity of this slurry be at a shear rate of 500 s^{-1} as predicted by each of these models?

11. You must analyze the flow behavior of blood. Its measured viscosity is 6.49 cP at a shear rate of 10 s^{-1}, and 4.66 cP at a shear rate of 80 s^{-1}.
    (a) How would you describe the viscous properties of blood?
    (b) What would the viscosity of the blood be if it were subjected to a shear stress of 50 dyn/cm^2 if it is described by: (1) the power law model; (2) the Bingham plastic model? Which answer do you think would be better, and why?

12. Write equations which define each of the following laws: Fick's, Fourier's, Newton's, and Ohm's. What is the conserved quantity in each of these laws? Can you represent all of these laws by one general expression? If so, does this mean that all of the processes represented by these laws are always analogous? If they aren't, why not?

13. The following data were measured for the viscosity of a 500 ppm polyacrylamide solution in distilled water:

Shear rate (s^{-1})	Viscosity (cP)	Shear rate (s^{-1})	Viscosity (cP)
0.015	300	15	30
0.02	290	40	22
0.05	270	80	15
0.08	270	120	11
0.12	260	200	8
0.3	200	350	6
0.4	190	700	5
0.8	180	2,000	3.3
2	100	4,500	2.2
3.5	80	7,000	2.1
8	50	20,000	2

Find the model which best represents these data, and determine the values of the model parameters by fitting the model to the data. (This can most easily be done by trial and error, using a spreadsheet.)

14. Determine the viscosity model which best represents the following data, and the values of the parameters in the model. Show a plot of

the data together with the line which represents the model, to show how well the model works. (Hint: The easiest way to do this is by trial and error, fitting the model equation to the data using a spreadsheet.)

Shear rate (s^{-1})	Viscosity (poise)	Shear rate (s^{-1})	Viscosity (poise)
0.007	7745	20	270
0.01	7690	50	164
0.02	7399	100	113
0.05	6187	200	77.9
0.07	5488	500	48.1
0.1	4705	700	40.4
0.2	3329	1,000	33.6
0.5	2033	2,000	23.8
0.7	1692	5,000	15.3
1	1392	7,000	13.2
2	952	10,000	11.3
5	576	20,000	8.5
7	479	50,000	6.1
10	394	70,000	5.5

15. You would like to determine the pressure drop in a slurry pipeline. In order to do this, you need to know the rheological properties of the slurry. To evaluate these properties, you test the slurry by pumping it through a 1/8 in. I.D. tube which is 10 ft long. You find that it takes a 5 psi pressure drop to produce a flow rate of 100 cm³/s in the tube, and that a pressure drop of 10 psi results in a flow rate of 300 cm³/s. What can you deduce about the rheological characteristics of the slurry from these data? If it is assumed that the slurry can be adequately described by the power law model, what would be the values of the appropriate fluid properties (i.e. the flow index and consistency parameter) for the slurry?

16. A film of paint, 3 mm thick, is applied to a flat surface which is inclined to the horizontal by an angle θ. If the paint is a Bingham plastic, with a yield stress of 150 dyn/cm², a limiting viscosity of 65 cP, and a specific gravity of 1.3, how large would the angle θ have to be before the paint would start to run? At this angle, what would the shear rate be if the paint follows the power law model instead, with a flow index of 0.6 and a consistency coefficient of 215 (in cgs units)?

17. A thick suspension is tested in a Couette (cup-and-bob) viscometer that has a cup radius of 2.05 cm, a bob radius of 2.00 cm, and a bob length of 15 cm. The following data were obtained:

Cup speed (rpm)	Torque on bob (dyn cm)
2	2,000
4	6,000
10	19,000
20	50,000
50	1,500,000

What can you deduce about the viscous properties of this material and the best model to use to represent these data?

18. A sample of a viscous fluid is tested in a cup-and-bob viscometer that has a cup radius of 2.1 cm, a bob radius of 2.0 cm, and a bob length of 5 cm. When the cup is rotated at 10 rpm, the torque measured at the bob is 6000 dyn cm, and at 100 rpm the torque is 15,000 dyn cm.
    (a) What is the viscosity of this sample?
    (b) What can you conclude about the viscous properties of the sample?
    (c) If the cup is rotated at 500 rpm, what would be the torque on the bob and the fluid viscosity? Clearly explain any assumptions you make to answer this question, and tell how you might check the validity of these assumptions.

19. Acrylic latex paint can be described by the Bingham plastic model, with a yield stress of 112 dyn/cm^2, a limiting viscosity of 80 cP, and a density of 0.95 g/cm^3. What is the maximum thickness of this paint that can be applied to a vertical wall without running?

20. Santa Claus and his loaded sleigh are sitting on your roof, which is covered with snow. The sled's two runners each have a length L and width W, and the roof is inclined at an angle θ to the horizontal. The thickness of the snow between the runners and the roof is H. If the snow has properties of a Bingham plastic, derive an expression for the total mass (M) of the loaded sleigh at which it will just start to slide on the roof, if it is pointed straight downhill. If the actual mass is twice this minimum mass, determine an expression for the speed at which the sled will slide.

21. Using data from the Appendix for the viscosity of water and SAE 10 lube oil as a function of temperature, plot the data in a form which is consistent with each of the following equations:

$$\mu = Ae^{B/T}$$

and

$$\log \mu = a + bT$$

where T is absolute temperature. Use regression analysis (a spreadsheet works well for this) to determine the best values of A, B, a, and b, and which of these equations best fits the data. Plot the data and the equation on the same plot (for each equation), to illustrate the goodness of fit.

22. You must design a piping system to handle a sludge waste product. However, you don't know the properties of the sludge, so you test it in a cup-and-bob viscometer with a cup diameter of 10 cm, a bob diameter of 9.8 cm, and a bob length of 8 cm. When the cup is rotated at 2 rpm, the torque on the bob is 2.4×10^4 dyn cm, and at 20 rpm it is 6.5×10^4 dyn cm.
    (a) If you use the power law model to describe the sludge, what would the values of the flow index and consistency be?
    (b) If you use the Bingham plastic model instead, what would be the values of the yield stress and limiting viscosity?

23. You must determine the horsepower required to pump a coal slurry through an 18 in. diameter pipeline, 300 mi long, at a rate of 5 million tons/yr. The slurry can be described by the Bingham plastic model, with a yield stress of 75 dyn/cm², a limiting viscosity of 40 cP, and a density of 1.4 g/cm³. For non-Newtonian fluids, the flow is not sensitive to the wall roughness.
    (a) Determine the dimensionless groups that characterize this system. You want to use these to design a lab experiment, from which you can scale up measurements to find the desired horsepower.
    (b) Can you use the same slurry in the lab as in the pipeline?
    (c) If you use a slurry in the lab that has a yield stress of 150 dyn/cm², a limiting viscosity of 20 cP, and a density of 1.5 g/cm³, what size pipe and what flow rate (in gpm) should you use in the lab?
    (d) If you run the lab system as designed and measure a pressure drop of ΔP (psi) over a 100 ft length of pipe, show how you

would use this information to determine the required horsepower for the pipeline.

24. You want to determine how fast a rock will settle in mud, which behaves like a Bingham plastic. The first step is to perform a dimensional analysis of the system.
    (a) List the important variables which have an influence on this problem, with their dimensions (when listing these variables, give careful attention to the factors which cause the rock to fall), and determine the appropriate dimensionless groups.
    (b) Design an experiment in which you measure the velocity of a solid sphere falling in a Bingham plastic in the lab, and use the dimensionless variables to scale the answer to find the velocity of a 2 in. diameter rock, with a density of 3.5 g/cm^3, falling in a mud with a yield stress of 300 dyn/cm^2, a limiting viscosity of 80 cP, and a density of 1.6 g/cm^3. Should you use this same mud in the lab, or can you use a different material which is also a Bingham plastic but has a different yield stress and limiting viscosity?
    (c) If you use a suspension in the lab with a yield stress of 150 dyn/cm^2, a limiting viscosity of 30 cP, and a density of 1.3 g/cm^3, and a solid sphere, how big should the sphere be, and how much should it weigh?
    (d) If the sphere in the lab falls at a rate of 4 cm/s, how fast will the 2 in. diameter rock fall in the other mud?

25. A pipeline has been proposed to transport a coal slurry 1200 mi from Wyoming to Texas, at a rate of 50 million tons/year, through a 36 in. diameter pipeline. The coal slurry has the properties of a Bingham plastic, with a yield stress of 150 dyn/cm^2, a limiting viscosity of 40 cP, and a specific gravity of 1.5. You must conduct a lab experiment in which the measured pressure gradient can be used to determine the total pressure drop in the pipeline.
    (a) Perform a dimensional analysis of the system to determine an appropriate set of dimensionless groups to use (you may neglect the effect of wall roughness for this fluid).
    (b) For the lab test fluid, you have available a sample of the above coal slurry, and three different muds with the following properties:

	Yield stress (dyn/cm²)	Limiting viscosity (cP)	Density (g/cm³)
Mud #1	50	80	1.8
Mud #2	100	20	1.2
Mud #3	250	10	1.4

Which of these would be the best to use in the lab, and why?

(c) What size pipe and what flow rate (in lb_m/min) should you use in the lab?

(d) If the measured pressure gradient in the lab is 0.016 psi/ft, what is the total pressure drop in the pipeline?

26. A fluid sample is subjected to a "sliding plate" (simple shear) test. The area of the plates is 100 ± 0.01 cm², and the spacing between them is 2 ± 0.1 mm. When the moving plate travels at a speed of 0.5 cm/s, the force required to move it is measured to be 150 dyn, and at a speed of 3 cm/s the force is 1100 dyn. The force transducer has a sensitivity of 50 dyn. What can you deduce about the viscous properties of the sample?

27. Blood is a non-Newtonian fluid. At a shear rate of 10 s^{-1} it has a viscosity of 6.49 cP, and at 80 s^{-1} the viscosity is 4.66 cP. What is the viscosity of blood when it is subjected to a shear stress of 50 dyn/cm²? Determine the answer by assuming that blood follows: (a) the power law; (b) the Bingham plastic model. Which model do you think gives the more realistic answer, and why?

28. You want to predict how fast a glacier which is 200 ft thick will flow down a slope inclined 25° to the horizontal. Assume that the glacier ice can be described by the Bingham plastic model, with a yield stress of 50 psi, a limiting viscosity of 840 poise, and a specific gravity of 0.98. The following materials are available to you in the lab, which also may be described by the Bingham plastic model:

	Yield stress (dyn/cm²)	Limiting viscosity (cP)	SG
Mayonnaise	300	130	0.91
Shaving cream	175	15	0.32
Catchup	130	150	1.2
Paint	87	95	1.35

You want to set up a lab experiment to measure the velocity at which the model fluid flows down an inclined plane, and scale this value to find the velocity of the glacier.

(a) Determine the appropriate set of dimensionless groups.
(b) Which of the above materials would be the best to use in the lab? Why?
(c) What is the film thickness that you should use in the lab, and at what angle should the plane be inclined?
(d) What would be the scale factor between the measured velocity in the lab and the glacier velocity?
(e) What problems might you encounter when conducting this experiment?

4

Fluid Statics

I. STRESS AND PRESSURE

The forces that exist within a fluid at any point may arise from various sources. These may be gravity or the "weight" of the fluid, some external driving force such as a pump or compressor, the internal resistance to relative motion between fluid elements, or inertial effects resulting from variation in local velocity and the mass of the fluid.

Any or all of these forces may result in local stresses within the fluid, which may be defined as a (local) "concentration of force", or force per unit area of fluid. Now both force and area are vectors, the direction of the area being defined by its normal vector, which points outward relative to the volume bounded by the area. Thus, each stress component has a magnitude and *two* directions associated with it, which are the characteristics of a "second order tensor" or "dyad". If the direction in which the local force acts is designated by a subscript j (e.g. j = x, y, or z in Cartesian coordinates), and the orientation of the local area element upon which it acts is designated by a subscript i, then the corresponding stress (σ_{ij}) is given by

$$\sigma_{ij} = \frac{F_j}{A_i}, \text{ where i, j = 1, 2, or 3 (e.g., x, y, or z)} \qquad (4\text{-}1)$$

Note that, since i and j each represent any of three possible directions, there are a total of nine possible components of the stress tensor (at any given point in a fluid). However, it turns out that the stress tensor is symmetric (i.e. the ij components are the same as the ji components) so that there are, at most, six independent stress components.

Because of the various origins of these forces, as mentioned above, there are different "types" of stresses. For example, in a fluid at rest the only stress is the pressure, which may arise from gravity (e.g. hydrostatic head) or compressional energy of the fluid. Although this pressure is a stress (e.g. a force per unit area), it is *isotropic*, that is, the force acts uniformly in all directions normal to any local area at a given point in the fluid. Such a stress consequently has no directional character and is thus a scalar. (Any isotropic tensor component is, by definition, a scalar). However, the stress components which arise from the fluid motion *do* have directional characteristics (determined by the relative motion in the fluid). These stresses may be associated with the local resistance to motion due to viscous or inertial properties and are *anisotropic* because of their directional character. We shall designate them by τ_{ij}, where the i and j have the same significance as in Eqn (4-1).

Thus the total stress at any point within a fluid is composed of both the isotropic pressure and anisotropic stress components, as follows:

$$\sigma_{ij} = -P \, \delta_{ij} + \tau_{ij} \qquad (4-2)$$

where P is the (isotropic) pressure. Pressure is considered a "negative" stress, by convention, since it is compressive, whereas a tensile stress is positive (i.e. a positive F_j acting on a positive A_i or a negative F_j on a negative A_i). The term δ_{ij} in Eqn (4-2) is a "unit tensor" (or Kronecker delta), which has a value of zero if $i \neq j$, and unity if $i = j$. This is required because the isotropic pressure acts only normal to local areas (e.g. $i = j$).

As mentioned above, the shear stresses τ_{ij} are associated with relative motion within the fluid and are therefore zero in any fluid at rest. Thus, the only stresses that can exist in a fluid at rest, or in a state of uniform motion, in which there is no *relative* motion between fluid elements, are those associated with the pressure. It is this situation with which we will be concerned in this chapter.

II. THE BASIC EQUATION OF FLUID STATICS

Consider a cylindrical region of arbitrary size and shape within a fluid, as shown in Fig. 4-1. We will apply a momentum balance to a "slice" of the fluid which has an area A and a thickness Δz and is located a vertical distance z above some horizontal reference plane. The density of the fluid in the slice is ρ, and the force of gravity (g) acts in the -z direction. A momentum balance on a "closed" system (e.g. the slice) is equivalent to Newton's second law of motion, i.e.,

$$\sum F_z = m \, a_z \qquad (4-3)$$

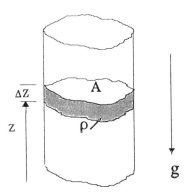

FIGURE 4-1 Arbitrary region within a fluid.

Since this is a vector equation, it is applied to the z vector components. ΣF_z is the sum of all of the forces acting on the system (the "slice") in the z direction, m is the mass of the system, and a_z is the acceleration in the z direction. Since the fluid is not moving, $a_z = 0$, and the momentum balance reduces to a force balance. The z forces acting on the system include the pressure on the bottom (at z) times the area, the pressure on the top (at z + Δz) times the area, and gravity, which is the "weight" of the fluid ($\rho g A \Delta z$). The latter two forces are negative, since they act in the -z direction. The momentum (force) balance thus becomes

$$(AP)_z - (AP)_{z+\Delta z} - \rho \, g \, A \, \Delta z = 0 \qquad (4\text{-}4)$$

If we divide through by $A\Delta z$, then take the limit as the slice shrinks to zero ($\Delta z \rightarrow 0$), the result is

$$\frac{dP}{dz} = -\rho \, g \qquad (4\text{-}5)$$

which is the *Basic Equation of Fluid Statics*. This equation states that at any point within a given fluid the pressure decreases as the elevation (z) increases, at a rate which is equal to the product of the fluid density and gravitational acceleration at that point. This equation is valid at all points within a given static fluid, regardless of the nature of the fluid. We shall now show how this equation can be applied in various special situations.

A. CONSTANT-DENSITY FLUIDS

If the fluid density (ρ) is constant, the fluid is termed "isochoric" (i.e. a given mass occupies a constant volume), although the somewhat more restrictive term "incompressible" is commonly used for this situation (liquids are normally considered to be incompressible or isochoric fluids). If gravity (g) is also constant, the only variables in the equation are pressure and elevation. Equation (4-5) can thus be integrated between any two points (1 and 2) in a given fluid to give

$$P_1 - P_2 = \rho \, g \, (z_2 - z_1) \tag{4-6}$$

which can also be written

$$\Phi_1 = \Phi_2 = \text{constant} \qquad \text{where } \Phi = P + \rho \, g \, z \tag{4-7}$$

This says that the sum of the local pressure and static head (Φ) (which may be called the *potential*) is a constant at all points within a given isochoric (incompressible) fluid. This is an important result for such fluids and can be applied directly to determine how the pressure varies with elevation in a static liquid, as illustrated by the following example.

Example 4-1: **Manometer**—The pressure difference between two points in a fluid (flowing or static) can be measured by using a manometer. The manometer contains an incompressible liquid (density ρ_m) which is immiscible with the flowing fluid (density ρ_f) in the pipe, and the legs of the manometer are connected to taps on the pipe where the pressure difference is desired (see Fig. 4-2). Equation (4-7) can be applied to any two points within either of the fluids, as follows:

$$(\Phi_1 = \Phi_3, \ \Phi_2 = \Phi_4)_f \ (\Phi_3 = \Phi_4)_m \tag{4-8}$$

or

$$P_1 + \rho_f g z_1 = P_3 + \rho_f g z_3$$
$$P_3 + \rho_m g z_3 = P_4 + \rho_m g z_4 \tag{4-9}$$
$$P_4 + \rho_f g z_4 = P_2 + \rho_f g z_2$$

When these three equations are added, P_3 and P_4 cancel out and, after collecting terms, the result is

$$\Delta \Phi = -\Delta \rho \, g \, \Delta h \tag{4-10}$$

where $\Delta\Phi = \Phi_2 - \Phi_1$, $\Delta\rho = \rho_m - \rho_f$, $\Delta h = z_4 - z_3$, and $\Phi = P + \rho g z$. This is the basic *manometer equation* and may be applied to a manometer in any orientation. Note that the manometer reading is a direct measure of the *potential difference* ($\Phi_2 - \Phi_1$), which is identical to the pressure difference ($P_2 - P_1$) only if the pipe is horizontal (i.e. $z_2 = z_1$).

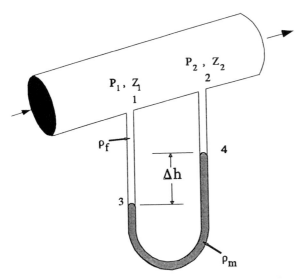

FIGURE 4-2 Manometer attached to pressure taps on pipe carrying a flowing fluid.

B. IDEAL GAS—ISOTHERMAL

If the fluid can be described by the ideal gas law (e.g. air, under atmospheric conditions), then

$$\rho = \frac{PM}{RT} \tag{4-11}$$

and Eqn (4-5) becomes

$$\frac{dP}{dz} = -\frac{PMg}{RT} \tag{4-12}$$

Now if the temperature is constant over all z (i.e. isothermal), Eqn (4-12) can be integrated from (P_1, z_1) to (P_2, z_2) to give the pressure as a function of elevation:

$$P_2 = P_1 \exp\left(-\frac{Mg\,\Delta z}{RT}\right) \tag{4-13}$$

where $\Delta z = z_2 - z_1$. Notice that for this case the pressure falls exponentially with elevation instead of linearly as for the incompressible fluid.

C. IDEAL GAS—ISENTROPIC

If there is no heat transfer or energy dissipated in the gas when going from state 1 to state 2, the process is adiabatic and reversible, i.e. *isentropic*. For an ideal gas under these conditions,

$$\frac{P}{\rho^k} = \text{constant} = \frac{P_1}{\rho_1^k} \tag{4-14}$$

where $k = c_p/c_v$ is the specific heat ratio for the gas (for an ideal gas, $c_p = c_v + R/M$). If the density is eliminated from Eqns (4-14) and (4-11), the result is:

$$\frac{T}{T_1} = \left(\frac{P}{P_1}\right)^{\frac{k-1}{k}} \tag{4-15}$$

which gives the relation between temperature and pressure in an isentropic ideal gas. If Eqn (4-15) is used to eliminate T from Eqn (4-12), the latter can be integrated to give the pressure as a function of elevation:

$$P_2 = P_1 \left[1 - \left(\frac{k-1}{k}\right)\frac{gM\,\Delta z}{RT_1}\right]^{\frac{k}{k-1}} \tag{4-16}$$

Equation (4-15) can be used to eliminate P_2/P_1 from this equation to give an expression for the temperature as a function of elevation in an isentropic system:

$$T_2 = T_1\left[1 - \left(\frac{k-1}{k}\right)\frac{gM\,\Delta z}{RT_1}\right] \tag{4-17}$$

D. THE STANDARD ATMOSPHERE

Neither Eqn (4-13) nor Eqn (4-16) would be expected to provide a very good representation of the pressure and temperature in the real atmosphere, which is neither isothermal nor isentropic. Thus, we must resort to the use of observations (i.e. empiricism) to describe such a system. In fact, atmospheric conditions vary considerably from time to time and from place to place over the earth. However, a reasonable representation of atmospheric conditions "averaged" year round and all over the earth is provided by the following expressions, which are based on observations:

$$\text{For } 0 < z < 36{,}150 \text{ ft: } \frac{dT}{dz} = -3.56°F/1000 \text{ ft} = -G \tag{4-18}$$

$$\text{For } z > 36{,}150 \text{ ft: } \quad T = -69.7°F$$

and the average temperature at ground level ($z = 0$) is assumed to be 59°F (519°R). These relations describe what is known as the "standard atmosphere", which represents an average state. Using Eqn (4-18) for the temperature as a function of elevation, and incorporating this into Eqn (4-12), gives

$$\frac{dP}{dz} = -\frac{PMg}{R\,(T_0 - Gz)} \tag{4-19}$$

where $T_0 = 519°R$. Integrating Eqn (4-19) gives the pressure as a function of elevation:

$$P_2 = P_1 \left[1 - \frac{G\,\Delta z}{T_0 - Gz_1} \right]^{\frac{Mg}{RG}} \tag{4-20}$$

III. MOVING SYSTEMS

We have stated that the only stress that can exist in a fluid at rest is *pressure*, since the shear stresses (which resist motion) are zero when the fluid is at rest. This conclusion may also apply to fluids in motion, as long as there is no *relative* motion within the fluid, since the shear stresses are determined by the velocity *gradients* (i.e. the shear rate). However, if the motion involves an acceleration, this can contribute an additional component to the pressure, as illustrated by the examples below.

A. VERTICAL ACCELERATION

Consider the vertical column of fluid illustrated in Fig. 4-1, but now imagine it to be on an elevator that is accelerating upward with an acceleration a_z, as shown in Fig. 4-3. Application of our momentum balance to the "slice" of fluid, as

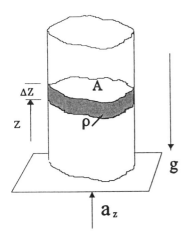

FIGURE 4-3 Vertically accelerating column of fluid.

before, gives

$$\sum F_z = ma_z \tag{4-21}$$

which is the same as Eqn (4-3) except that now $a_z \neq 0$. The same procedure that led to Eqn (4-5) now gives

$$\frac{dP}{dz} = -\rho \, (g + a_z) \tag{4-22}$$

which shows that the effect of a superimposed vertical acceleration is equivalent to increasing the acceleration due to gravity by a factor of a_z. In fact, this result may be generalized to any direction; that is, an acceleration in the i direction will result in a pressure gradient within the fluid in the $-i$ direction, of magnitude ρa_i:

$$\frac{\partial P}{\partial x_i} = -\rho a_i \tag{4-23}$$

Two applications of this result are illustrated below.

B. HORIZONTALLY ACCELERATING FREE SURFACE

Consider a pool of water in the bed of your pickup truck. If you accelerate from rest, the water will slosh toward the rear, and you want to know how fast you can accelerate (a_x) without sloshing the water out of the back of the truck (see Fig. 4-4). That is, you must determine the slope of the water surface as a function of the rate of acceleration a_x.

Now at any point within the liquid, there is both a vertical pressure gradient [due to gravity, Eqn (4-5)] and a horizontal pressure gradient [due to the acceleration a_x, Eqn (4-23)]. At any place in the liquid, the total differential pressure dP between two points separated by dx in the horizontal direction and

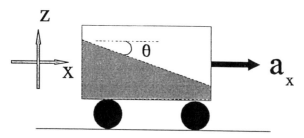

FIGURE 4-4 Horizontally accelerating tank.

dz in the vertical direction is given by

$$dP = \frac{\partial P}{\partial x} dx + \frac{\partial P}{\partial z} dz$$

$$= -\rho\, a_x\, dx - \rho\, g\, dz \tag{4-24}$$

Since the surface of the water is open to the atmosphere, the condition P = constant (1 atm) applies at the surface, so that

$$(dP)_s = 0 = -\rho\, g\, (dz)_s - \rho\, a_x\, (dx)_s \tag{4-25}$$

or

$$\left(\frac{dz}{dx}\right)_s = -\frac{a_x}{g} = \tan\theta \tag{4-26}$$

This is the slope of the surface, and it is seen to be independent of fluid properties. A knowledge of the initial position of the surface plus the surface slope determines the elevation at the rear of the truck bed and hence whether or not the water will spill out.

C. ROTATING FLUID

Now consider an open bucket of water resting on a turntable, rotating at an angular velocity of ω (see Fig. 4-5). Because of the (inward) radial acceleration that exists due to the rotation ($\omega^2 r$), there is a corresponding radial pressure gradient at all points in addition to the vertical pressure gradient due to gravity. Thus, the pressure differential between two points separated by dr and dz anyplace within the fluid is

$$dP = \left(\frac{\partial P}{\partial z}\right) dz + \left(\frac{\partial P}{\partial r}\right) dr = \rho\,(-g\, dz + \omega^2 r\, dr) \tag{4-27}$$

As in the case of the accelerating tank, we can determine the shape of the free surface from the surface condition that the pressure is constant:

$$(dP)_s = 0 = -g\,(dz)_s + \omega^2 r\,(dr)_s \tag{4-28}$$

which can be integrated to give the equation for the shape of the surface:

$$z = z_0 + \frac{\omega^2 r^2}{2g} \tag{4-29}$$

This shows that the rotating surface will achieve a parabolic shape.

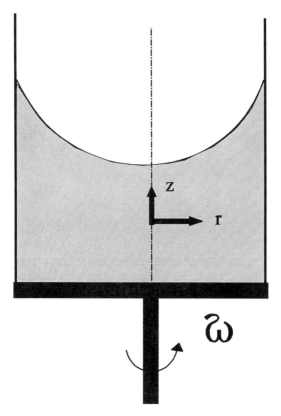

FIGURE 4-5   Rotating fluid.

IV. BUOYANCY

By Archimedes' principle, the buoyant force exerted on a submerged body is equal to the weight of the displaced fluid and acts in a direction opposite to the acceleration vector. Thus the effective ''net weight'' of a submerged body is its actual weight less the weight of an equal volume (V_s) of the fluid. This is equivalent to replacing the density of the body (ρ_s) in the expression for the weight of the body ($\rho_s g V_s$) by the difference between the density of the body and that of the fluid ($\Delta\rho g V_s$, where $\Delta\rho = \rho_s - \rho_f$).

   An equivalent result applies to a body submerged in a fluid that is subjected to any acceleration. For example, a solid particle submerged in a fluid within a centrifuge at a point where the angular velocity is ω is subjected to a net radial *inward* force equal to $\Delta\rho\omega^2 r V_s$. Thus, the effect of buoyancy is to effectively reduce the density of the body by an amount equal to the density of the surrounding fluid.

V. STATIC FORCES ON SOLID BOUNDARIES

The force exerted on a solid boundary by a static pressure can be expressed as

$$\vec{F} = \int_A P \, d\vec{A} \tag{4-30}$$

Note that both force and area are vectors, whereas pressure is a scalar. Hence, the directional character of the force is determined by the orientation of the surface on which the pressure acts. That is, the component of force acting in a given direction on a surface is the integral of the pressure over the *projected* area of the surface, where the surface vector (normal to the surface) is the direction of the force.

Example 4-2: Consider the pressure exerted by a fluid contained within a pipe, on the wall of the pipe, as illustrated in Fig. 4-6. The pressure P acts equally in all directions on the inside of the pipe, and the resulting force exerted within the pipe wall normal to a plane through the pipe centerline is simply the product of the pressure and the projected area of the wall on this plane, e.g. $F_x = PA_x = 2PRL$. This (disruptive) force, which acts to pull apart the metal within the wall, is resisted by the (restorative) internal stress within the metal holding it together, as measured by the effective *working stress*, σ, of the particular metal of which the pipe is made. If we assume a thin-walled pipe (i.e. we neglect

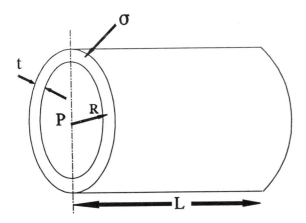

FIGURE 4-6   Fluid pressure inside pipe.

the variation of the stress from point to point within the wall), a force balance between the "disruptive" pressure force and the "restorative" stress force gives

$$2PRL = 2 \sigma t L \tag{4-31}$$

or

$$t \cong RP/\sigma \tag{4-32}$$

This relation determines the pipe wall thickness required to contain a fluid at a pressure P in a pipe of radius R, made of a material with a working stress σ. The dimensionless pipe wall thickness (times 1000) is known as the *Schedule Number* of the pipe:

$$\text{Schedule No.} \cong \frac{1000t}{R} = \frac{1000P}{\sigma} \tag{4-33}$$

This expression is only approximate, as it does not make any allowance for the effects of such things as pipe threads or corrosion. To compensate for these factors, an additional allowance is made in the wall thickness for the working definition of the "schedule thickness", t_s:

$$\text{Schedule No.} = \frac{1000P}{\sigma} = \frac{1750t_s - 200}{D_0} \tag{4-34}$$

where both t_s and D_0 (the pipe outside diameter) are measured in inches. This relation between schedule number and pipe dimensions can be compared with the actual dimensions of commercial pipe for various schedule pipe sizes, as tabulated in Appendix F.

PROBLEMS

STATICS

1. The manometer equation is $\Delta\Phi = -\Delta\rho \, g \, \Delta h$, where $\Delta\Phi$ is the difference in the total pressure plus static head $(P + \rho gz)$ between the two points to which the manometer is connected, $\Delta\rho$ is the difference in the densities of the two fluids in the manometer, Δh is the manometer reading, and g is the acceleration due to gravity. If $\Delta\rho$ is 12.6 g/cm^3, and Δh is 6 in. for a manometer connected to two points on a horizontal pipe, calculate the value of ΔP in the following units:
   (a) dyn/cm^2; (b) psi; (c) Pascals; (d) atmospheres.

2. A manometer containing an oil with SG = 0.92 is connected across an orifice plate in a horizontal pipeline carrying seawater (SG = 1.1). If the manometer reading is 16.8 cm, what is the pressure drop across the orifice in psi? What is it in inches of water?

3. A mercury manometer is used to measure the pressure drop across an orifice which is mounted in a vertical pipe. A liquid with a density of 0.87 g/cm^3 is flowing upward through the pipe and the orifice. The distance between the manometer taps is 1 ft. If the pressure in the pipe at the upper tap is 30 psig, and the manometer reading is 15 cm, what is the pressure in the pipe at the lower manometer tap in psig?

4. A mercury manometer is connected between two points in a piping system which contains water. The downstream tap is 6 ft higher than the upstream tap, and the manometer reading is 16 in. If a pressure gauge in the pipe at the upstream tap reads 40 psia, what would a pressure gauge at the downstream tap read in (a) psia, (b) dyn/cm^2, (c) Pa, (d) kg$_f$/m^2?

5. An inclined tube manometer with a reservoir is used to measure the pressure gradient in a large pipe carrying oil (SG = 0.91) (see the Figure). The pipe is inclined at an angle of 60° to the horizontal,

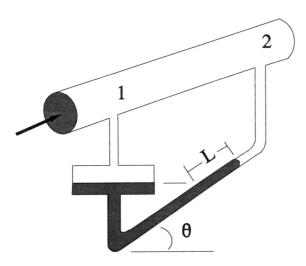

PROBLEM 5

and flow is uphill. The manometer tube is inclined at an angle of 20° to the horizontal, and the pressure taps on the pipe are 5 in. apart. The manometer reservoir diameter is eight times as large as the manometer tube diameter, and the manometer fluid is water. If the manometer reading is 3 in., and the displacement of the interface in the reservoir is neglected, what is the pressure drop in the pipe in (a) psi, (b) Pa, (c) inches of water? What is the percentage error introduced by neglecting the change of elevation of the interface in the reservoir?

6. Water is flowing downhill in a pipe which is inclined 30° to the horizontal. A mercury manometer is attached to pressure taps 5 cm apart on the pipe. The interface in the downstream manometer leg is 2 cm higher than the interface in the upstream leg. What is the pressure gradient ($\Delta P/L$) in the pipe in Pa/m, dyn/cm^3, in. H$_2$O/ft, and psi/mi?

7. Repeat Problem 6 for the case in which the water in the pipe is flowing uphill instead of downhill, all other conditions remaining the same.

8. Two horizontal pipelines are parallel, with one carrying salt water ($\rho = 1.988$ slugs/ft^3) and the other carrying fresh water ($\rho = 1.937$ slugs/ft^3). An inverted manometer using linseed oil ($\rho = 1.828$ slugs/ft^3) as the manometer fluid is connected between the two pipelines. The interface between the oil and the fresh water in the manometer is 38 in. above the centerline of the freshwater pipeline, and the oil-salt water interface in the manometer is 20 in. above the centerline of the salt water pipeline. If the manometer reading is 8 in., determine the difference in the pressures in the pipelines in Pa and in psi.

9. Two identical tanks are 3 ft in diameter and 3 ft high and are both vented to the atmosphere. The top of tank B is level with the bottom of tank A, and they are connected by a line from the bottom of A to the top of B with a valve in it. Initially A is full of water, and B is empty. The valve is opened for a short time, letting some of the water drain into B. An inverted manometer having an oil with SG = 0.7 is connected between taps on the bottom of each tank. The manometer reading is 6 in., and the oil-water interface in the leg connected to tank A is higher. What is the water level in each of the tanks?

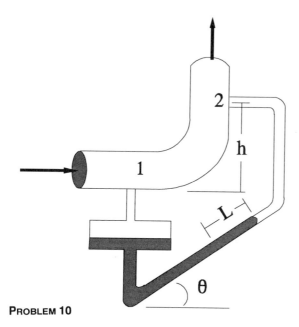

PROBLEM 10

10. An inclined tube manometer is used to measure the pressure drop in an elbow through which water is flowing (see Figure). The manometer fluid is an oil with SG = 1.15. The distance L is the distance along the inclined tube that the interface has moved from its equilibrium (no pressure differential) position. If h = 6 in., L = 3 in., θ = 30°, the reservoir diameter is 2 in., and the tubing diameter is 0.25 in., calculate the pressure drop ($P_1 - P_2$) in (a) atm; (b) Pa; (c) cm of water; (d) dyn/cm². What would be the percentage error in pressure difference as read by the manometer if the change in level in the reservoir is neglected?

11. The three-fluid manometer illustrated in the Figure is used to measure a very small pressure difference, $P_1 - P_2$. The cross sectional area of each of the reservoirs is A, and that of the manometer legs is a. The three fluids have densities ρ_a, ρ_b, and ρ_c, and the difference in elevation of the interfaces in the reservoir is x. Derive the equation which relates the manometer reading h to the pressure difference $P_1 - P_2$. How could you simplify this equation if A >> a?

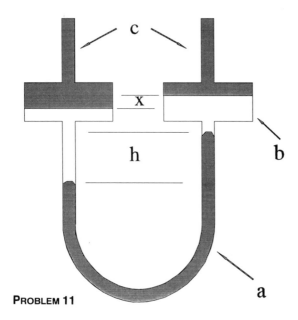

PROBLEM 11

12. A tank that is vented to the atmosphere contains a liquid with a density of 0.9 g/cm³. A dip tube is inserted into the top of the tank, which extends to a point 1 ft from the bottom of the tank. Air is bubbled slowly through the dip tube, and the air pressure in the tube is measured with a mercury (SG = 13.6) manometer. One leg of the manometer is connected to the air line feeding the dip tube, and the other leg is open to the atmosphere. If the manometer reading is 5 in., what is the depth of the liquid in the tank?

13. An inclined manometer is used to measure the pressure drop between two taps on a pipe carrying water, as shown in the Figure. The manometer fluid is an oil with SG = 0.92, and the manometer reading is 8 in. The manometer reservoir is 4 in. in diameter, the tubing is 1/4 in. in diameter, and the tube is inclined at an angle of 30° to the horizontal. The pipe is inclined at 20° to the horizontal and the pressure taps are 40 in. apart.
    (a) What is the pressure difference between the two pipe taps that would be indicated by the difference in readings of two pressure gauges attached to the taps, in psi, Pa, and inches of water?
    (b) Which way is the water flowing?
    (c) What will the manometer reading be if the valve is closed?

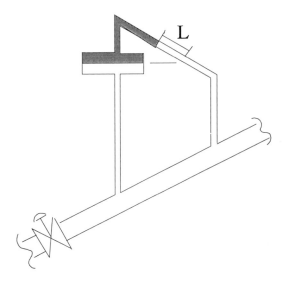

L

PROBLEM 13

14. The pressure gradient required to force water through a straight horizontal 1/4 in. ID tube at a rate of 2 gpm is 1.2 psi/ft. Consider this same tubing coiled in an expanding helix with a vertical axis. Water enters the bottom of the coil and flows upward at a rate of 2 gpm. A mercury manometer is connected between two pressure taps on the coil, one near the bottom where the coil radius is 6 in., and the other near the top where the coil radius is 12 in. The taps are 2 ft apart in the vertical direction, and there is a total of 5 ft of tubing between the two taps. Determine the manometer reading, in cm.

15. It is possible to achieve a weightless condition for a limited time in an airplane by flying in a circular arc above the earth (like a rainbow). If the plane flies at 650 mph, what should the radius of the flight path be (in miles) to achieve weightlessness?

16. Water is flowing in a horizontal pipe bend at a velocity of 10 ft/s. The radius of curvature of the inside curve of the bend is 4 in., and the pipe I.D. is 2 in. A mercury manometer is connected to taps located radially on the inside and outside of the bend. Assuming that the water velocity is uniform over the pipe cross section, what would be the manometer reading in centimeters? What would it be if the

water velocity were 5 ft/s? Convert the manometer reading to equivalent pressure difference in psi and Pa.

17. Calculate the atmospheric pressure at an elevation of 3000 m, assuming: (a) the air is incompressible, at a temperature of 59°F; (b) the air is isothermal at 59°F and is an ideal gas; (c) the pressure distribution follows the model of the standard atmosphere, with a surface temperature of 59°F.

18. One pound mass of air (MW = 29) at sea level and 70°F is contained in a balloon, which is then carried to an elevation of 10,000 ft in the atmosphere. If the balloon offers no resistance to expansion of the gas, what is its volume at this elevation?

19. A gas well contains hydrocarbon gases with an average molecular weight of 24, which can be assumed to be an ideal gas with a specific heat ratio of 1.3. The pressure and temperature at the top of the well are 250 psig and 70°F, respectively. The gas is being produced at a slow rate, so conditions in the well can be considered to be isentropic.
    (a) What are the pressure and temperature at a depth of 10,000 ft?
    (b) What would the pressure be at this depth if the gas were assumed to be isothermal?
    (c) What would the pressure be at this depth if the gas were assumed to be incompressible?

20. The adiabatic atmosphere obeys the following equation:

    $$PV^k = \text{constant}$$

    where k is a constant and V is volume. If the temperature decreases 0.3°C for every 100 ft increase in altitude, what is the value of k? [Note: Air is an ideal gas; g = 32.2 ft/s^2; R = 1544 ft lb$_f$/°R lb mol).]

21. Using the actual dimensions of commercial steel pipe from Appendix F, plot the pipe wall thickness versus the pipe diameter for both Schedule 40 and Schedule 80 pipe, and fit the plot with a straight line by linear regression analysis. Rearrange your equation for the line in a form consistent with the given equation for the schedule number as a function of wall thickness and diameter:

    $$\text{Sched No} = (1750t_s - 200)/D$$

    and use the values of the slope and intercept of the linear data fit to calculate values equivalent to the 1750 and 200 in this equation. Do this using (for D): (a) the nominal pipe diameter; (b) the outside pipe diameter. Offer an explanation for any discrepancies or differences in the numerical values determined from the data fit as compared to those in the equation.

22. The "yield stress" for carbon steel is 35,000 psi, and the "working stress" is half of this value. What schedule number would you recommend for a pipe carrying ethylene at a pressure of 2500 psi, if the pipeline design calls for a pipe of 2 in. I.D.? Give the dimensions of the pipe that you would recommend. What would be a safe maximum pressure to recommend for this pipe?

23. Consider a 90° elbow in a 2 in. pipe (all of which is in the horizontal plane). A pipe tap is drilled through the wall of the elbow on the inside curve of the elbow, and another through the outer wall of the elbow directly across from the inside tap. The radius of curvature of the inside of the bend is 2 in., and that of the outside of the bend is 4 in. The pipe is carrying water, and a manometer containing an immiscible oil with SG = 0.90 is connected across the two taps on the elbow. If the reading of the manometer is 7 in., what is the average velocity of the water in the pipe, assuming that the flow is uniform across the pipe inside the elbow?

24. A pipe carrying water is inclined at an angle of 45° to the horizontal. A manometer containing a fluid with SG = 1.2 is attached to taps on the pipe, which are 1 ft apart. If the liquid interface in the manometer leg which is attached to the lower tap is 3 in. below the interface in the other leg, what is the pressure gradient in the pipe ($\Delta P/L$), in units of psi/ft and Pa/m? Which direction is the water flowing?

25. A vertical section of pipe contains a valve, across which a manometer is connected. Water is flowing upward through the pipe, and the manometer fluid is an oil with a specific gravity of 0.85. The distance between the manometer taps on the pipe is 3 ft, and the manometer reading is 6 in.
    (a) What is the friction loss in the valve, in units of ft lb_f/lb_m, psi, feet of water, and kcal/kg? (Hint: Use Bernoulli's equation, Eqn 5-35.)
    (b) If the manometer is replaced by a transducer that measures $P_1 - P_2$ directly, what would it read, in psi?

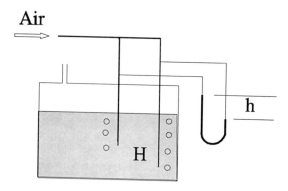

Air

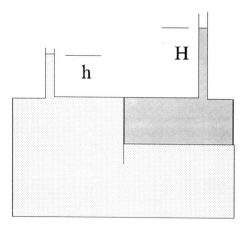

H

h

PROBLEM 26

26. A tank contains a liquid of unknown density (see the Figure). Two dip tubes are inserted into the tank, each to a different level in the tank, and air is bubbled very slowly through them into the liquid. A manometer is used to measure the difference in pressure between the two dip tubes. If the difference in level of the ends of the dip tubes (H) is 1 ft, and the manometer reads 1.5 ft (h) with water as the manometer fluid, what is the density of the liquid in the tank?

27. The tank shown in the Figure has a partition which separates two immiscible fluids. Most of the tank contains water, and oil is floating above the water to the right of the partition. The height of the water

H

h

PROBLEM 27

in the standpipe (h) is 10 cm, and the interface between the oil and water is 20 cm below the top of the tank and 25 cm above the bottom of the tank. If the specific gravity of the oil is 0.82, what is the height of the oil in the standpipe (H)?

28. A manometer which is open to the atmosphere contains water, with a layer of oil floating on the water in one leg (see Figure). If the level of the water in the left leg is 1 cm above the center of the leg, the interface between the water and oil is 1 cm below the center in the right leg, and the oil layer on the right extends 2 cm above the center, what is the density of the oil?

29. An open cylindrical drum, with a diameter of 2 ft and a length of 4 ft, is turned upside down in the atmosphere and then submerged in a liquid so that it floats partially submerged upside down, with air trapped inside. If the drum weighs 150 lb$_f$, and it floats with 1 ft above the surface of the liquid, what is the density of the liquid? How much additional weight must be added to the drum to make it sink to the point where it floats just level with the liquid?

30. A solid spherical particle, with a radius of 1 mm and a density of 1.3 g/cm^3, is immersed in water in a centrifuge. If the particle is 10 cm from the axis of the centrifuge, which is rotating at a rate of 100 rpm, what direction will the particle be traveling relative to a horizontal plane?

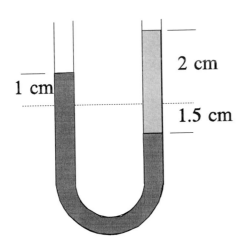

2 cm

1 cm

1.5 cm

PROBLEM 28

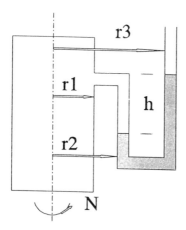

31. A manometer with mercury as the manometer fluid is attached to the wall of a closed tank containing water (see the Figure). The entire system is rotating about the axis of the tank at N rpm. The radius of the tank is r_1, the distances from the tank centerline to the manometer legs are r_2 and r_3 (as shown), and the manometer reading is h. If N = 30 rpm, r_1 = 12 cm, r_2 = 15 cm, r_3 = 18 cm, and h = 2 cm, determine the gauge pressure at the wall of the tank and also at the centerline.

5

Conservation Principles

I. THE "SYSTEM"

The basic principles that govern the analysis and solution of flow problems include the conservation of mass, energy, and momentum. For flow problems, these conservation laws will be applied to a "system", which is defined as any clearly specified region or volume of fluid, with either macroscopic or microscopic dimensions (this is sometimes referred to as a "control volume"), as illustrated in Fig. 5-1. The general conservation law is

(Rate of "X" into the system) − (Rate of "X" out of the system)

= (Rate of accumulation of "X" in the system)

where "X" is the conserved quantity, i.e. mass, energy, or momentum. In the case of momentum, since a "rate of momentum" is equivalent to a force (by Newton's second law), the "rate in" term must also include any (net) forces acting *on* the system. It is emphasized that the system is *not* the "containing vessel" (a pipe, tank, pump, etc.) but is the *fluid* contained within the designated boundary. We will show how this generic expression applies for each of the three conserved quantities.

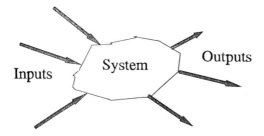

FIGURE 5-1 A system with inputs and outputs.

II. CONSERVATION OF MASS

A. MACROSCOPIC BALANCE

For a given system (e.g. Fig. 5-1), each entering stream will carry mass into the system (at rate $\dot{m}_i$) and each exiting stream will carry mass out of the system (at rate $\dot{m}_o$). Hence, the conservation of mass, or continuity, equation for the system is

$$\sum_{in} \dot{m}_i - \sum_{out} \dot{m}_o = \frac{dm_s}{dt} \tag{5-1}$$

where m_s is the mass of the system, and

$$\dot{m} = \int_A \rho \vec{V} \cdot d\vec{A} = \rho \vec{V} \cdot \vec{A} \tag{5-2}$$

That is, the mass flow rate through a given surface is the integrated value of the local mass flow rate over the area of the surface. Note that mass flow rate is a scalar, whereas velocity and area are vectors. Thus it is the scalar (or dot) product of the velocity and area that is required (the "direction" or orientation of the surface is that of the unit vector that is normal to the surface). The corresponding definition of the average velocity through the conduit is

$$V = \frac{1}{A} \int \vec{V} \cdot d\vec{A} = \frac{Q}{A} \tag{5-3}$$

where $Q = \dot{m}/\rho$ is the volumetric flow rate and the area A is the projected component of $\vec{A}$ that is normal to $\vec{V}$. For a system at steady state, Eqn (5-1) reduces to

$$\sum_{in} \dot{m}_i = \sum_{out} \dot{m}_o \tag{5-4}$$

or

$$\sum_{in} (\rho VA)_i = \sum_{out} (\rho VA)_o \qquad (5\text{-}5)$$

Example 5-1: Water is flowing at a velocity of 7 ft/s in both 1 in. and 2 in. I.D. pipes, which are joined together and feed into a 3 in. I.D. pipe, as shown in Figure 5-2. Determine the water velocity in the 3 in. pipe.

Solution—Since the system is at steady state, Eqn (5-5) applies:

$$(\rho VA)_1 + (\rho VA)_2 = (\rho VA)_3$$

Since the fluid density is constant, this may be solved for V_3:

$$V_3 = V_1 \frac{A_1}{A_3} + V_2 \frac{A_2}{A_3}$$

Since $A = \pi D^2/4$, this gives

$$V_3 = 7 \frac{ft}{s} \left[\frac{1}{9} + \frac{4}{9} \right] = 3.89 \frac{ft}{s}$$

B. MICROSCOPIC BALANCE

The conservation of mass can be applied to an arbitrarily small fluid element to derive the "microscopic continuity" equation, which must be satisfied at all points within a continuous fluid. This can be done by considering an arbitrary

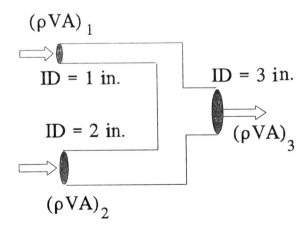

$(\rho VA)_1$

ID = 1 in.

ID = 3 in.

ID = 2 in.

$(\rho VA)_3$

$(\rho VA)_2$

FIGURE 5-2   Continuity.

(cubical) differential element of dimensions dx, dy, dz, with mass flow components into or out of each surface, e.g:

$$\dot{m}_{out} - \dot{m}_{in} = dy\ dz[(\rho v_x)_{x+dx} - (\rho v_x)_x]$$
$$+ dx\ dz[(\rho v_y)_{y+dy} - (\rho v_y)_y]$$
$$+ dx\ dy[(\rho v_z)_{z+dz} - (\rho v_z)_z]$$
$$= -\frac{\partial \rho}{\partial t} dx\ dy\ dz \tag{5-6}$$

Dividing by the volume of the element (dx dy dz), and taking the limit as the size of the element shrinks to zero, gives

$$\frac{\partial(\rho v_x)}{\partial x} + \frac{\partial(\rho v_y)}{\partial y} + \frac{\partial(\rho v_z)}{\partial z} = -\frac{\partial \rho}{\partial t} \tag{5-7}$$

This is the microscopic (or local) continuity equation and must be satisfied at all points within a flowing fluid. If the fluid is incompressible (i.e. constant ρ), Eqn (5-7) reduces to

$$\frac{\partial v_x}{\partial x} + \frac{\partial v_y}{\partial y} + \frac{\partial v_z}{\partial z} = 0 \tag{5-8}$$

We will make use of this equation in subsequent chapters.

III. CONSERVATION OF ENERGY

Energy can take a wide variety of forms, such as internal (thermal), mechanical work, kinetic, potential, surface, electrostatic, electromagnetic, and nuclear. Also, for nuclear reactions or velocities approaching the speed of light, the interconversion of mass and energy can be significant. However, we will not be concerned with situations involving nuclear reactions or speeds near that of light, and some other possible forms of energy will usually be negligible as well. Our purposes will be adequately served if we consider only internal (thermal), kinetic, potential (due to gravity), mechanical (work), and heat forms of energy. For the system illustrated in Fig. 5-1, a unit mass of fluid in each inlet and outlet stream may contain a certain amount of internal energy (u) by virtue of its temperature, kinetic energy (v^2) by virtue of its velocity, potential energy (gz) due to its position in the (gravitational) field, and "pressure" energy (P/ρ). The "pressure" energy is sometimes called the *flow work*, because it is associated with the amount of work or energy required to "inject" or "eject" the fluid element into or out of the system at the appropriate pressure. In addition, energy can cross the boundaries of the system other than with the flow streams, in the form of heat (Q) resulting from a temperature difference and "shaft" work (W).

Shaft work is so named because it is normally associated with work done by a pump, compressor, mixer, etc., that is driven by a shaft.

The sign conventions for heat (Q) and work (W) are arbitrary and consequently vary from one reference to another. Heat is usually taken to be positive when it is added to the system, so it would seem to be consistent to use this same convention for work (as is done in most scientific references). However, engineers, being pragmatic, use a sign convention that is directly associated with "value". That is, if work can be *extracted from* the system, then it is *positive*, since a positive asset can be sold to produce revenue. However, if work must be *put into* the system, then it is *negative*, since it must be purchased (a negative asset). This convention is also more consistent with the "driving force" interpretation of the terms in the energy balance, as we shall see.

With this introduction, we may write the conservation of energy expression for any system as a rate equation, as follows:

$$\sum_{in} \left(h + gz + \frac{V^2}{2} \right)_i \cdot \dot{m}_i - \sum_{out} \left(h + gz + \frac{V^2}{2} \right)_o \cdot \dot{m}_o + \dot{Q} - \dot{W}$$

$$= \frac{d}{dt} \left[\left(u + gz + \frac{V^2}{2} \right) m_s \right]_s \tag{5-9}$$

Here, $h = u + P/\rho$ is the *enthalpy* per unit mass of fluid. Note that the inlet and exit streams include enthalpy (i.e. both internal energy u and flow work P/ρ), whereas the "system energy" includes the internal energy but no P/ρ flow work (for obvious reasons). If there is only one inlet and one exit stream, then $\dot{m}_i = \dot{m}_o = \dot{m}$, and if the system is at steady state, the energy balance becomes

$$\Delta h + g\Delta z + \frac{1}{2} \Delta V^2 = q - w \tag{5-10}$$

where $\Delta = ("\text{out}" - "\text{in}")$, and $q = \dot{Q}/\dot{m}$, $w = \dot{W}/\dot{m}$ are the heat added to the system and work done by the system, respectively, per unit mass of fluid. This expression also applies to a system composed of the fluid between any two points along a streamline (or a "stream tube") within a flow field. Specifically, if these two points are only an infinitesimal distance apart, the result is the differential form of the energy balance:

$$dh + gdz + vdv = \delta q - \delta w \tag{5-11}$$

where $dh = du + d(P/\rho)$. The "d( )" notation represents a total or "exact" differential and applies to those quantities which are determined only by the state (T, P) of the system and are thus "point" properties. The "δ( )" notation represents quantities which are inexact differentials and depend upon the path taken from one point to another.

Note that the energy balance contains several different forms of energy, which may be generally classified as either "mechanical" energy, associated with motion or position, or "thermal" energy, associated with temperature. Mechanical energy is the more useful form, in that it is a higher level energy that can be converted directly into useful work and includes the potential energy, kinetic energy, and shaft work terms. The thermal energy terms, i.e. enthalpy, internal energy, and heat, are not directly available to do useful work unless they are transformed into mechanical energy, in which case it is the mechanical energy that does the work.

In fact, the total amount of energy represented by a relatively small temperature change is often equivalent to a relatively large amount of "mechanical energy". For example, 1 Btu of "thermal energy" is equivalent to 778 ft lb_f of "mechanical energy". This means that the amount of energy required to raise the temperature of 1 lb of water by 1 degree Fahrenheit (the definition of the Btu) is equivalent to the amount of energy required to raise the elevation of that same pound of water by 778 ft! Thus, for systems that involve significant temperature changes, the mechanical energy terms (e.g. pressure, potential and kinetic energy, and work) may be negligible compared with the thermal energy terms (heat transfer, internal energy, etc.). In such cases, the energy balance equation reduces to a "heat balance", i.e. $\Delta h = q$. However, the reader should be warned that "heat" is not a conserved quantity and that the inherent assumption when writing a "heat balance", namely, that other forms of energy are negligible, should always be confirmed. Before proceeding further, we will take a closer look at the significance of enthalpy and internal energy, since these cannot be measured directly but are determined indirectly by measuring other properties such as temperature and pressure.

A. INTERNAL ENERGY

An infinitesimal change in internal energy is an exact differential and is uniquely determined by a change in temperature and pressure (for a given composition). Since the density is also uniquely determined by temperature and pressure (e.g. by an equation of state for the material), internal energy may be expressed as a function of any two of the three T, P, ρ (or $\overline{V} = 1/\rho$). Hence, we may write

$$du = \left(\frac{\partial u}{\partial T}\right)_{\overline{V}} dT + \left(\frac{\partial u}{\partial \overline{V}}\right)_{T} d\overline{V} \tag{5-12}$$

By making use of thermodynamic identities, this is equivalent to

$$du = c_v dT + \left[T\left(\frac{\partial P}{\partial T}\right)_{\overline{V}} - P\right] d\overline{V} \tag{5-13}$$

where

$$c_v = \left(\frac{\partial u}{\partial T}\right)_{\bar{v}} \tag{5-14}$$

is the heat capacity (specific heat) of the material at constant volume (e.g. constant density). We will now consider several special cases of these equations for various materials.

1. Ideal Gas

For an ideal gas,

$$\rho = \frac{PM}{RT} \quad \text{so that} \quad T\left(\frac{\partial P}{\partial T}\right)_{\bar{v}} = P \tag{5-15}$$

Thus Eqn (5-13) reduces to

$$du = c_v dT \quad \text{or} \quad \Delta u = \int_{T_1}^{T_2} c_v dT = \bar{c}_v(T_2 - T_1) \tag{5-16}$$

which shows that the internal energy for an ideal gas is a function of temperature only.

2. Non-Ideal Gas

For a non-ideal gas, Eqn (5-15) is not valid, so

$$T\left(\frac{\partial P}{\partial T}\right)_{\bar{v}} \neq P \tag{5-17}$$

Thus the last term in Eqn (5-13) does not cancel as it did for the ideal gas, and

$$\Delta u = fn(T, P) \tag{5-18}$$

The form that the implied function fn(T, P) takes may be mathematical if the material can be described by a non-ideal equation of state, or it could be empirical, such as for steam properties, which are described by data tabulated in steam tables.

3. Solids and Liquids

For solids and liquids, $\rho \simeq$ constant (or $d\bar{V} = 0$). Thus,

$$du = c_v dT \quad \text{or} \quad \Delta u = \int_{T_1}^{T_2} c_v dT = \bar{c}_v(T_2 - T_1) \tag{5-19}$$

which shows that the internal energy depends upon temperature only (as for the ideal gas, but for an entirely different reason).

B. ENTHALPY

We may choose to express the enthalpy as a function of temperature and pressure, in which case

$$dh = \left(\frac{\partial h}{\partial T}\right)_P dT + \left(\frac{\partial h}{\partial P}\right)_T dP \qquad (5\text{-}20)$$

which, from thermodynamic identities, is equivalent to

$$dh = c_p dT + \left[\bar{V} - T\left(\frac{\partial \bar{V}}{\partial T}\right)_P\right] dP \qquad (5\text{-}21)$$

Here,

$$c_p = \left(\frac{\partial h}{\partial T}\right)_P \qquad (5\text{-}22)$$

is the heat capacity (specific heat) of the material at constant pressure. We again consider three special cases.

1.   Ideal Gas

For an ideal gas,

$$T\left(\frac{\partial \bar{V}}{\partial T}\right)_P = \bar{V} \text{ and } c_P = c_v + \frac{R}{M} \qquad (5\text{-}23)$$

Thus Eqn (5-21) for the enthalpy becomes

$$dh = c_p dT \text{ or } \Delta h = \int_{T_1}^{T_2} c_p dT = \bar{c}_P (T_2 - T_1) \qquad (5\text{-}24)$$

which shows that the enthalpy for an ideal gas is a function of temperature only (as is the internal energy).

2.   Non-Ideal Gas

For a non-ideal gas,

$$T\left(\frac{\partial \bar{V}}{\partial T}\right)_P \neq \bar{V} \quad \text{so that} \quad \Delta h = fn (T, P) \qquad (5\text{-}25)$$

which, as for Δu, may be either an analytical or an empirical function. All gases behave as ideal gases under appropriate conditions (i.e. far enough from the critical point) and become more non-ideal as the critical point is approached. Thus, for a given gas, if the temperature and pressure are sufficiently far from the critical point that the enthalpy is essentially independent of pressure at a given temperature, the gas can probably be adequately described by the ideal gas law (see, for example, the pressure-enthalpy diagrams for various gases in Appendix D).

3. Solids and Liquids

For solids and liquids, $\rho \simeq$ constant, so that $(\partial \overline{V}/\partial T)_P = 0$ and $c_P \simeq c_v$. Therefore,

$$dh = c_p dT + \overline{V} dP \qquad (5\text{-}26)$$

so that

$$\Delta h = \int_{T_1}^{T_2} c_p dT + \int_{P_1}^{P_2} \frac{dP}{\rho} = \overline{c}_p (T_2 - T_1) + \frac{P_2 - P_1}{\rho} \qquad (5\text{-}27)$$

This shows that for solids and liquids the enthalpy depends upon both temperature and pressure. This is in contrast to the internal energy, which depends on temperature only.

IV. IRREVERSIBLE EFFECTS

We have noted that if there is a significant change in temperature, the "thermal" energy terms (i.e. q and u) may represent much more energy than the "mechanical" terms (i.e. pressure, potential and kinetic energy, and work). On the other hand, if no significant temperature difference exists between the system and its surroundings, the only source of "heat" (i.e. temperature change) is the internal (irreversible) dissipation of mechanical energy into thermal energy, or "friction". The origin of this "friction loss" is the irreversible work required to overcome intermolecular forces (i.e. the forces between the "fluid elements"). This can be quantified as follows.

For a system at equilibrium (i.e. in a reversible or "static" state), thermodynamics tells us that

$$du = Tds - Pd(1/\rho) \text{ and } Tds = \delta q \qquad (5\text{-}28)$$

That is, the total change in entropy comes from the heat transferred across the system boundary (δq). However, a flowing fluid is in a "dynamic" or irreversible state, so that

$$Tds > \delta q \text{ or } Tds = \delta q + \delta e_f \qquad (5\text{-}29)$$

in which case

$$du = \delta q + \delta e_f - Pd(1/\rho) \qquad (5\text{-}30)$$

Here, δe_f represents the "irreversible energy" associated with the departure of the system from equilibrium. The farther from equilibrium (e.g. the faster the motion), the greater is the associated irreversible energy. The origin of this energy (or "extra entropy") is the mechanical energy which drives the system, which is thus reduced by the amount e_f. This energy ultimately appears as an increase in the thermal energy of the system (Eqn 5-30). This mechanism of transfer of useful mechanical energy to low-grade (non-useful) thermal energy is referred to as "energy dissipation". Although e_f is often referred to as the energy "loss", it is seen that this energy is not really lost but is transformed (dissipated) from the useful mechanical energy form to non-useful thermal energy.

When Eqn (5-30) is introduced into the definition of enthalpy, we get

$$dh = du + d\left(\frac{P}{\rho}\right) = \delta q + \delta e_f + \frac{dP}{\rho} \qquad (5\text{-}31)$$

Substituting this for the enthalpy in the differential energy balance, Eqn (5-11), gives

$$\frac{dP}{\rho} + gdz + VdV + \delta w + \delta e_f = 0 \qquad (5\text{-}32)$$

This can be integrated along a streamline from the "inlet" to the "outlet" of the system to give

$$\int_{P_i}^{P_o} \frac{dP}{\rho} + g(z_o - z_i) + \frac{1}{2}(V_o^2 - V_i^2) + e_f + w = 0 \qquad (5\text{-}33)$$

where, from Eqn (5-30),

$$e_f = (u_o - u_i) - q + \int_{\rho_i}^{\rho_o} Pd\left(\frac{1}{\rho}\right) \qquad (5\text{-}34)$$

Equation (5-33) is simply a rearrangement of the steady-state energy balance equation (Eqn 5-10) and is just as general. Without the "friction loss" (e_f) term (which includes all of the thermal energy effects) it reduces to a "mechanical energy" balance, which is often referred to as the *Bernoulli equation*. Although some purists may argue, we will refer to Eqn (5-33) as the *generalized Bernoulli equation*, or just the Bernoulli equation, for short. Equation (5-33), along with Eqn (5-34), accounts for all of the possible thermal and mechanical energy terms and is the form of the energy balance that is most convenient when mechanical energy terms dominate and thermal effects are minor.

If the fluid is incompressible (constant density), Eqn (5-33) becomes

$$\frac{\Delta\Phi}{\rho} + \frac{1}{2}\Delta(V^2) + e_f + w = 0 \qquad (5\text{-}35)$$

where $\Phi = P + \rho gz$. For a fluid at rest, $e_f = V = w = 0$, and Eqn (5-35) reduces to the basic equation of fluid statics for an incompressible fluid (we will consider compressible fluids in more detail later). For gases, however, if the pressure change is such that the density does not change more than about 40%, the incompressible equation can be applied with reasonable accuracy by assuming the fluid density to be constant at a value equal to the average density in the system.

Note that if each term of Eqn (5-35) is divided by g, then each term will have the dimension of length. The result is called the "head" form of the Bernoulli equation, in which each term represents the equivalent amount of energy in a column of the fluid of the specified height. For example, the pressure term becomes the "pressure head $(-\Delta P/\rho g = H_P)$", the potential energy term becomes the "static head $(-\Delta z = H_z)$", the kinetic energy term becomes the "velocity head $(\Delta V^2/2g = H_v)$", the friction loss becomes the "head loss $(e_f/g = H_f)$", and the work term is, typically, the "pump head $(-w/g = H_w)$".

A. KINETIC ENERGY CORRECTION

In the above equations, we have assumed that the fluid velocity (v) at a given point in the system (e.g. in a tube) is uniform over the cross section of the flow stream. However, such is not the case in conduits, since the fluid velocity is zero at a non-moving boundary or wall and thus must increase with distance from the wall. Now the total rate at which kinetic energy is transported by a fluid element $(\dot{m})$ with velocity $\vec{v}$ through a differential area $\vec{A}$ is $(v^2 d\dot{m}2)$, where $d\dot{m} = \rho \vec{v} \cdot d\vec{A}$. Thus, the total rate of transport of kinetic energy through the cross section A is

$$\int \frac{1}{2} v^2 \, d\dot{m} = \frac{\rho}{2} \int v^3 \, dA \qquad (5\text{-}36)$$

If the fluid velocity is everywhere uniform at a value equal to the average velocity $(\overline{V})$, the rate of kinetic energy transport would be

$$\frac{1}{2} \rho \, \overline{V}^3 \, A \qquad (5\text{-}37)$$

Therefore, a *kinetic energy correction factor*, α, may be defined as the ratio of the true rate of kinetic energy transport relative to that which would occur if the fluid velocity is everywhere equal to the average velocity (e.g. plug flow):

$$\alpha = \frac{\text{True KE transport rate}}{\text{Plug flow KE transport rate}} = \frac{1}{A} \int_A \left(\frac{v}{\overline{V}}\right)^3 dA \qquad (5\text{-}38)$$

Thus, the Bernoulli equation should include this kinetic energy correction factor, i.e.

$$\frac{\Delta\Phi}{\rho} + \frac{1}{2} \Delta \left(\alpha \overline{V}^2\right) + e_f + w = 0 \qquad (5\text{-}39)$$

As will be shown later, the velocity profile for a Newtonian fluid in laminar flow in a circular tube is parabolic. When this is introduced into Eqn (5-38), the result is $\alpha = 2$. For highly turbulent flow, the profile is much flatter and $\alpha \approx 0.85$, although for practical applications it is usually assumed that $\alpha = 1$ for turbulent flow.

Example 5-2:   **Kinetic Energy Correction Factor for Laminar Flow of a Newtonian Fluid**—We will show later that the velocity profile for the laminar flow of a Newtonian fluid in a circular tube is parabolic. Since the velocity is zero at the wall of the tube and maximum in the center, the profile can be represented by

$$v(r) = v_{max} \left(1 - \frac{r^2}{R^2} \right)$$

This can be used to calculate the kinetic energy correction factor from Eqn (5-38) as follows. First we must calculate the average velocity, $\overline{V}$, using Eqn (5-3):

$$\overline{V} = \frac{1}{\pi R^2} \int_0^R v(r) 2\pi r \ dr$$

$$= 2 v_{max} \int_0^1 (1 - x^2) x \ dx$$

$$= v_{max}/2$$

This shows that the average velocity is simply one-half of the maximum (centerline) velocity. Thus, replacing $\overline{V}$ in Eqn (5-38) by $v_{max}/2$ and then integrating the parabolic velocity profile over the tube cross section leads to $\alpha = 2$ (the details of the manipulation are left as an exercise for the reader).

Example 5-3:   **Diffuser**—A diffuser is a section in a conduit over which the flow area increases gradually from upstream to downstream, as illustrated in Fig. 5-3. If the inlet and outlet areas (A_1 and A_2) are known and the upstream pressure and velocity (P_1 and V_1) are given, we would like to find the downstream pressure and velocity (P_2 and V_2). If the fluid is incompressible, the continuity equation gives V_2:

$$(\rho VA)_1 = (\rho VA)_2 \qquad \text{or} \qquad V_2 = V_1 \frac{A_1}{A_2}$$

The pressure (P_2) is determined by Bernoulli's equation. If the diffuser is horizontal, there is no work done between the inlet and outlet, and the friction loss is very small (which will be a good assumption for a well-designed diffuser

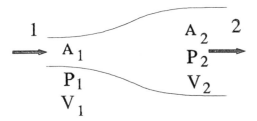

FIGURE 5-3 Diffuser.

with a gradually varying area), the Bernoulli equation gives

$$P_2 = P_1 + \frac{\rho}{2}(V_1^2 - V_2^2) - \rho\, e_f \cong P_1 + \frac{\rho V_1^2}{2}\left(1 - \frac{A_1^2}{A_2^2}\right)$$

Since $A_1 < A_2$, this shows that $P_2 > P_1$, i.e. the pressure increases because the kinetic energy decreases, and the losses are small. Such a diffuser is said to have a "high pressure recovery".

Example 5-4: **Sudden Expansion**—We now consider an incompressible fluid flowing from a small conduit, through a sudden expansion into a larger conduit, as illustrated in Fig. 5-4. The objective, as in the previous example, is to determine the exit pressure and velocity (P_2 and V_2), given the upstream conditions and the dimensions of the ducts. Since the conditions are all identical to those of the above diffuser example, the continuity and Bernoulli equations are also identical. The major difference is that the friction loss is not as small as for the diffuser. Because of inertia, the fluid cannot follow the sudden 90° change in direction of the boundary, so considerable turbulence is generated after the fluid

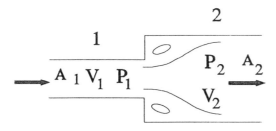

FIGURE 5-4 Sudden expansion.

leaves the small duct and before it can expand to fill the large duct. The equation for P_2 is the same as before:

$$P_2 = P_1 + \frac{\rho \, V_1^2}{2} \left(1 - \frac{A_1^2}{A_2^2} \right) - \rho \, e_f$$

The "pressure recovery" is reduced by the friction loss, which is relatively high for the sudden expansion. The pressure recovery is therefore relatively low.

Example 5-5: **The Torricelli Problem**—Consider an open vessel with diameter D_1 containing a fluid at a depth of h, which is draining out of a hole of diameter D_2 in the bottom of the tank. We would like to determine the velocity of the fluid flowing out of the hole in the bottom. As a first approximation, we neglect the friction loss in the tank and through the hole. Point 1 is taken at the surface of the fluid in the tank, point 2 is taken at the exit of the hole, and the pressure is atmospheric at both points. The velocity in the tank is related to that through the hole by the continuity equation:

$$(\rho VA)_1 = (\rho VA)_2$$

or

$$V_1 = V_2 \, \frac{A_2}{A_1} = V_2 \, \beta^2$$

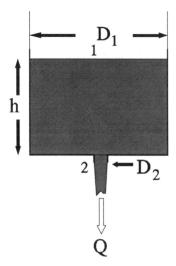

FIGURE 5-5   Draining tank—the Torricelli problem.

where $\beta = D_2/D_1$. The Bernoulli equation for an incompressible fluid between points 1 and 2 is

$$\frac{P_2 - P_1}{\rho} + g\,(z_2 - z_1) + \frac{1}{2}\,(\alpha_2 V_2^2 - \alpha_1 V_1^2) + w + e_f = 0$$

Since points 1 and 2 are both at atmospheric pressure, $P_2 = P_1$. We assume that $\alpha = 1$ and neglect friction so that $e_f = 0$ and $w = 0$. Setting $z_2 - z_1 = -h$, eliminating V_1 from these two equations, and solving for V_2 gives

$$V_2 = \sqrt{\frac{2gh}{1 - \beta^4}}$$

This is known as the *Torricelli equation*. Now consider what happens as the hole gets larger. Specifically, as $\beta \to 1$ (i.e. as $D_2 \to D_1$), then $V_2 \to \infty$! This is obviously an unrealistic limit, so there must be something wrong. Of course, our assumption that friction is negligible may be valid at low velocities, but as the velocity increases it becomes less valid, and it is obviously invalid long before this condition is reached.

Upon examining the equation for V_2, we see that it is independent of the properties of the fluid in the tank. We might suspect that this is not accurate, since if the tank were to be filled with CO_2 we intuitively expect that it would drain more slowly than if it were filled with water. So, what is wrong? In this case, it is our assumption that $P_2 = P_1$. Of course, the pressure is atmospheric at both points 1 and 2, but we have neglected the static head of air between these points, which represents a difference in the pressure. This results in a buoyant force due to the air and has a significant effect on the drainage of CO_2, although it will be negligible for water. Thus, if we account for the static head of air, i.e. $P_2 - P_1 = \rho_a g h$ in the Bernoulli equation, and then solve for V_2, we get

$$V_2 = \sqrt{\frac{2gh(1 - \rho_a/\rho)}{1 - \beta^4}}$$

where ρ is the density of the fluid in the tank. This now shows that as $\rho \to \rho_a$, the velocity goes to zero, as we would expect.

These examples illustrate the importance of knowing what can and cannot be neglected in a given problem and the need to match the appropriate assumptions to the specific problem conditions in order to arrive at a valid solution.

V. CONSERVATION OF MOMENTUM

A macroscopic momentum balance for a flow system must include all equivalent forms of momentum. In addition to the rate of momentum carried into and out of the system by the entering and leaving streams, we must include the sum of all of the forces that act *on* the system (the system being a specifically defined volume of *fluid*). This follows from Newton's second law, which shows the equivalence between force and the rate of momentum. The macroscopic conservation of momentum thus becomes

$$\sum_{\text{on fluid}} \vec{F} + \sum_{\text{in}} (\dot{m}\,\vec{V})_i - \sum_{\text{out}} (\dot{m}\,\vec{V})_o = \frac{d}{dt}(m\,\vec{V})_{\text{sys}} \qquad (5\text{-}40)$$

Note that this is a vector equation, since momentum is a vector. For three-dimensional space it can be represented by three component equations, one for each direction. If there is only one entering and one leaving stream, then $\dot{m}_i = \dot{m}_o = \dot{m}$. In addition, if the system is at steady state the momentum balance becomes

$$\sum_{\text{on fluid}} \vec{F} = \sum_{\text{out}} (\dot{m}\,\vec{V})_o - \sum_{\text{in}} (\dot{m}\,\vec{V})_i \qquad (5\text{-}41)$$

Note that the vector (directional) character of the ''rate of momentum'' terms (i.e. $\dot{m}\,\vec{V}$) is that of the velocity, since $\dot{m}$ is a scalar (that is, $\dot{m} = \rho\,\vec{V} \cdot \vec{A}$ is a scalar product).

A. ONE-DIMENSIONAL FLOW IN A TUBE

We will apply the steady-state momentum balance to a fluid flowing in a tube, as illustrated in Fig. 5-6 (the ''stream tube'' may be bounded by either solid or imaginary boundaries—the only condition is that no fluid crosses the boundaries other that through the ''inlet'' and ''outlet'' planes). The cross section does

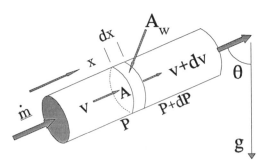

FIGURE 5-6 Momentum balance on a "slice" in a stream tube.

not have to be circular—it can be any shape. The fluid element in the "slice" of thickness dx is our system, and the momentum balance equation on this system is

$$\sum_{\text{on fluid}} F_x + \dot{m}V_x - \dot{m}(V_x + dV_x) = \sum_{\text{on fluid}} F_x - \dot{m}\, dV_x \tag{5-42}$$

$$= \frac{d}{dt}(\rho V_x A\, dx) = 0$$

The forces acting on the fluid include those due to pressure (dF_p), gravity (dF_g), wall drag (dF_w), and external shaft work (δW, not shown on Fig. 5-3):

$$\sum_{\text{on fluid}} F_x = dF_p + dF_s + F_{ext} + dF_w$$

where

$$dF_p = A_x[P - (P + dP)] = - A_x\, dP$$
$$dF_g = \rho g_x A_x\, dx = -\rho A_x\, dx\, g\, \cos\theta = -\rho g A_x\, dz$$
$$dF_w = -\tau_w\, dA_w = -\tau_w W_p\, dx$$
$$-\delta W = F_{ext}\, dx \quad \text{or} \quad F_{ext} = -\frac{\delta W}{dx} \tag{5-43}$$

Here, τ_w is the stress exerted *by* the fluid *on* the wall, and W_p is the perimeter of the wall in the cross section that is wetted by the fluid (the "wetted perimeter"). After substituting the expressions for the forces from Eqn (5-43) into the momentum balance equation, Eqn (5-42), and dividing the result by $-\rho A$, where $A = A_x$, the result is

$$\frac{dP}{\rho} + g\, dz + \frac{\tau_w W_p}{\rho A}\, dx + \delta w + V\, dV = -\frac{dV}{dt}\, dx \tag{5-44}$$

where $\delta w = \delta W/(\rho A\, dx)$ is the work done per unit mass of fluid. Integrating this expression from the inlet to the outlet, and assuming steady state, gives

$$\int_{P_i}^{P_o} \frac{dP}{\rho} + g(z_o - z_i) + \frac{1}{2}(V_o^2 - V_i^2) + \int_L \frac{\tau_w W_p}{\rho A}\, dx + w = 0 \tag{5-45}$$

If this is compared with the Bernoulli equation (Eqn 5-33), they are seen to be identical, provided

$$e_f = \int_L \left(\frac{\tau_w W_p}{\rho A}\right) dx \tag{5-46}$$

That is,

$$e_f \cong \frac{\tau_w W_p L}{\rho A} = \frac{\tau_w}{\rho}\left(\frac{4L}{D_h}\right) \tag{5-47}$$

where

$$D_h = 4\left(\frac{A}{W_p}\right) \tag{5-48}$$

is called the *hydraulic diameter*. Note that this applies to a conduit of any cross-sectional shape. For a circular tube, D_h is identical to the tube diameter.

We see that there are several ways of interpreting the term e_f. From the Bernoulli equation, it represents the "lost" (or "dissipated") energy associated with the irreversibilities of the flow. From the momentum balance, it is also seen to be a consequence of the stress between the fluid and the tube wall (τ_w), i.e. it can be interpreted as the work required to overcome the resistance to flow in the conduit. These interpretations are both correct and are equivalent.

Although the energy and momentum balances lead to equivalent results for this special case of one-dimensional fully developed flow in a tube, this is an exception and not a rule. In general, the momentum balance gives additional information relative to the forces exerted on and/or by the fluid in the system, which is not given by the energy balance or Bernoulli equation. This will be illustrated subsequently.

B. THE LOSS COEFFICIENT

From observation of the Bernoulli equation, we see that the friction loss (e_f) may be made dimensionless by dividing by the kinetic energy per unit mass of fluid. The result is the dimensionless *loss coefficient*, K_f:

$$K_f = \frac{e_f}{V^2/2} \tag{5-49}$$

A loss coefficient may be associated with any element that offers resistance to flow (i.e. in which energy is dissipated), such as a length of conduit, a valve, pipe fitting, contraction, expansion, etc. This will be discussed in detail later.

As can be seen from Eqn (5-46), the pipe wall stress may be made dimensionless by dividing by the kinetic energy per unit volume of fluid. The result is known as the *Fanning friction factor, f*:

$$f = \frac{\tau_w}{\rho V^2/2} \tag{5-50}$$

Although $\rho V^2/2$ is kinetic energy per unit volume, ρV^2 is also the *flux of momentum* carried by the fluid along the conduit. The latter interpretation is more logical in Eqn (5-50), since the wall stress is equivalent to a flux of momentum from the fluid to the wall. However, the "conventional" definition includes the (arbitrary) factor of 1/2. Other definitions of the pipe friction factor are in use, which are some multiple of the Fanning friction factor. For example, the Darcy friction factor, which is equal to 4f, is used frequently by mechanical and civil engineers. Thus, it is important to know which definition is implied when data for friction factors are used.

Since the friction loss and wall stress are related by Eqn (5-47), the loss coefficient for pipe flow and the Fanning friction factor are related as follows:

$$K_f = \frac{4\,fL}{D_h}\ (\text{pipe}) \tag{5-51}$$

Example 5-6: **Friction Loss in a Sudden Expansion**—Figure 5-7 shows the flow in a sudden expansion from a small conduit to a larger one. We assume that the conditions upstream of the expansion (point 1) are known, as well as the areas A_1 and A_2. We desire to find the velocity and pressure downstream of the expansion (V_2 and P_2) as well as the loss coefficient K_f. As before, V_2 is determined from the mass balance, or continuity equation, applied to the system (which is the fluid in the shaded area). Assuming constant density:

$$V_2 = V_1\,\frac{A_1}{A_2}$$

The Bernoulli equation for this system is

$$\frac{P_2 - P_1}{\rho} + \frac{1}{2}\,(V_2^2 - V_1^2) + e_f = 0$$

This equation contains two unknowns, P_2 and e_f. Thus, we need another equation, which is the momentum balance:

$$\sum F_x = \dot{m}(V_{2x} - V_{1x})$$

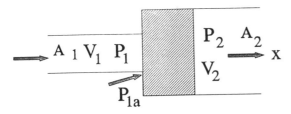

FIGURE 5-7 Sudden expansion.

where $V_{1x} = V_1$ and $V_{2x} = V_2$, since the flow is in only one direction. Accounting for all of the forces that can act on the system through each section of the boundary, this becomes

$$P_1 A_1 + P_{1a}(A_2 - A_1) - P_2 A_2 + F_{wall} = \rho V_1 A_1 (V_2 - V_1)$$

where P_{1a} is the pressure on the left boundary of the system (i.e. the washer-shaped surface), and F_{wall} is the force due to the drag of the wall on the fluid at the horizontal boundary of the system. Since the fluid pressure cannot change discontinuously, $P_{1a} \cong P_1$. Also, since the contact area with the wall of the system is relatively small, we may neglect F_{wall} with no serious consequences. The result is

$$(P_1 - P_2)A_2 = \rho V_1^2 A_1 \left(\frac{A_1}{A_2} - 1 \right)$$

This can be solved for P_2, which, when inserted into the Bernoulli equation, allows us to solve for e_f:

$$e_f = \frac{V_1^2}{2} \left(1 - \frac{A_1}{A_2} \right)^2 = \frac{K_f V_1^2}{2}$$

Thus,

$$K_f = \left(1 - \frac{A_1}{A_2} \right)^2 = (1 - \beta^2)^2$$

where

$$\beta = D_1/D_2$$

The loss coefficient is seen to be a function only of the geometry of the system (we have implicitly assumed that the flow is plug flow, i.e. highly turbulent). For most systems (flow in valves, fittings, etc.), the loss coefficient cannot be determined accurately from simple theoretical concepts (as in this case) but must be determined empirically.

Example 5-7: **Flange Forces on a Pipe Bend**—Consider an incompressible fluid flowing through a pipe bend, as illustrated in Fig. 5-8. We would like to determine the forces in the bolts in the flanges which hold the bend in the pipe, knowing the geometry of the bend, the flow rate through the bend, and the exit pressure (P_2) from the bend. Taking the system to be the fluid within the pipe bend, a steady-state balance on "x momentum" is

$$\sum (F_x)_{on\ sys} = \dot{m}(V_{2x} - V_{1X})$$

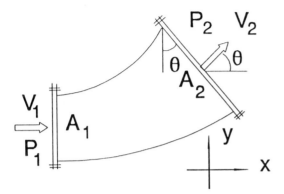

Figure 5-8 Flange forces in pipe bend.

Various factors contribute to the forces on the left side of this equation, namely:

$$\sum (F_x)_{\text{on fluid}} = P_1 A_{1x} + P_2 A_{2x} - (F_x)_{\text{on wall by fluid}}$$

$$= P_1 A_1 - P_2 A_2 \cos\theta - (F_x)_{\text{on bolts}}$$

(Note that each "area" has two sides, and it is the side in contact with the "system" fluid that is the one we want, e.g. the A_{x2} normal vector may point in either the $+x$ or $-x$ direction, but the side of A_2 in contact with our system is on the $-x$ side, so that $A_{2x} = -A_2 \cos\theta$). Also, the right side of the momentum balance becomes

$$\dot{m}(V_{2x} - V_{1x}) = \dot{m}(V_2 \cos\theta - V_1)$$

Solving these equations for $(F_x)_{\text{on wall}}$ gives

$$(F_x)_{\text{on wall}} = (F_x)_{\text{on bolts}} = P_1 A_1 - P_2 A_2 \cos\theta - \dot{m}(V_2 \cos\theta - V_1)$$

Similarly, the "y momentum" balance is

$$\sum (F_y)_{\text{on sys}} = \dot{m}(V_{2y} - V_{1y})$$

which becomes

$$(F_y)_{\text{on wall}} = (F_y)_{\text{on bolts}} = -P_2 A_2 \sin\theta - \dot{m}V_2 \sin\theta$$

This assumes that the xy plane is horizontal. If the y direction is vertical, the total weight of the bend including the fluid inside could be included as an additional (negative) force component. The magnitude and direction of the net force are

$$\overline{F} = \sqrt{F_x^2 + F_y^2}, \quad \phi = \tan^{-1}\left(\frac{F_y}{F_x}\right)$$

where ϕ is the direction of the net force, measured counterclockwise from the $+x$-axis. Note that either P_1 or P_2 must be known, but that the other is determined by the Bernoulli equation if the loss coefficient is known:

$$\frac{P_2 - P_1}{\rho} + g(z_2 - z_1) + \frac{1}{2}(V_2^2 - V_1^2) + e_f = 0 \text{ where } e_f = \frac{1}{2}K_fV_1^2$$

Methods for evaluating the loss coefficient K_f will be discussed later.

It should be noted that, when evaluating the forces acting on the system due to the pressure, the effects of the pressure transmitted to the system from the surrounding atmosphere have not been included. Although this pressure does transmit forces through the system boundaries to the fluid, these forces all cancel out so that it is the *net* pressure (e.g. the gauge pressure) in the system which should be used in the momentum balance equation.

C. CONSERVATION OF ANGULAR MOMENTUM

In addition to the conservation of linear momentum, an expression for the conservation of angular momentum (or the moment of momentum) can also be written. For a fixed mass (m) moving in the x direction with a velocity of V_x, the linear x momentum (M_x) is mV_x. Likewise, a mass m rotating counterclockwise about a center of rotation at an angular velocity $\omega = d\theta/dt$ has an angular momentum (L_θ) equal to $mV_\theta R = m\omega R^2$, where R is the distance from the center of rotation to m. Note that the angular momentum has dimensions of length times momentum and is often referred to as the "moment of momentum". If the mass is not a point but is a rigid distributed mass (M) rotating at a given angular velocity, the total angular momentum is given by

$$L_\theta = \int_M \omega r^2 \, dm = \omega \int_M r^2 dm = \omega I \qquad (5\text{-}52)$$

where I is the moment of inertia of the body with respect to the center of rotation.

For a fixed mass, the conservation of linear momentum is equivalent to Newton's second law:

$$\sum \vec{F} = m\vec{a} = \frac{d(m\vec{V})}{dt} = m\frac{d\vec{V}}{dt} \qquad (5\text{-}53)$$

and the corresponding expression for the conservation of angular momentum is

$$\sum \Gamma_\theta = \sum F_\theta R = \frac{d(I\omega)}{dt} = I\frac{d\omega}{dt} = I\alpha \qquad (5\text{-}54)$$

where Γ_θ is the moment (torque) acting on the system, and $d\omega/dt = \alpha$ is the angular acceleration.

For a flow system in which streams with curved streamlines may carry angular momentum into and out of the system, the general macroscopic angular momentum balance is

$$\sum_{in} (\dot{m}RV_\theta)_i - \sum_{out} (\dot{m}RV_\theta)_o + \sum \Gamma_\theta = \frac{d(I\omega)}{dt} = I\alpha \qquad (5\text{-}55)$$

For a steady-state system with only one inlet and one outlet stream, this becomes

$$\sum \Gamma_\theta = \dot{m}[(RV_\theta)_o - (RV_\theta)_i] = \dot{m}[(R^2\omega)_o - (R^2\omega)_i] \qquad (5\text{-}56)$$

This equation is also known as the *Euler turbine equation*, since it applies directly to turbines and all rotating fluid machinery. We will find it useful later in the analysis of the performance of centrifugal pumps.

D. MOVING BOUNDARY SYSTEMS—RELATIVE MOTION

We sometimes encounter a system which is in contact with a moving boundary, such that the fluid which composes the system is carried along with the boundary while streams carrying momentum and/or energy flow into and/or out of the system. Examples of this include the flow impinging on a turbine blade in which the system is the fluid in contact with the moving blade or flow of exhaust gases from a moving rocket motor. In such cases we generally have information concerning the velocity of the fluid relative to the boundary (i.e. relative to the system) V_r and we must also consider the velocity of the system V_s in order to determine the absolute velocity of the fluid which is required for the conservation equations. For example, consider a system which is moving in the x direction with a velocity of V_s, a fluid stream entering the system with a velocity in the x direction relative to the system of V_{ri}, and a stream which leaves the system with a velocity of V_{ro} relative to the system. The absolute velocity in the x direction is related to the relative velocity and the system velocity by

$$V_x = V_{sx} + V_{rx} \qquad (5\text{-}57)$$

and the linear momentum balance equation becomes

$$\sum \vec{F} = \dot{m}_o \vec{V}_o - \dot{m}_i \vec{V}_i \dot{m}_i^{\rightarrow}{}_i + \frac{d(m\vec{V}_s)}{dt}$$

$$= \dot{m}_o(\vec{V}_{ro} + \vec{V}_s) - \dot{m}_i(\vec{V}_{ri} + \vec{V}_s) + \frac{d(m\vec{V})}{dt} \qquad (5\text{-}58)$$

Example 5-8: **Turbine Blade**—Consider a fluid stream impinging on a turbine blade which is moving with a velocity V_s. We would like to know what the velocity of the impinging stream should be in order to transfer the maximum energy to the blade. The impinging steam velocity is V_i, the stream leaves the blade at V_o, and the system is the fluid in contact with the blade, which is moving at velocity V_s. Since $V_o = V_{ro} + V_s$ and $V_i = V_{ri} + V_s$, the system velocity cancels out of the momentum equation

$$F_x = \dot{m}(V_o - V_i) = \dot{m}(V_{ro} - V_{ri})$$

If the friction loss is negligible, the work done by the fluid on the blade is equal to the change in kinetic energy of the fluid, by Bernoulli's equation:

$$-w = \frac{1}{2}(V_i^2 - V_o^2)$$

which shows that the maximum work is done when $V_o = 0$. Now the continuity equation for the fluid stream is

$$\dot{m}_i = \dot{m}_o, \text{ or}$$
$$V_{ri} = -V_{ro}, \text{ or}$$
$$(V_i - V_s) = (V_o - V_s)$$

since the density doesn't change. Solving for V_s gives

$$V_s = \frac{1}{2}(V_i - V_o)$$

Since the maximum work is done when $V_o = 0$, this corresponds to

$$V_i = 2 V_s$$

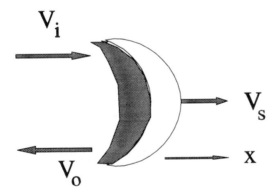

FIGURE 5-9   Turbine blade.

for the condition at which the maximum work will be transferred from the fluid to the blade.

E. MICROSCOPIC MOMENTUM BALANCE

We may apply the conservation of momentum principle to a system composed of an arbitrary (differential) cubical volume within any flow field. Accounting for convection of momentum through all of the surfaces, all possible stress components on any and all of the surfaces, and any body forces (e.g. gravity), a general microscopic form of the momentum equation can be derived which is valid at all points within any fluid. This is done in a manner similar to the derivation of the microscopic mass balance (continuity) equation, by dividing the balance equation by the (differential) volume, then taking the limit as the dimensions of the cube shrink to zero. (This procedure can be found in various textbooks and is left as an exercise for the reader.) The result is a vector equation which can be represented in general form as

$$\rho\left(\frac{\partial \vec{v}}{\partial t} + \vec{v} \cdot \vec{\nabla}\vec{v}\right) = -\vec{\nabla}P + \vec{\nabla} \cdot \vec{\tau} + \rho\vec{g} \qquad (5\text{-}59)$$

The three component equations of this momentum equation are given in detail in the Appendix for Cartesian, cylindrical, and spherical coordinates. When coupled with appropriate equations relating the shear stress components $(\vec{\tau})$ to the velocity components, the result is a set of differential equations which can be solved (in principle) with the appropriate boundary conditions for the velocity components as a function of time and space. In laminar flows, the shear stress components depend only upon the velocity gradients through an appropriate rheological constitutive equation. For example, for Newtonian fluids, the constitutive equation, generalized from the one-dimensional form (i.e. $\tau = \mu\dot{\gamma}$), is

$$\vec{\tau} = \mu[(\vec{\nabla}\vec{v}) + (\vec{\nabla}\vec{v})'] \qquad (5\text{-}60)$$

The components of this equation are given in Appendix D for Cartesian, cylindrical, and spherical coordinate systems. When these are used to eliminate the stress components from the momentum equation, the result is called the *Navier-Stokes equations*. These equations apply to the laminar flow of any Newtonian fluid in any system and are the starting point for the detailed solution of many fluid flow problems. Similar equations may be developed for non-Newtonian fluids, based upon the appropriate rheological model for the fluid. However, the number of flow problems for which analytical solutions are possible is rather limited, so that numerical computer techniques are required for many problems of practical interest. These procedures are beyond the scope of this book, but we will illustrate the application of the momentum equations to the solution of a simple example problem.

Example 5-9:   **Flow Down an Inclined Plane**—Consider the flow of a thin layer or film of liquid down a flat plate which is inclined at an angle θ to the vertical, as illustrated in Fig. 5-10. The width of the plate is W (normal to the plane of the figure). Flow is only in the x direction (parallel to the surface), and the velocity will vary only in the y direction (normal to the surface). These prescribed conditions constitute the definition of the problem to be solved. The objective is to determine the film thickness δ as a function of the flow rate per unit width of plate (Q/W), the fluid properties, and other parameters in the problem. Since $v_y = v_z = 0$, the microscopic mass balance (continuity equation) reduces to

$$\frac{\partial v_x}{\partial x} = 0$$

which tells us that the velocity v_x is independent of x. Hence, the only independent variable is y. Considering the x component of the momentum equation, and discarding all y and z velocity and stress components and all derivatives other than those with respect to the y direction, the result is

$$0 = \frac{\partial \tau_{yx}}{\partial y} + \rho g \cos\theta$$

The pressure gradient term has been discarded because the system is open to the atmosphere and therefore the pressure is constant (or, at most, hydrostatic)

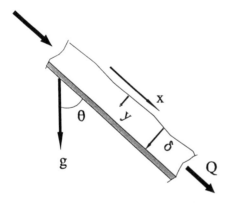

FIGURE 5-10   Flow down inclined plane.

everywhere. This equation can be integrated to give the shear stress distribution in the film:

$$\tau_{yx} = -\rho g y \cos\theta$$

where the constant of integration is zero, since there is zero (negligible) stress at the free surface of the film ($y = 0$). Note that this result is valid for any fluid under any flow conditions (laminar or turbulent), since it is simply a statement of the conservation of momentum. If the fluid is Newtonian and the flow is laminar, the shear stress is

$$\tau_{yx} = \mu \frac{\partial v_x}{\partial y}$$

Eliminating the stress between the last two equations gives an equation which can be integrated to give the velocity distribution:

$$v_x = \frac{\rho g \delta^2 \cos\theta}{2\,\mu} \left(1 - \frac{y^2}{\delta^2} \right)$$

where the boundary condition that $v_x = 0$ at $y = \delta$ (the wall) has been used to evaluate the constant of integration. The volumetric flow rate can now be determined:

$$Q = W \int_0^\delta v_x \, dy = \frac{W\rho g \delta^3 \cos\theta}{3\mu}$$

The film thickness is seen to be proportional to both the cube root of the flow rate and the fluid viscosity. The wall stress on the plate is seen to be

$$\tau_w = (-\tau_{yx})_{y=\delta} = \rho g \delta \cos\theta$$

which is just the component of the weight of the fluid on the plate acting parallel to the plate.

The above procedure can be used to solve a variety of steady, fully developed laminar flow problems, such as flow in a tube or in a slit between parallel walls, for Newtonian or non-Newtonian fluids. However, if the flow is turbulent, the turbulent eddies transport momentum in three dimensions within the flow field, which contributes additional momentum flux components to the shear stress terms in the momentum equation. The resulting equations cannot be solved exactly for such flows, and the methods for analyzing them will be discussed in the following chapter.

PROBLEMS

CONSERVATION OF MASS AND ENERGY

1. Water is flowing into the top of a tank at a rate of 200 gpm. The tank is 18 in. in diameter and has a 3 in. diameter hole in the bottom, through which the water flows out. If the inflow rate is adjusted to match the outflow rate, what will the height of the water be in the tank if friction is negligible?

2. A vacuum pump operates at a constant volumetric flow rate of 10 liters/min, based upon the pump inlet conditions. How long will it take to pump down a 100 liter tank containing air from 1 atm to 0.01 atm, assuming the temperature is constant?

3. Air is flowing at a constant mass flow rate into a tank having a volume of 3 ft^3. The temperature of the tank and air is constant at 70°F. If the pressure in the tank is observed to increase at a rate of 5 psi/min, what is the flow rate of air into the tank?

4. A tank contains water initially at a depth of 3 ft. The water flows out of a hole in the bottom of the tank, and air at a constant pressure of 10 psig is admitted to the top of the tank. If the water flow rate is directly proportional to the gauge pressure at the bottom of the tank, derive expressions for the water flow rate and air flow rate as a function of time. Be sure to define the symbols you use in your equations.

5. The flow rate of a hot coal-oil slurry in a pipeline is measured by injecting a small side stream of cool oil and measuring the resulting temperature change downstream in the pipeline. The slurry is initially at 300°F and has a density of 1.2 g/cm^3 and a specific heat of 0.7 BTU/(lb$_m$ °F). With no side stream injected, the temperature downstream of the mixing point is 298°F. With a side stream at 60°F and a flow rate of 1 lb$_m$/s, the temperature at this point is 295°F. The side stream has a density of 0.8 g/cm^3 and a C$_p$ of 0.6 BTU/(lb$_m$ °F). What is the mass flow rate of the slurry?

6. A gas enters a horizontal 3 in. sch 40 pipe at a constant rate of 0.5 lb$_m$/s, a temperature of 70°F, and pressure of 1.15 atm. The pipe is wrapped with an 20 kW heating coil covered with a layer of thick insulation. At the point where the gas is discharged, the pressure is

1.05 atm. What is the gas temperature at the discharge point, assuming it to be ideal with a MW of 29 and a C_p of 0.24 BTU/(lb$_m$ °F)?

7. Water is flowing into the top of an open cylindrical tank (diameter D) at a rate of Q_i and out of a hole in the bottom at a rate of Q_o. The tank is made of wood which is very porous, and the water is leaking out through the wall uniformly at a rate of q per unit area of wetted surface area. The initial depth of water in the tank is z_1. Derive an equation for the depth of water in the tank at any time. If Q_i = 10 gpm, Q_o = 5 gpm, D = 5 ft, q = 0.1 gpm/ft^2 and z_1 = 3 ft, is the level in the tank rising or falling?

8. Air is flowing steadily through a horizontal tube at a constant temperature of 32°C and a mass flow rate of 1 kg/s. At one point upstream, where the tube diameter is 50 mm, the pressure is 345 kPa. At another point downstream, the diameter is 75 mm, and the pressure is 359 kPa. What is the value of the friction loss (e_f) between these two points? [C_p = 1005 J/(kg K)]

9. Steam is flowing through a horizontal nozzle. At the inlet, the velocity is 1000 ft/s, and the enthalpy is 1320 BTU/lb$_m$. At the outlet, the enthalpy is 1200 BTU/lb$_m$. If heat is lost through the nozzle at a rate of 5 BTU/lb$_m$ of steam, what is the outlet velocity?

10. Oil is being pumped from a large storage tank, where the temperature is 70°F, through a 6 in. I.D. pipeline. The oil level in the tank is 30 ft above the pipe exit. If a 25 hp pump is required to pump the oil at a rate of 600 gpm through the pipeline, what would the temperature of the oil at the exit be if no heat is transferred across the pipe wall? State any assumptions that you make. Oil properties: SG = 0.92, μ = 35 cP, C_p = 0.5 BTU/(lb$_m$ °F).

11. Freon-12 enters a 1 in. sch 80 pipe at 170°F, 100 psia, and a velocity of 10 ft/s. At a point somewhere downstream, the temperature has dropped to 140°F and the pressure to 15 psia. Calculate the velocity at the downstream conditions and the Reynolds number at both the upstream and downstream conditions.

12. Number 3 fuel oil (30° API) is transferred from a storage tank at 60°F to a feed tank in a power plant at a rate of 2000 bbl/day. Both tanks are open to the atmosphere and are connected by a pipeline containing 1200 ft equivalent length of 1 1/2 in. sch 40 steel pipe

and fittings. The level in the feed tank is 20 ft higher than that in the storage tank, and the transfer pump is 60% efficient.

(a) What horsepower motor is required to drive the pump?

(b) If the heat capacity of the oil is 0.5 BTU/(lb$_m$ °F) and the pump and transfer line are perfectly insulated, what is the temperature of the oil entering the feed tank?

13. You must pump fuel oil from a storage tank (which is vented to the atmosphere) to a burner nozzle in a small boiler, through a 1 in. sch 40 pipe, 200 ft long, at a rate of 25 gpm. The oil level in the tank is 15 ft above the nozzle in the boiler, and the line pressure at the end of the pipe at the nozzle must be 90 psig.

(a) What is the value of the Reynolds number in the pipe and the horsepower of the motor required to drive the pump, if it is 60% efficient?

(b) If the pump and line are insulated, how much will the temperature of the oil change across the pump, and also across the entire line, from the tank to the burner?

[Oil properties: density = 0.87 g/cm^3, viscosity = 65 cP, specific heat = 0.5 BTU/(lb$_m$ °F)]

14. The fuel oil in Problem 12 is contained in an open cylindrical tank, which is 6 in. in diameter and 2 ft deep. A 1/4 in. I.D. vertical tube which is 3 ft long is attached to the bottom of the tank. The oil drains by gravity from the tank, down through the tube, and is collected in an open bucket. Calculate the flow rate of the oil (in gpm) and the Reynolds number in the tube. (Neglect any friction loss in the tank and the entrance to the tube.)

15. Water is pumped at a rate of 90 gpm by a centrifugal pump driven by a 10 hp motor. The water enters the pump through a 3 in. sch 40 pipe at 60°F and 10 psig, and leaves through a 2 in. sch 40 pipe at 100 psig. If the water loses 0.1 BTU/lb$_m$ while passing through the pump, what is the water temperature leaving the pump?

16. A pump, driven by a 7.5 hp motor, takes water in at 75°F and 5 psig and discharges it at 60 psig, at a flow rate of 600 lb$_m$/min. If no heat is transferred to or from the water while it is in the pump, what will the temperature of the water be leaving the pump?

17. A high-pressure pump takes water in at 70°F, 1 atm, through a 1 in. I.D. suction line and discharges it at 1000 psig through a 1/8 in. I.D.

line. The pump is driven by a 20 hp motor and is 65% efficient. If the flow rate is 500 g/s, and the temperature of the discharge is 73°F, how much heat is transferred between the pump casing and the water, per pound of water? Does it go into or out of the water?

BERNOULLI'S EQUATION

18. Water is flowing from one large tank to another through a 1 in. diameter pipe. The level in tank A is 40 feet above the level in tank B. The pressure above the water is 5 psig in tank A, and 20 psig in tank B. Which direction is the water flowing?

19. A pump which is driven by a 7.5 hp motor takes water in at 75°F and 5 psig and discharges it at 60 psig, at a flow rate of 600 lb$_m$/min. If no heat is transferred between the water in the pump and the surroundings, what will be the temperature of the water leaving the pump?

20. A 90% efficient pump driven by a 50 hp motor is used to transfer water at 70°F from a cooling pond to a heat exchanger through a 6 in. sch 40 pipeline. The heat exchanger is located 25 ft above the level of the cooling pond, and the water pressure at the discharge end of the pipeline is 40 psig. With all valves in the line wide open, the water flow rate is 650 gpm. What is the rate of energy loss in the pipeline due to friction, in kilowatts?

21. A pump takes water from the bottom of a large tank where the pressure is 50 psig, and delivers it through a hose to a nozzle which is 50 ft above the bottom of the tank, at a rate of 100 lb$_m$/s. The water exits the nozzle into the atmosphere at a velocity of 70 ft/s. If a 10 hp motor is required to drive the pump, which is 75% efficient, find (a) the friction loss in the pump and (b) the friction loss in the rest of the system. Express your answers in units of ft lb$_f$ /lb$_m$.

22. You have purchased a centrifugal pump to transport water at a maximum rate of 1000 gpm from one reservoir to another, through an 8 in. sch 40 pipeline. The total pressure drop through the pipeline is 50 psi. If the pump has an efficiency of 65% at maximum flow conditions, and there is no heat transferred across the pipe wall or the pump casing, calculate:
(a) The temperature change of the water through the pump.
(b) The horsepower of the motor that would be required to drive the pump.

23. The hydraulic turbines at Boulder Dam power plant are rated at 86,000 kW, when water is supplied at a rate of 66.3 m³/s. The water enters at a head of 145 m at 20°C and leaves through a 6 m diameter duct.
    (a) Determine the efficiency of the turbines.
    (b) What would be the rating of these turbines if the dam power plant was on Jupiter (g = 26 m/s²)?

24. Water is draining from an open conical funnel. The diameter at the top of the funnel is 1 cm and that at the bottom is 0.5 cm, and the funnel is 5 cm high. The friction loss in the funnel per unit mass of fluid is given by 0.4V², where V is the velocity leaving the funnel. What is the volumetric flow rate of the water and the value of the Reynolds number entering and leaving the funnel?

25. Water is being transferred between two open tanks (from A to B) at a rate of 100 gpm by a pump. The pump receives the water from the bottom of tank A through a 3 in. sch 40 pipe and discharges it into the top of tank B through a 2 in. sch 40 pipe. The point of discharge into B is 75 ft higher than the surface of the water in A. The friction loss in the piping system is 8 psi, and both tanks are 50 ft in diameter. What is the head (in feet) which must be delivered by the pump to move the water at the desired rate? If the pump is 70% efficient, what horsepower motor is required to drive the pump?

26. A 4 in. diameter open can has a 1/4 in. diameter hole in the bottom. The can is immersed bottom down in a pool of water, to a point where its bottom is 6 in. below the water surface, and is held there while the water flows through the hole into the can. How long will it take for the water in the can to rise to the same level as that outside the can? Neglect friction, and assume a "pseudo steady state", i.e. time changes are so slow that at any instant the steady-state Bernoulli equation applies.

27. Carbon tetrachloride (SG = 1.6) is being pumped at a rate of 2 gpm through a pipe which is inclined upward at an angle of 30°. An inclined tube manometer (with a 10° angle of inclination) using mercury as the manometer fluid (SG = 13.6) is connected between two taps on the pipe, which are 2 ft apart. The manometer reading is 6 in. If no heat is lost through the tube wall, what is the temperature rise of the CCl₄ over a 100 ft length of the tube?

28. A pump which is taking water at 50°F from an open tank at a rate of 500 gpm is located directly over the tank. The suction line entering the pump is a nominal 6 in. sch 40 straight pipe 10 ft long and extends 6 ft below the surface of the water in the tank. If friction in the suction line is neglected, what is the pressure at the pump inlet (in psi)?

29. A pump is transferring water from tank A to tank B, both of which are open to the atmosphere, at a rate of 200 gpm. The surface of the water in tank A is 10 ft above ground level, and that in tank B is 45 ft above ground level. The pump is located at ground level, and the discharge line which enters tank B is 50 ft above ground level at its highest point. All piping is 2 in. I.D., and the tanks are 20 ft in diameter. If friction is neglected, what would be the required pump head rating for this application (in feet), and what size motor (horsepower) would be needed to drive the pump if it is 60% efficient? (Assume the temperature is constant at 77°F.)

30. A surface effect (air cushion) vehicle measures 10 ft by 20 ft and weighs 6000 lb_f. The air is supplied by a blower mounted on top of the vehicle, which must supply sufficient power to lift the vehicle 1 in. off the ground. Calculate the required blower capacity in scfm (standard ft^3/min), and the horsepower of the motor required to drive the blower if it is 80% efficient. Neglect friction, and assume the air to be an ideal gas at 80°F with properties evaluated at an average pressure.

31. If the air cushion car in Problem 21 is equipped with a 2 hp blower which is 70% efficient, what would be (a) the clearance between the skirt of the car and the ground and (b) the air flow rate in scfm.

32. An ejector pump operates by injecting a high-speed fluid stream into a slower stream to increase its pressure. Consider water flowing at a rate of 50 gpm through a 90° elbow in a 2 in. I.D. pipe. A stream of water is injected at a rate of 10 gpm through a 1/2 in. I.D. pipe through the center of the elbow in a direction parallel to the downstream flow in the larger pipe. If both streams are at 70°F, determine the increase in pressure in the larger pipe at the point where the two streams mix.

33. A large tank has a 51 mm diameter hole in the bottom. When the depth of the liquid is 15 m above the hole, the flow rate through the

hole is found to be 0.0324 m^3/s. What is the head loss due to friction in the hole?

34. Water at 68°F is pumped through a 1000 ft length of 6 in. sch 40 pipe. The discharge end of the pipe is 100 ft above the suction end. The pump is 90% efficient and is driven by a 25 hp motor. If the friction loss in the pipe amounts to 70 ft lb$_f$/lb$_m$ of water, what is the flow rate through the pipe, in gpm? (P_{in} = P_{out} = 1 atm)

35. You want to siphon water out of a large tank using a 5/8 in. I.D. hose. The highest point of the hose is 10 ft above the water surface in the tank, and the hose exit is 5 ft below this level. If friction is neglected, what would be the flow rate through the hose (in gpm), and what is the minimum pressure in the hose (in psi)?

36. It is desired to siphon a volatile liquid out of a deep open tank. If the liquid has a vapor pressure of 200 mm Hg and a density of 45 lb$_m$/ft^3, and the surface of the liquid is 30 ft below the top of the tank, is it possible to siphon the liquid? If so, what would be the velocity be through a frictionless siphon 1/2 in. in diameter if the exit of the siphon tube is 3 ft below the level in the tank?

37. The propeller of a speedboat is 1 ft in diameter and 1 ft below the surface of the water. At what speed (rpm) will cavitation occur? The vapor pressure of the water is 18.65 mm Hg at 70°F.

38. A conical funnel is full of liquid. The diameter of the top (mouth) is D$_1$, and that of the bottom (spout) is D$_2$ (where D$_2$ << D$_1$), and the depth of the fluid above the bottom is H$_o$. Derive an expression for the time required for the fluid to drain by gravity to a level of H$_o$/2, assuming frictionless flow.

39. An open cylindrical tank, which is 2 ft in diameter and 4 ft high, is full of water. If the tank has a 2 in. diameter hole in the bottom, how long will it take for half of the water to drain out, if friction is neglected?

40. A window is left slightly open while the air conditioning system is operating. The air conditioning blower develops a pressure of 2 in. of water (gauge) inside the house, and the window opening measures 1/8 in. x 20 in. Neglecting friction, what is the flow rate of air

through the opening in cfm (ft^3/min), and how much horsepower is required to move this air?

41. A large tank has a 5.1 mm diameter hole in the bottom. When the depth of liquid in the tank is 1.5 m above the hole, the flow rate through the hole is found to be 324 cm^3/s. What is the head loss due to friction in the hole?

42. A window is left slightly open while the air conditioning system is running. The air conditioning blower develops a pressure of 2 in. of water (gauge) inside the house, and the window opening measures 1/8 in. by 20 in. Neglecting friction, what is the flow rate of air through the opening, in scfm (standard ft^3/min; use 60°F, 1 atm as standard conditions)? How much horsepower is required to move this air?

43. Water at 68°F is pumped through a 1000 ft length of 6 in. sch 40 pipe. The discharge end of the pipe is 100 ft above the suction end. The pump is 90% efficient and is driven by a 25 hp motor. If the friction loss in the pipe amounts to 70 ft lb$_f$/lb$_m$ of water, what is the flow rate through the pipe, in gpm?

44. The plumbing in your house is 3/4 in. sch 40 galvanized pipe, and it is connected to an 8 in. sch 80 water main in which the pressure is 15 psig. When you turn on a faucet in your bathroom (which is 12 ft higher than the water main), the water flows out at a rate of 20 gpm.
    (a) How much energy is lost due to friction in the plumbing?
    (b) If the water temperature in the water main is 60°F, and the pipes are well insulated, what would the temperature of the water be leaving the faucet?
    (c) If there were no friction loss in the plumbing, what would the flow rate be (in gpm)?

45. A 60% efficient pump, driven by a 10 hp motor, is used to transfer bunker C fuel oil from a storage tank to a boiler through a well-insulated line. The pressure in the tank is 1 atm, and the temperature is 100°F. The pressure at the burner in the boiler is 100 psig, and it is 100 ft above the level in the tank. If the temperature of the oil entering the burner is 102°F, what is the oil flow rate, in gpm? Oil properties: SG = 0.8, C$_p$ = 0.5 BTU/(lb$_m$°F)

FLUID FORCES, MOMENTUM TRANSFER

46. You have probably noticed that when you turn on the garden hose, it will whip about uncontrollably if it is not restrained. This is because of the unbalanced forces developed by the change of momentum in the tube. If a 1/2 in. I.D. hose carries water at a rate of 50 gpm, and the open end of the hose is bent at an angle of 30° to the rest of the hose, calculate the force components (magnitude and direction) exerted by the water on the bend in the hose. Assume that the loss coefficient in the hose is 0.25.

47. Repeat Problem 46 above, for the case in which a nozzle is attached to the end of the hose, and the water exits the nozzle through a 1/4 in. opening. The loss coefficient for the nozzle is 0.3.

48. A 4 in. I.D. fire hose discharges water at a rate of 1500 gpm through a nozzle which has a 2 in. I.D. exit. The nozzle is conical and converges through a total included angle of 30°. What is the total force transmitted to the bolts in the flange where the nozzle is attached to the hose? Assume that the loss coefficient in the nozzle is 3.0 based on the velocity in the hose.

49. A 90° horizontal reducing bend has an inlet diameter of 4 in. and an outlet diameter of 2 in. If water enters the bend at a pressure of 40 psig and a flow rate of 500 gpm, calculate the force (net magnitude and direction) exerted on the supports which hold the bend in place. The loss coefficient for the bend may be assumed to be 0.75 based on the highest velocity in the bend.

50. A 90° threaded elbow is attached to the end of a 3 in. sch 40 pipe, and a reducer with an inside diameter of 1 in. is threaded into the elbow. If water is pumped through the pipe at a rate of 500 gpm, calculate the forces exerted on the pipe at the point where the elbow is attached. The loss coefficient for the elbow and reducer can be taken to be 1.5, based on the inlet velocity.

51. A fireman is holding the nozzle of a fire hose which he is using to put out a fire. The hose is 3 in. in diameter, and the nozzle is 1 in. in diameter. The water flow rate is 200 gpm, and the loss coefficient for the nozzle is 0.25 (based on the exit velocity). How much force must the fireman use to restrain the nozzle? Must he push or pull on

the nozzle to apply the force? What is the pressure at the end of the hose, where the water enters the nozzle?

52. Water flows through a 30° pipe bend at a rate of 200 gpm. The diameter of the entrance to the bend is 2.5 in., and that of the exit is 3 in. The pressure in the pipe is 30 psig, and the pressure drop in the bend is negligible. What is the total force (magnitude and direction) exerted by the fluid on the pipe bend?

53. A nozzle with a 1 in. I.D. outlet is attached to a 3 in. I.D. fire hose. Water pressure inside the hose is 100 psig, and the flow rate is 100 gpm. Calculate the force (magnitude and direction) required to hold the nozzle at an angle of 45° relative to the axis of the hose.

54. Water flows through a 45° expansion pipe bend at a rate of 200 gpm, exiting into the atmosphere. The inlet to the bend is 2 in. I.D., the exit is 3 in. I.D., and the loss coefficient for the bend is 0.3. Calculate the force (magnitude and direction) exerted by the fluid on the bend, relative to the direction of the entering stream.

55. A patrol boat is powered by a water jet engine, which takes water in at the bow through a 1 ft diameter duct and pumps it out the stern through a 3 in. diameter exhaust jet. If the water is pumped at a rate of 5000 gpm, determine:
    (a) The thrust rating of the engine;
    (b) The maximum speed of the boat, if the drag coefficient is 0.005 based on an underwater area of 600 ft^2;
    (c) The horsepower required to operate the motor (neglecting friction in the motor, pump, and ducts)

56. A patrol boat is powered by a water jet pump engine. The engine takes water in through a 3 ft diameter duct in a bow and discharges it through a 1 ft diameter duct in the stern. The drag coefficient of the boat has a value of 0.01 based on a total underwater area of 1500 ft^2. Calculate the pump capacity in gpm and engine horsepower required to achieve a speed of 35 mph, neglecting friction in the pump and ducts.

57. A piece of rubber hose is attached to a tank containing air at 5 psig. The hose is 10 ft long and 1/4 in. I.D., and the free end is open to the atmosphere. If the hose is bent at a 90° angle, calculate the net

force (magnitude and direction) exerted by the air on the hose. Assume that the hose is hydraulically smooth and T = 32°F.

58. Water is flowing through a 45° pipe bend at a rate of 200 gpm. The inlet to the bend is 1 $^1/_2$ in. inside diameter, and the exit is 1 in. in diameter, with the exit into the atmosphere. The fiction loss in the bend can be characterized by a loss coefficient of 0.3 (based on the inlet velocity). Calculate the net force (magnitude and direction) transmitted to the flange holding the pipe section in place.

59. The arms of a lawn sprinkler are 8 in. long and 3/8 in. I.D. Nozzles at the end of each arm direct the water in a direction which is 45° from the arms. If the total flow rate is 10 gpm, determine:
    (a) The moment developed by the sprinkler if it is held stationary and not allow to rotate.
    (b) The angular velocity, in rpm, of the sprinkler if there is no friction in the bearings.
    (c) The trajectory of the water from the end of the rotating sprinkler (i.e. the radial and angular velocity components)

60. A water sprinkler contains two 1/4 in. I.D. jets at the ends of a rotating hollow (3/8 in. I.D.) tube, which direct the water 90° to the axis of the tube. If the water leaves at 20 ft/s, what torque would be necessary to hold the sprinkler in place?

61. An open container 8 in. high with an inside diameter of 4 in. weighs 5 lb$_f$ when empty. The container is placed on a scale, and water flows into the top of the container through a 1 in. diameter tube at a rate of 40 gpm. The water flows horizontally out through two 1/2 in. holes on opposite sides of the container. Under steady conditions, the height of the water in the tank is 7 in. The pressure is atmospheric across all openings.
    (a) Determine the reading on the scale.
    (b) Determine how far the holes in the sides of the container should be from the bottom so that the level in the container is constant.

62. A boat is tied to a dock by a line from the stern of the boat to the dock. A pump inside the boat takes water in through the bow and discharges it out the stern at rate of 3 ft^3/s through a pipe running through the hull. The pipe inside area is 0.25 ft^2 at the bow and 0.15 ft^2 at the stern. Calculate the tension on the line, assuming that inlet and outlet pressures are equal.

63. An open cylindrical tank is mounted on wheels. The tank is 6 in. in diameter and contains water to a depth of 3 ft. On the side of the tank near the bottom is a 1.5 in. I.D. outlet, to which is attached a ball valve, which has a loss coefficient of 1.2. When the valve is opened, the water flows out in a horizontal stream. Calculate:
    (a) The flow rate of the water, in gpm
    (b) The thrust exerted on the tank by the escaping water, and the direction that the tank will move
    If the diameter of the outlet and valve are increased, will the thrust on the tank increase or decrease? Why?

64. A jet ejector pump is shown in the Figure. A high-speed stream (Q_a) is injected at a rate of 50 gpm, through a small tube 1 in. in diameter, into a stream (Q_b) in a larger, 3 in. diameter, tube. The energy and momentum are transferred from the small stream to the larger stream, which increases the pressure in the pump. The fluids come into contact at the end of the small tube and become perfectly mixed a short distance downstream (the flow is turbulent). The energy dissipated in the system is significant, but the wall force between the end of the small tube and the point where mixing is complete can be neglected. If both streams are water at 60°F, and Q_b = 100 gpm, calculate the pressure rise in the pump.

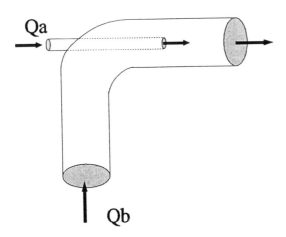

PROBLEM 64

LAMINAR FLOW

65. Use the microscopic equations of motion in Appendix E as a starting point to derive a relation between the volumetric flow rate and the pressure gradient for a Newtonian fluid in a pipe which is valid for any orientation of the pipe axis. (Hint: The critical starting point requires that you identify which velocity and velocity gradients are non-zero, and hence the corresponding non-zero stress components, for this problem. This allows you to "tailor" the differential equations to suit the problem, and the resulting equations can be integrated, with appropriate boundary conditions, to get the answer.)

66. A viscous molten polymer is pumped through a thin slit between two flat surfaces. The slit has a depth H, width W, and length L and is inclined upward at an angle θ to the horizontal (H<<W). The flow is laminar, and the polymer is non-Newtonian, with properties that can be represented by the power law model.
    (a) Derive an equation relating the volume flow rate of the polymer (Q) to the applied pressure difference along the slit, the slit dimensions, and the fluid properties.
    (b) Using the definition of the Fanning friction factor (f), solve your equation for f in terms of the remaining quantities.
    The corresponding solution for a Newtonian fluid can be written f = $24/N_{Re}$. Use your solution to obtain an equivalent expression for the power law Reynolds number (i.e. N_{RePL} = 24/f). Use the hydraulic diameter as the length scale in the Reynolds number. (Note: It is easiest to take the origin of your coordinates at the center of the slit, then calculate the flow rate for one-half the slit and double this to get the answer. Why is this the easiest way?)

67. Acrylic latex paint can be described as a Bingham plastic, with a yield stress of 200 dyn/cm^2, a limiting viscosity of 50 cP, and a density of 0.95 g/cm^3.
    (a) What is the maximum thickness at which a film of this paint could be spread on a vertical wall without running?
    (b) If the power law model were used to describe this paint, such that the apparent viscosity predicted by both the power law and Bingham plastic models is the same at shear rates of 1 and 100 s^{-1}, what would the flow rate be of the film having the thickness predicted in (a)?

68. A vertical belt is moving upward continuously through a liquid bath, at a velocity V. A film of the liquid adheres to the belt that tends to drain downward due to gravity. The equilibrium thickness of the film is determined by the steady-state condition at which the downward drainage velocity of the surface of the film is exactly equal to the upward velocity of the belt. Derive an equation for the film thickness if the fluid is (a) Newtonian; (b) a Bingham plastic.

69. Water at 70°F is draining by gravity down the outside of a 4 in. O.D. vertical tube, at a rate of 1 gpm.
    (a) Determine the thickness of the film. Is the flow laminar or turbulent?
    (b) Would you expect a hydraulic jump to occur in this film? If so, what would the thickness of the film be downstream of the jump?

70. For laminar flow of a Newtonian fluid in a tube:
    (a) Show that the average velocity over the cross section is half the maximum velocity in the tube.
    (b) Derive the kinetic energy correction factor (i.e. $\alpha = 2$).

71. A slider bearing can be described as one plate moving with a velocity v parallel to a stationary plate, with a viscous lubricant in between the plates. The force applied to the moving plate is F, and the distance between the plates is H. If the lubricant is a grease with properties that can be described by the power law model, derive an equation relating the velocity v to the applied force F and the gap clearance H, starting with the general microscopic continuity and momentum equations. If the area of the plate is doubled, with everything else staying the same, how will the velocity v change?

72. Consider a fluid flowing in a conical section, as illustrated in the Figure. The mass flow rate is the same going in (through point 1) as it is coming out (point 2), but the velocity changes because the

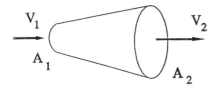

PROBLEM 72   Flow through a conical section.

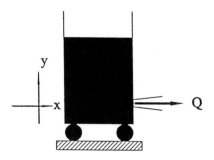

PROBLEM 73   Flow from hole in tank.

area changes. They are related by

$$(\rho \, V \, A)_1 = (\rho \, V \, A)_2$$

where ρ is the fluid density (assumed to be constant here). Because the velocity changes, the transport of momentum will be different going in than going out, which results in a net force in the fluid. Derive an expression for the magnitude of this force associated with the change in momentum, and determine the direction in which the force transmitted from the fluid to the cone will act.

73.   Consider the tank on wheels shown in the Figure. Water is draining out of a hole in the side of the open tank, at a rate of 10 gpm. If the tank diameter is 2 ft and the diameter of the hole is 2 in., determine the magnitude and direction of the force transmitted from the water to the tank.

6

Pipe Flow

I. FLOW REGIMES

In 1883, Osborn Reynolds conducted a classical experiment, illustrated in Fig. 6-1, in which he measured the pressure drop as a function of flow rate for water in a tube. He found that at low flow rates the pressure drop was directly proportional to the flow rate, but as the flow rate was increased, a point was reached where the relation was no longer linear and the "noise" or scatter in the data increased considerably. At still higher flow rates the data became more reproducible, but the relation between pressure drop and flow rate became almost quadratic instead of linear.

To investigate this phenomenon further, Reynolds introduced a trace of dye into the flow to observe what was happening. At the low flow rates where the linear relation was observed, the dye was seen to remain a coherent, rather smooth, thread throughout most of the tube. However, where the data scatter occurred, the dye trace was seen to be rather unstable, and it broke up after a short distance. At still higher flow rates, where the quadratic relation was observed, the dye dispersed almost immediately, forming a uniform "cloud" throughout the tube. The stable flow observed initially was termed *laminar* because it was observed that the fluid elements moved in smooth layers or lamella relative to each other with no mixing. The unstable flow pattern, characterized by a high degree of mixing between the fluid elements, was termed *turbulent*. Although the transition from laminar to turbulent flow occurred rather abruptly,

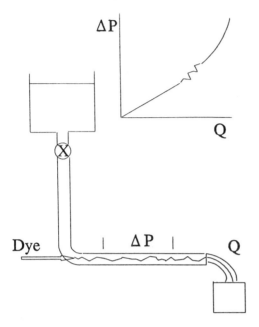

FIGURE 6-1   Reynolds' experiment.

there is nevertheless a *transition* region where the flow is unstable but not thoroughly mixed.

Careful study of various fluids in tubes of different sizes has indicated that laminar flow in a tube persists up to a point where the value of the Reynolds number ($N_{Re} = DV\rho/\mu$) is about 2000, and turbulent flow occurs when N_{Re} is greater than about 4000, with a transition region in between. Actually, unstable flow (turbulence) occurs when perturbations in the flow are amplified, whereas laminar flow occurs when these perturbations are damped out. Since turbulent flow cannot occur unless there are perturbations, studies have been conducted on systems in which extreme care has been taken to eliminate any perturbations due to irregularities in the boundary surfaces, sudden changes in direction, vibrations, etc. Under these conditions, it has been possible to sustain laminar flow in a tube to a Reynolds number of the order of 100,000 or more. However, under all but the most unusual conditions there are sufficient natural perturbations in all practical systems such that turbulence begins in a pipe at a Reynolds number of about 2000.

The physical significance of the Reynolds number can be appreciated better if it is rearranged as follows:

$$N_{Re} = \frac{DV\rho}{\mu} = \frac{\rho V^2}{\mu V/D} \tag{6-1}$$

The numerator is the flux of "inertial" momentum carried by the fluid along the tube in the axial direction. The denominator is proportional to the viscous shear stress in the tube, which is equivalent to the flux of "viscous" momentum normal to the flow direction, i.e. in the radial direction. Thus, the Reynolds number is a ratio of the momentum flux due to inertia (in the flow direction) to the momentum flux due to viscous stresses (in the radial direction). Since viscous forces are a manifestation of intermolecular attractive forces, they are stabilizing, whereas inertial forces tend to pull the fluid elements apart and are destabilizing. It is thus quite logical that stable (laminar) flow should occur at low Reynolds numbers where viscous forces dominate, whereas unstable (turbulent) flow occurs at high Reynolds numbers where inertial forces dominate. Also, laminar flows are dominated by viscosity and are independent of the fluid density, whereas fully turbulent flows are dominated by the fluid density and are essentially independent of the fluid viscosity at high turbulence levels. For fluids flowing near solid boundaries (e.g. inside conduits), viscous forces dominate in the immediate vicinity of the boundary, whereas for turbulent flows (high Reynolds numbers), inertial forces dominate in the region far from the boundary. We will consider both the laminar and turbulent flow of Newtonian and non-Newtonian fluids in pipes in this chapter.

II. GENERAL RELATIONS FOR PIPE FLOWS

For steady, uniform, fully developed flow in a pipe (or any conduit), the conservation of mass, energy, and momentum equations can be arranged in specific forms which are most useful for the analysis of such problems. These general expressions are valid for both Newtonian and non-Newtonian fluids, in either laminar or turbulent flow.

A. ENERGY BALANCE

Consider a section of uniform cylindrical pipe, of length L and radius R, inclined upward at an angle θ to the horizontal, as shown in Fig. 6-2. The energy balance (or Bernoulli equation) as applied to the fluid flowing in the pipe can be written

$$\frac{-\Delta\Phi}{\rho} = e_f = K_f \frac{V^2}{2} \tag{6-2}$$

where $\Phi = P + \rho gz$, $K_f = 4fL/D$, and f is the Fanning friction factor.

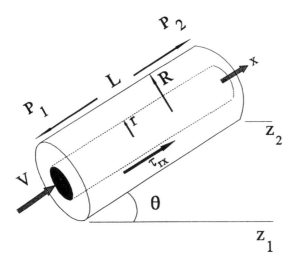

FIGURE 6-2   Pipe flow.

B.   MOMENTUM BALANCE

We can write a momentum balance on a cylindrical volume of fluid of radius r, length L, centered on the pipe centerline, as follows:

$$\sum F_x = (P_1 - P_2)\pi r^2 - \pi r^2 L\rho g \sin\theta + 2\pi r L\tau_{rx} = 0 \qquad (6\text{-}3)$$

Solving Eqn (6-3) for τ_{rx} gives

$$\tau_{rx} = \frac{\Delta\Phi r}{2L} = -\tau_w \frac{r}{R} \qquad (6\text{-}4)$$

where $\Delta\Phi = \Delta P + \rho g\Delta z = \Delta P + \rho g L \sin\theta$, and τ_w is the stress exerted by the fluid on the tube wall. Equation (6-4) is equivalent to Eqn (6-2), since

$$f = \frac{\tau_w}{\frac{1}{2}\rho v^2} = \frac{K_f}{4L/D} \qquad (6\text{-}5)$$

Note that, from Eqn (6-4), the shear stress is negative in the fluid (i.e. the fluid outside the cylindrical ''system'' of radius r is moving slower than that inside the system and hence exerts a force in the $-x$ direction on the fluid inside the ''system'', which is bounded by the ''r'' surface). However, the stress at the wall (τ_w) is defined as the force exerted *by the fluid on the wall* (which is a $-r$ surface), in the $+x$ direction.

C. CONTINUITY

Continuity provides a relation between the volumetric flow rate (Q) passing through a given cross section in the pipe and the local velocity (v_x), i.e.

$$Q = \int_A v_x \, dA = \pi \int_0^R 2rv_x \, dr = \pi \int_A v_x \, d(r^2) \tag{6-6}$$

This can be integrated by parts, as follows:

$$Q = -\pi \int_A r^2 \, dv_x = -\pi \int_0^R r^2 \frac{dv_x}{dr} \, dr \tag{6-7}$$

All of these equations are general, since they apply to any fluid (Newtonian or non-Newtonian) in any type of flow (laminar or turbulent) in steady, fully developed flow in a uniform cylindrical tube with any orientation.

III. NEWTONIAN FLUIDS

A. LAMINAR FLOW

For a Newtonian fluid in laminar flow,

$$\tau_{rx} = \mu \frac{dv_x}{dr} \quad \text{or} \quad \frac{dv_x}{dr} = \frac{\tau_{rx}}{\mu} \tag{6-8}$$

When the velocity gradient from Eqn (6-8) is substituted into Eqn (6-7), and Eqn (6-4) is used to eliminate the shear stress, Eqn (6-7) becomes

$$Q = -\frac{\pi}{\mu} \int_0^R r^2 \tau_{rx} \, dr = \frac{\pi \tau_w}{\mu R} \int_0^R r^3 \, dr \tag{6-9}$$

or

$$Q = \frac{\pi \tau_w R^3}{4\mu} = -\frac{\pi \Delta \Phi R^4}{8\mu L} = -\frac{\pi \Delta \Phi D^4}{128 \, \mu L} \tag{6-10}$$

Equation (6-10) is known as the *Hagen-Poiseuille equation*. It can also be derived by equating the shear stress for a Newtonian fluid, Eqn (6-8), to that from the momentum balance for tube flow, Eqn (6-4), and integrating to obtain the velocity profile:

$$v_x(r) = \frac{\tau_w R}{2\mu} \left(1 - \frac{r^2}{R^2} \right) \tag{6-11}$$

Inserting this into Eqn (6-6) and integrating over the tube cross section gives Eqn (6-10) for the volumetric flow rate. If the wall stress (τ_w) in Eqn (6-6) is expressed in terms of the Fanning friction factor (i.e. $\tau_w = f\rho v^2/2$) and the result solved for f, what results is the dimensionless form of the Hagen-Poiseuille equation:

$$f = \frac{4\pi D\mu}{Q\rho} = \frac{16\mu}{DV\rho} = \frac{16}{N_{Re}} \qquad (6\text{-}12)$$

It should be recalled that dimensional analysis shows that the laminar flow of a Newtonian fluid in a tube can be characterized by a single dimensionless group which is equivalent to the product fN_{Re} (this group is independent of the fluid density, which cancels out). It follows that the value of this group must be constant for all such flows, regardless of the fluid viscosity or density, the size of the tube, the flow rate, etc. Although the magnitude of this constant could not be obtained from dimensional analysis, we have now shown from theoretical principles that the value is 16, which is also in agreement with experimental observations. Equation (6-12) is valid for $N_{Re} < 2000$, as previously discussed.

B. TURBULENT FLOW

As previously noted, if the Reynolds number in the tube is larger than about 2000 the flow will no longer be laminar. Since fluid elements in contact with a stationary solid boundary are also stationary (i.e. the fluid sticks to the wall), the velocity increases from zero at the boundary to a maximum value at some distance from the boundary. For uniform flow in a symmetrical duct, the maximum velocity occurs at the centerline of the duct. The region of flow in which the velocity varies with the distance from the boundary, as illustrated in Fig. 6-3 is called the *boundary layer*.

1. The Boundary Layer

Since the fluid velocity at the boundary is zero, there will always be a region adjacent to the wall which is laminar. This is called the *laminar sublayer* and

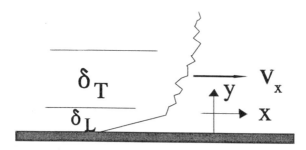

FIGURE 6-3 Boundary layer.

is designated δ_L in Fig. 6-3. Note that for tube flow, if $N_{Re} < 2000$, the entire flow is laminar and $\delta_L = R$. The turbulent boundary layer (δ_T) includes the region in the vicinity of the wall which is turbulent, in which the velocity varies with the distance from the wall (y). Beyond this region the fluid is almost completely mixed in what is often called the *turbulent core*, and the velocity is independent of y. The transition from the laminar sublayer to the turbulent boundary layer is gradual, not abrupt, and the transition region is called the *buffer zone*.

2. Turbulent Momentum Flux

The velocity field in turbulent flow may be described by a local "mean" (or time-averaged) velocity, upon which is superimposed a time-dependent fluctuating component, or turbulent "eddy". Even in a "one-dimensional" flow, in which the mean velocity has only one component (as illustrated in Fig. 6-3), the turbulent eddies have a three-dimensional structure. Thus, for the flow illustrated in Fig. 6-3, the local velocity components are

$$v_x(y, t) = \bar{v}_x(y) + v'_x(y, t)$$
$$v_y(y, t) = 0 + v'_y(y, t) \tag{6-13}$$
$$v_z(y, t) = 0 + v'_z(y, t)$$

The time-averaged velocity $(\bar{v})$ obviously has zero components in the y and z directions, but the eddy velocity components are nonzero in all three directions. The time-averaged velocity is defined as

$$\bar{v}_x = \frac{1}{T} \int_0^T v_x \, dt \qquad \text{so that} \qquad \int_0^T v'_v \, dt = 0 \tag{6-14}$$

The average is taken over the time T, which is long compared to the period of the eddy fluctuation.

Now the eddy motion transports momentum, and the corresponding momentum flux components are equivalent to (negative) shear stress components, as follows:

$$\tau'_{xx} = -\rho(v'_x)^2 \qquad \tau'_{xy} = -\rho \, v'_x v'_y \qquad \tau'_{xz} = -\rho v'_x v'_z$$
$$\tau'_{yx} = \tau'_{xy} \qquad \tau'_{yy} = -\rho(v'_y)^2 \qquad \tau'_{yz} = -\rho v'_y v'_z \tag{6-15}$$
$$\tau'_{zx} = \tau'_{xz} \qquad \tau'_{zy} = \tau'_{yz} \qquad \tau'_{zz} = -\tau(v'_z)^2$$

These "turbulent momentum flux components" are also called *Reynolds stresses*. Thus, the total stress in a Newtonian fluid in turbulent flow is composed of both viscous and turbulent (Reynolds) stresses:

$$\tau_{ij} = \mu\left(\frac{\partial \bar{v}_i}{\partial x_j} + \frac{\partial \bar{v}_j}{\partial x_i}\right) - \rho v'_i v'_j \tag{6-16}$$

Although Eqn (6-16) can be used to eliminate the stress components from the general microscopic equations of motion, a solution for the turbulent flow field still cannot be obtained unless the spatial dependence of the eddy velocities or turbulent stresses is known. A classic (simplified) model for the turbulent stresses, attributed to Prandtl, is outlined below.

3. Mixing Length Theory

The turbulent eddies (with velocity components v_x', v_y', v_z') are continuously being generated, grow, and die out. During this process, there is an exchange of momentum between the eddies and the mean flow. Considering a two-dimensional turbulent field in a smooth tube, Prandtl assumed (incorrectly) that $v_x' \approx v_y'$, so that

$$\tau_{yx}' = -\rho v_x' v_y' \cong -\rho(v_x')^2 \qquad (6\text{-}17)$$

He also assumed that each eddy moves a distance l (the "mixing length") while exchanging its momentum with the mean flow, i.e.

$$\frac{v_x'}{l} \cong \frac{d\bar{v}_x}{dy} \qquad (6\text{-}18)$$

Using Eqn (6-18) to eliminate the eddy velocity from Eqn (6-17) gives

$$\tau_{yx}' = \mu_e \frac{d\bar{v}_x}{dy} \qquad (6\text{-}19)$$

where

$$\mu_e = \rho l^2 \left| \frac{d\bar{v}_x}{dy} \right| \qquad (6\text{-}20)$$

is called the *eddy viscosity*. Note that the eddy viscosity is *not* a fluid property but is a function of the eddy characteristics (e.g. the mixing length or the degree of turbulence) and the mean velocity gradient. The only fluid property involved is the density, since turbulent momentum transport is an inertial (i.e. mass-dominated) effect. Since turbulence (and all motion) is zero at the wall, Prandtl assumed that the mixing length should be proportional to the distance from the wall, i.e.

$$l = \kappa y \qquad (6\text{-}21)$$

Since these relations apply only in the vicinity of the wall, Prandtl further assumed that the eddy (Reynolds) stress must be of the same order as the wall stress, i.e.

$$\tau_{yx}' \cong \tau_w = \rho \kappa^2 y^2 \left(\frac{d\bar{v}_x}{dy} \right)^2 \qquad (6\text{-}22)$$

Integrating Eqn (6-22) over the turbulent boundary layer (from y_1, the edge of the buffer layer, to y) gives

$$\bar{v}_x - \bar{v}_{x1} = \frac{1}{\kappa} \left(\frac{\tau_w}{\rho} \right)^{1/2} \ln \left(\frac{y}{y_1} \right) \qquad (6\text{-}23)$$

This equation is called the *von Karman equation* (or, sometimes, the "law of the wall"), and can be written in dimensionless form as

$$v^+ = \frac{1}{\kappa} \ln y^+ + A \qquad (6\text{-}24)$$

where

$$v^+ = \frac{\bar{v}_x}{v_*} = \frac{\bar{v}_x}{V} \sqrt{\frac{2}{f}}, \qquad y^+ = \frac{yv_* \rho}{\mu} = \frac{yV\rho}{\mu} \sqrt{\frac{f}{2}} \qquad (6\text{-}25)$$

The term

$$v_* = \sqrt{\frac{\tau_w}{\rho}} = V \sqrt{\frac{f}{2}} \qquad (6\text{-}26)$$

is called the *friction velocity*, because it is a wall stress parameter but has dimensions of velocity. The parameters κ and A in the von Karman equation have been determined from experimental data on Newtonian fluids in smooth pipe to be $\kappa = 0.4$ and $A = 5.5$. Equation (6-24) applies only within the turbulent boundary layer (outside the buffer region), which has been found empirically to correspond to $y^+ \geq 26$.

Within the laminar sublayer the turbulent eddies are negligible, so that

$$\tau_{yx} \cong \tau_w = \mu \frac{d\bar{v}_x}{dy} \qquad (6\text{-}27)$$

In dimensionless form, this is equivalent to

$$\frac{dv^+}{dy^+} = 1 \qquad (6\text{-}28)$$

or

$$v^+ = y^+ \qquad (6\text{-}29)$$

Equation (6-29) applies to the laminar sublayer region in a Newtonian fluid, which has been found to correspond to $0 \leq y^+ \leq 5$. The intermediate region, or "buffer zone" between the laminar sublayer and the turbulent boundary layer can be represented by the empirical equation

$$v^+ = -3.05 + 5.0 \ln y^+ \tag{6-30}$$

which applies for $5 < y^+ < 26$.

4. Friction Loss in Smooth Pipe

For a Newtonian fluid in a smooth pipe, these equations can be integrated over the pipe cross section to give the average fluid velocity, e.g.

$$V = \frac{2}{R^2} \int_0^R \bar{v}_x r \, dr = 2v_* \int_0^1 v^+(1 - x) \, dx \tag{6-31}$$

where $x = y/R = 1 - r/R$. If the von Karman equation (Eqn 6-24) for v^+ is introduced into this equation, and the laminar sublayer is neglected, the integral can be evaluated to give

$$\frac{1}{\sqrt{f}} = 4.1 \log(N_{Re} \sqrt{f}) - 0.60 \tag{6-32}$$

The constants in this equation were modified by Nikuradse from observed data taken in smooth pipe, as follows:

$$\frac{1}{\sqrt{f}} = 4.0 \log(N_{re} \sqrt{f}) - 0.40 \tag{6-33}$$

Equation (6-33) is also known as the *von Karman/Nikuradse equation* and agrees well with observations for friction loss in smooth pipe over the range $5 \times 10^3 < N_{Re} < 5 \times 10^6$.

An alternative equation for smooth tubes was derived by Blasius based on observations that the mean velocity profile in the tube could be represented approximately by

$$\bar{v}_x = v_{max} \left(1 - \frac{r}{R}\right)^{1/7} \tag{6-34}$$

A corresponding expression for the friction factor can be obtained by writing this expression in dimensionless form and substituting the result into Eqn (6-31):

$$f = \frac{0.0791}{N_{Re}^{1/4}} \tag{6-35}$$

Equation (35) represents the friction factor for Newtonian fluids in smooth tubes quite well over a range of Reynolds numbers up to about 10^5. The Prandtl mixing length theory and the von Karman and Blasius equations are referred to as "semiempirical" models. That is, even though these models result from a process of logical reasoning, the results cannot be deduced solely from first principles because they require the introduction of certain parameters which can only be evaluated empirically.

5. Friction Loss in Rough Tubes

Since all models for turbulent flows are semiempirical in nature, it is necessary to rely upon empirical observations (e.g. data) for a quantitative description of friction loss in such flows. For Newtonian fluids in long tubes, we have shown by dimensional analysis that the friction factor should be a unique function of the Reynolds number and the relative roughness of the tube wall. This result has been used to correlate a wide range of measurements in various sizes of tubes, a variety of fluids, and a wide range of flow rates in terms of a generalized plot of f vs N_{Re}, with ε/D as a parameter. This correlation, shown in Fig. 6-4, is called a *Moody diagram.*

The laminar region (for $N_{Re} < 2000$) is described by the theoretical Hagen-Poiseuille equation (Eqn 6-12), which is plotted on Fig. 6-4. In this region, the only fluid property that influences friction loss is the viscosity (since the density cancels out). Furthermore, the roughness has a negligible effect in laminar flow,

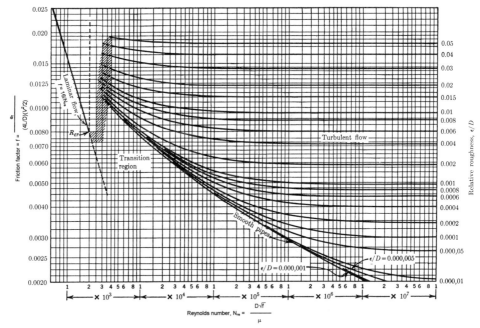

FIGURE 6-4 Moody diagram.

as will be explained shortly. The "critical zone" is the range of transition from laminar to turbulent flow, which corresponds to values of N_{Re} from about 2000 to 4000. Data are not very reproducible in this range, and correlations are unreliable. The so-called transition zone on Fig. 6-4 is the region where the friction factor depends strongly on both the Reynolds number and relative roughness. In the region labeled "complete turbulence, rough pipes", the lines are horizontal, which means that the friction factor is independent of Reynolds number (i.e. independent of viscosity) and is a function of the relative roughness only.

For turbulent flow in smooth tubes, the semiempirical models of Prandtl, von Karman/Nikuradse, or Blasius represent the friction factor quite well. Whether a tube is hydraulically "smooth" or "rough" depends upon the size of the wall roughness elements relative to the thickness of the laminar sublayer. Since laminar flow is stable, if the flow perturbations due to the roughness elements lie entirely within the laminar region, the disturbances will be damped out and will not affect the rest of the flow field. However, if the roughness elements protrude through the laminar sublayer into the turbulent region, which is unstable, the disturbance will grow throughout the turbulent flow, thus enhancing the Reynolds stresses and consequently the energy dissipation (friction loss). Since the thickness of the laminar sublayer decreases as the Reynolds number increases, a tube with a given roughness may be hydraulically smooth at a low Reynolds number but hydraulically rough at a high Reynolds number.

For rough tubes in turbulent flow ($N_{Re} > 4000$), the von Karman equation was modified empirically by Colebrook to include the effect of wall roughness, as follows:

$$\frac{1}{\sqrt{f}} = -4 \log \left[\frac{\varepsilon/D}{3.7} + \frac{1.255}{N_{Re} \sqrt{f}} \right] \tag{6-36}$$

It is noted that the term $N_{Re}\sqrt{f}$ is, by definition,

$$N_{Re} \sqrt{f} = \left(\frac{e_f D^3 \rho^2}{2L\mu^2} \right)^{1/2} \tag{6-37}$$

which is independent of flow rate. Thus the dimensionless groups in the Colebrook equation are in a form which is convenient if the flow rate is to be found and the allowable friction loss (e.g. driving force), tube size, and fluid properties are known.

6. Roughness Factors

The actual size of the roughness elements on the conduit wall obviously varies from one material to another, with age and usage, and with the amount of dirt, scale, etc. Characteristic values of wall roughness have been determined for various materials, as shown in Table 6-1. The most common pipe material—

TABLE 6-1   Equivalent Roughness of Various Surfaces

Material	Condition	Roughness Range	Recommended
Drawn brass, copper, stainless steel	New	0.01-0.0015 mm 0.0004-0.00006 in.	0.002 mm 0.00008 in.
Commercial steel	New	0.1-0.02 mm 0.004-0.0008 in.	0.045 mm 0.0018 in.
	Light rust	1.0-0.15 mm 0.04-0.006 in.	0.3 mm 0.015 in.
	General rust	3.0-1.0 mm 0.1-0.04 in.	2.0 mm 0.08 in.
Iron	Wrought, new	0.045 mm 0.002 in.	0.045 mm 0.002 in.
	Cast, new	1.0-0.25 mm 0.04-0.01 in.	0.30 mm 0.025 in.
	Galvanized	0.15-0.025 mm 0.006-0.001 in.	0.15 mm 0.006 in.
	Asphalt-coated	1.0-0.1 mm 0.04-0.004 in.	0.15 mm 0.006 in.
Sheet metal	Ducts Smooth joints	0.1-0.02 mm 0.004-0.0008 in.	0.03 mm 0.0012 in.
Concrete	Very smooth	0.18-0.025 mm 0.007-0.001 in.	0.04 mm 0.0016 in.
	Wood floated, brushed	0.8-0.2 mm 0.03-0.007 in.	0.3 mm 0.012 in.
	Rough, visible form marks	2.5-0.8 mm 0.1-0.03 in.	2.0 mm 0.08 in.
Wood	Stave, used	1.0-0.25 mm 0.035-0.01 in.	0.5 mm 0.02 in.
Glass or plastic	Drawn tubing	0.01-0.0015 mm 0.0004-0.00006 in.	0.002 mm 0.0008 in.
Rubber	Smooth tubing	0.07-0.006 mm 0.0003-0.00025 in.	0.01 mm 0.0004 in.
	Wire-reinforced	4.0-0.3 mm 0.15-0.01 in.	1.0 mm 0.04 in.

clean, new commercial steel or wrought iron—has been found to have an effective roughness of about 0.0018 in. (0.045 mm). Other surfaces, such as concrete, may vary by as much as several orders of magnitude, depending upon the nature of the surface finish.

These roughness values are not measured directly but have been determined indirectly. Surfaces artificially roughened by sand grains of various sizes were studied initially by Nikuradse. Measurements of f and N_{Re} were plotted to establish the reference curves for various known values of ε/D for these surfaces, which are represented on the Moody diagram. The equivalent roughness factors for other materials are determined from similar measurements in conduits made of the material, by plotting the data on the Moody diagram and comparing the results with the reference curves (or by using the Colebrook equation). For this reason, such roughness values are sometimes termed the *equivalent sand grain roughness*.

C. ALL FLOW REGIMES

The expressions for the friction factor in both laminar and turbulent flow were combined into a single expression by Churchill (1977) as follows:

$$f = 2\left[\left(\frac{8}{N_{Re}}\right)^{12} + \frac{1}{(A + B)^{3/2}}\right]^{1/12} \tag{6-38}$$

where

$$A = \left[2.457 \ln \left(\frac{1}{\left(\frac{7}{N_{Re}}\right)^{0.9} + \frac{0.27\varepsilon}{D}}\right)\right]^{16}$$

and

$$B = \left(\frac{37,530}{N_{Re}}\right)^{16}$$

Equation (6-38) adequately represents the Fanning friction factor over the entire range of Reynolds numbers within the accuracy of the data used to construct the Moody diagram, including a reasonable estimate for the intermediate or transition region between laminar and turbulent flow. Note that it is explicit in f.

IV. POWER LAW FLUIDS

We will consider the corresponding expressions for the friction loss in laminar and turbulent flow for non-Newtonian fluids in pipes, for the two simplest (two-parameter) models—the power law and Bingham plastic. The power law model is very popular for representing the viscosity of a wide variety of non-Newtonian fluids because of its simplicity and versatility. However, extreme care should be exercised in its application, since the range of shear stress (or shear rate) expected in the application should not extend beyond the range of the rheological data used to evaluate the model parameters for reliable results. Both laminar and turbulent pipe flow of highly loaded slurries of fine particles, for example, can often be adequately represented by either of these two models, as shown by Darby et al. (1992).

A. LAMINAR FLOW

Since the shear stress and shear rate are negative in pipe flow, the appropriate form of the power law model for laminar pipe flow is

$$\tau_{rx} = m\dot{\gamma}_{rx}^n = -m \left(-\frac{dv_x}{dr} \right)^n \tag{6-39}$$

By equating the shear stress from Eqns (6-39) and (6-4), solving for the velocity gradient, and introducing the result into Eqn (6-7) to determine the flow rate (as was done for the Newtonian fluid), we find

$$Q = \pi \left(\frac{\tau_w}{mR} \right)^{1/n} \int_0^R r^{2+1/n} \, dr = \pi \left(\frac{\tau_w}{mR} \right)^{1/n} \left(\frac{n}{3n+1} \right) R^{(3n+1)/n} \tag{6-40}$$

This is the equivalent of the Hagen-Poiseuille equation for a power law fluid. It can be written in dimensionless form by expressing the wall stress in terms of the friction factor using Eqn (6-5), solving for f, and equating the result to $16/N_{Re}$ (the equivalent Newtonian expression). The result is an expression which is identical to the dimensionless Hagen-Poiseuille equation and thus defines the equivalent Reynolds number for a power law fluid, i.e.

$$fN_{RePL} = 16 \tag{6-41}$$

where

$$N_{RePL} = \frac{8D^n V^{2-n} \rho}{m[2(3n+1)/n]^n} \tag{6-42}$$

is the power law Reynolds number. Note that, because n is an empirical parameter and can take on any value, the units in expressions for power law fluids can be very complex. Thus, the calculations are simplified if a scientific system of dimensions and units is used (e.g. SI or cgs), which avoids the necessity of introducing the conversion factor g_c. In fact, the evaluation of most dimensionless groups is usually simplified by the use of such units.

B. TURBULENT FLOW

Dodge and Metzner (1959) modified the von Karman equation to apply to power law fluids, with the following result:

$$\frac{1}{\sqrt{f}} = \frac{4}{n^{0.75}} \log[N_{RePL} f^{(1-n/2)}] - \frac{0.4}{n^{1.2}} \tag{6-43}$$

Like the von Karman equation, this equation is implicit in f. Equation (6-43) is applicable to any non-Newtonian fluid if the parameter n is interpreted to be the point slope of the shear stress versus shear rate from (laminar) viscosity measurements, at the wall shear stress (or shear rate) corresponding to the conditions of interest in turbulent flow. However, it is not a simple matter to acquire data over the appropriate range or to solve the equation for f for a given flow rate and pipe diameter in turbulent flow.

Note that there is no effect of pipe wall roughness in Eqn (6-43), in contrast to the case for Newtonian fluids. There are insufficient data in the literature to provide a reliable estimate of the effect of roughness on friction loss for non-Newtonian fluids in turbulent flow. However, the evidence that does exist suggests that the roughness is not as significant for non-Newtonian fluids as for Newtonian fluids. This is partly due to the fact that the majority of non-Newtonian turbulent flows lie in the low Reynolds number range, and partly due to the fact that the laminar boundary layer tends to be thicker for non-Newtonian fluids than for Newtonian fluids (i.e. the flows are most often in the "hydraulically smooth" range for common pipe materials).

C. ALL FLOW REGIMES

An expression which represents the friction factor for the power law fluid over the entire range of Reynolds numbers (laminar through turbulent) has been given by Darby et al. (1992), as follows:

$$f = (1 - \alpha)f_L + \frac{\alpha}{(f_T^{-8} + f_{Tr}^{-8})^{1/8}} \tag{6-44}$$

where

$$f_L = \frac{16}{N_{RePL}} \tag{6-45}$$

$$f_T = \frac{0.0682n^{-1/2}}{N_{RePL}^{1/(1.87+2.39n)}} \tag{6-46}$$

$$f_{Tr} = 1.79 \times 10^{-4} \exp(-5.24n)N_{RePL}^{(0.414+0.757n)} \tag{6-47}$$

The "critical" Reynolds number which corresponds to the onset of turbulence is given by

$$N_{RePLc} = 2100 + 875(1 - n) \tag{6-48}$$

The parameter α is given by

$$\alpha = \frac{1}{1 + 4^{-\Delta}} \tag{6-49}$$

where

$$\Delta = N_{RePL} - N_{RePLc} \qquad (6\text{-}50)$$

Equation (6-45) applies for $N_{RePL} < N_{RePLc}$, Eqn (6-46) applies for $4000 < N_{RePL} < 10^5$, and Eqn (6-47) applies for $N_{RePLc} < N_{RePL} < 4000$.

V. BINGHAM PLASTICS

The Bingham plastic model usually provides a good representation for the viscosity of concentrated slurries, suspensions, emulsions, foams, etc. Such materials often exhibit a yield stress which must be exceeded before the material will flow at a significant rate. Other examples include paint, shaving cream, and mayonnaise. There are also many "fluids" which may have a yield stress which is not as pronounced—blood, for example.

It is recalled that a "plastic" is really two materials. At low stresses, below the critical or yield stress (τ_0), the material behaves as a solid, whereas for stresses above the yield stress the material behaves as a fluid. The Bingham model for this behavior is

$$\text{For } |\tau| < \tau_0: \quad \dot{\gamma} = 0$$
$$\text{For } |\tau| > \tau_0: \quad \tau = \pm\tau_0 + \mu_\infty\dot{\gamma} \qquad (6\text{-}51)$$

Since the shear stress and shear rate can be either positive or negative, the plus is used in Eqn (6-51) in the former case, and the minus for the latter. For tube flow, since the shear stress and shear rate are both negative, the appropriate form of the model is

$$\text{For } |\tau_{rx}| < \tau_0: \quad \frac{dv_x}{dr} = 0$$
$$\text{For } |\tau_{rx}| > \tau_0: \quad \tau_{rx} = -\tau_0 + \mu_\infty\frac{dv_x}{dr} \qquad (6\text{-}52)$$

A. LAMINAR FLOW

Because the shear stress is always zero at the centerline in pipe flow and increases linearly with distance from the center toward the wall (Eqn 6-4), there will be a finite distance from the center over which the stress is always less than the yield stress. In this region, the material has solid-like properties and does not yield but moves as a rigid "plug". The radius of this plug (r_o) is, from Eqn (6-4),

$$r_o = R\frac{\tau_0}{\tau_w} \qquad (6\text{-}53)$$

Since the stress outside of this plug region exceeds the yield stress, the material will deform as a fluid between the plug and the wall. The flow rate must therefore be determined by combining the flow in the "plug" with that in the "fluid" region:

$$Q = \int_A v_x \, dA = Q_{plug} + \pi \int_{r_o^2}^{R^2} v_x \, dr^2 \tag{6-54}$$

Evaluating the integral by parts, and noting that the Q_{plug} term cancels with $\pi r^2 v_{plug}$, the result is

$$Q = -\pi \int_{r_o}^{R} r^2 \dot{\gamma} \, dr \tag{6-55}$$

When Eqn (6-52) is used to eliminate the shear rate and Eqn (6-4) is used for the shear stress, the integral can be evaluated to give

$$Q = \frac{\pi R^3 \tau_w}{4 \, \mu_\infty} \left[1 - \frac{4}{3}\left(\frac{\tau_o}{\tau_w}\right) + \frac{1}{3}\left(\frac{\tau_o}{\tau_w}\right)^4 \right] \tag{6-56}$$

This equation is known as the *Buckingham-Reiner equation*. It can be cast in dimensionless form and rearranged as follows:

$$f_L = \frac{16}{N_{Re}} \left[1 + \frac{1}{6}\left(\frac{N_{He}}{N_{Re}}\right) - \frac{1}{3}\left(\frac{N_{He}^4}{f^3 N_{Re}^7}\right) \right] \tag{6-57}$$

where the Reynolds number is given by

$$N_{Re} = \frac{DV\rho}{\mu_\infty} \tag{6-58}$$

and

$$N_{He} = \frac{D^2 \rho \tau_o}{\mu_\infty^2} \tag{6-59}$$

is the *Hedstrom number*. Note that the Bingham plastic reduces to a Newtonian fluid if $\tau_o = N_{He} = 0$, in which case Eqn (6-57) reduces to the Newtonian result, i.e. $f = 16/N_{Re}$ [see Eqn (6-12)]. Note that there are actually only two independent dimensionless groups in Eqn (6-57) (consistent with the results of dimensional analysis for a fluid with two rheological properties, τ_o and μ_∞), which are the combined groups fN_{Re} and N_{He}/N_{Re}. The ratio N_{He}/N_{Re} is also called the *Bingham number*, $N_{Bi} = D\tau_o/\mu V$. Equation (6-57) is implicit in f, so it must be solved by iteration for known values of N_{Re} and N_{He}. This is not difficult, however, since the last term in Eqn (6-57) is usually much smaller than the other terms, so that neglecting this term usually provides a very good first estimate for f. Inserting this first estimate into the neglected term to revise f, and repeating the procedure, usually results in rapid convergence.

B. TURBULENT FLOW

For a Bingham plastic, there is no abrupt transition from laminar to turbulent flow, as is observed for Newtonian fluids. Instead, a gradual deviation from purely laminar flow to fully turbulent flow occurs. For turbulent flow, the friction factor can be represented by the empirical expression of Darby and Melson (1982) (as modified by Darby et al., 1992):

$$f_T = \frac{10^a}{N_{Re}^{0.193}} \tag{6-60}$$

where:

$$a = -1.41 \, [1 + 0.146 \, \exp(-2.9 \times 10^{-5} \, N_{He})] \tag{6-61}$$

C. ALL REYNOLDS NUMBERS

The friction factor for a Bingham plastic can be calculated for any Reynolds number, from laminar through turbulent, from the equation:

$$f = (f_L^m + f_T^m)^{1/m} \tag{6-62}$$

where

$$m = 1.7 + \frac{40,000}{N_{Re}} \tag{6-63}$$

In Eqn (6-62), f_T is given by Eqn (6-60) and f_L is given by Eqn (6-57).

VI. PIPE FLOW PROBLEMS

There are three typical problems encountered in pipe flows, depending upon what is known and what is to be found. These are the "unknown driving force", "unknown flow rate", and "unknown diameter" problems, and we will outline here the procedure for the solution of each of these for both Newtonian and non-Newtonian (power law and Bingham plastic) fluids. A fourth problem, perhaps of even more practical interest for piping system design, is the "most economical diameter" problem, which will be considered in the next chapter.

We note first that the Bernoulli equation can be written

$$DF = e_f + \frac{1}{2} (\alpha_2 V_2^2 - \alpha_1 V_1^2) \tag{6-64}$$

where

$$e_f = \left(\frac{4fL}{D}\right)\left(\frac{V^2}{2}\right) = \frac{32fLQ^2}{\pi^2 D^5} \tag{6-65}$$

and

$$DF = -\left(\frac{\Delta\Phi}{\rho} + w\right) \tag{6-66}$$

DF represents the net energy input into the fluid per unit mass (or the "net driving force") and is the combination of static head, pressure difference, and pump work. When any of the terms in Eqn (6-66) are negative, they represent a positive "driving force" for moving the fluid through the pipe (positive terms represent forces resisting the flow, e.g. an increase in elevation, pressure, etc.). In most applications, the kinetic energy terms are negligible or cancel out, although this should be verified for each problem. We will use the Bernoulli equation in the form of Eqn (6-64) (usually without the kinetic energy terms) for analyzing pipe flows, and we will use the total volumetric flow rate (Q) as the flow variable instead of the velocity, since this is the usual measure of capacity in a pipeline. The expression for the Newtonian Reynolds number consistent with this is

$$N_{Re} = \frac{4Q\rho}{\pi D \mu} \tag{6-67}$$

A. UNKNOWN DRIVING FORCE

For this problem, we want to know the net driving force (DF) which is required to move a given fluid at a specified rate (Q) through a specified pipe (D, L, ε).

1. Newtonian Fluid

The "knowns" and "unknowns" in this case are

Given: Q, μ, ρ, D, L, ε *Find*: DF

All of the relevant variables and parameters are uniquely related through the three dimensionless variables f, N_{Re}, and ε/D by the Moody diagram or the Churchill equation. Furthermore, the unknown (DF = e_f) appears in only one of these groups (f). Thus the procedure is:

(1) Calculate the Reynolds number from Eqn (6-67).
(2) Calculate ε/D.
(3) Determine f from the Moody diagram or Churchill equation (Eqn 6-38) (if $N_{Re} < 2000$, use f = $16/N_{Re}$).
(4) Calculate e_f (hence DF) from Eqn (6-65).

This is a straightforward calculation. From the resulting value of DF, the required pump head can be determined, for example, from a knowledge of the upstream and downstream pressures and elevations, using Eqn (6-66).

2. Power Law Fluid

The equivalent problem statement is

Given: Q, m, n, ρ, D, L   *Find*: DF

Note that we have an additional fluid property (m and n instead of μ), but we also assume that pipe roughness has a negligible effect, so that the total number of variables is the same. The corresponding dimensionless variables are f, N_{RePL}, and n (which are related by Eqn 6-44), and the unknown (DF = e_f) appears in only one group (f). The above procedure may thus also be applied for the power law fluid, if the appropriate equations are used, as follows:

(1) Calculate the Reynolds number (N_{RePL}), using Eqn (6-42) and the volumetric flow rate instead of the velocity, i.e.

$$N_{RePL} = \frac{2^{7-3n}\rho Q^{2-n}}{m\pi^{2-n}D^{4-3n}} \left[\frac{n}{3n+1} \right]^n \tag{6-68}$$

(2) Calculate f from Eqn (6-44).
(3) Calculate e_f (hence DF) from Eqn (6-65).

3. Bingham Plastic

The problem statement is

Given: Q, μ_∞, τ_o, ρ, D, and L   *Find*: DF

The number of variables is the same as in the above problems, hence the number of groups relating these variables is the same. For the Bingham plastic, these are f, N_{Re}, and N_{He}, which are related by Eqn (6-62), along with Eqns (6-57), (6-60), and (6-63). The unknown (DF = e_f) appears in only f, as before. The solution procedure is similar to that above:

(1) Calculate the Reynolds number:

$$N_{Re} = \frac{4Q\rho}{\pi D\mu_\infty} \tag{6-69}$$

(2) Calculate the Hedstrom number:

$$N_{He} = \frac{D^2\rho\tau_o}{\mu_\infty^2} \tag{6-70}$$

(3) Determine f from Eqns (6-62), (6-60), and (6-57). [Note that an iteration is required to determine f_L from Eqn (6-57)].
(4) Calculate e_f, hence DF, from Eqn (6-65).

B. UNKNOWN FLOW RATE

In this case, the flow rate is to be determined when a given fluid is transported in a given pipe with a known net driving force (e.g. pump head, pressure head, and/or hydrostatic head). The same total variables are involved, and hence the dimensionless variables are the same and are related in the same way as for the unknown driving force problems. The main difference is that now the unknown (Q) appears in two of the dimensionless variables (f and N_{Re}), which requires a different solution strategy.

1. Newtonian Fluid

The problem statement is

Given: DF, D, L, ε, μ, ρ *Find*: Q

The strategy we use is to redefine the relevant dimensionless variables, by combining the original groups in such a way that the unknown variable appears in one group. For example, from the definition of f and N_{Re} the unknown (Q) will cancel if we combine these groups as follows:

$$fN_{Re}^2 = \left(\frac{DF\pi^2D^5}{32LQ^2}\right)\left(\frac{4Q\rho}{\pi D\mu}\right)^2 = \frac{DF\rho^2D^3}{2L\mu^2} \tag{6-71}$$

Thus, if we work with the three dimensionless variables fN_{Re}^2, N_{Re}, and ε/D, the unknown (Q) appears in only one of these (N_{Re}), which is now the (dimensionless) unknown.

There are various approaches that we can take to solve this problem. Since the Reynolds number is unknown, an explicit solution is not possible using the established relations between the friction factor and Reynolds number (e.g. the Moody diagram or Churchill equation). We can, however, proceed by a trial-and-error method which requires an initial "guess" for an unknown variable, using the basic relations to solve for this variable, revising the "guess" accordingly, and repeating the process (iterating) until agreement between calculated and "guessed" values is achieved. Note that in this context, either f or N_{Re} can be considered the "unknown" dimensionless variable, since they both involve the unknown Q. As an aid in making the choice between these, a glance at the Moody diagram shows that the practical range of possible values of f is approximately one order of magnitude, whereas the corresponding possible range of N_{Re} is over five orders of magnitude! Thus, the chances of our initial "guess" being close to the final answer are greatly enhanced if we choose to iterate on f instead of N_{Re}. Using this approach, the procedure is:

(1) Guess a value of f. A value of 0.005 is a good starting point, as this looks like a reasonable "dart throw" at the Moody diagram.

(2) Calculate the value of fN_{Re}^2 from given values.

(3) Calculate $N_{Re} = (fN_{Re}^2/f)^{1/2}$, using f = 0.005.

(4) Using the N_{Re} value from step 3, and the known value of ε/D, determine f from the Moody diagram or Churchill equation (if $N_{Re} <$ 2000, use f = $16/N_{Re}$).

(5) Insert the value of f from step 4 into step 3 to get a revised value of N_{Re}.

(6) Repeat steps 4 and 5 until N_{Re} doesn't change.

(7) Calculate $Q = \pi D\mu N_{Re}/4\rho$.

This procedure doesn't assume prior knowledge of whether the flow is laminar or turbulent, which, of course, cannot be determined until the Reynolds number is known. If the additional assumption of turbulent flow is made, the following simpler procedure can be followed. However, if the assumption of turbulent flow is not verified by the final answer, the procedure will be incorrect, as illustrated below.

(1) Calculate fN_{Re}^2 from known quantities, as in step 2 above, and ε/D.

(2) Assuming turbulent flow, the Colebrook equation (Eqn 6-36) gives $1/\sqrt{f}$ explicitly from the known values of $(fN_{Re}^2)^{1/2}$ and ε/d.

(3) Calculate $N_{Re} = (fN_{Re}^2)^{1/2} (1/\sqrt{f})$

(4) If $N_{Re} > 4000$, calculate $Q = \pi D\mu N_{Re}/4\rho$.

(5) If $N_{Re} < 2000$, the flow is laminar and $fN_{Re} = 16$ should be used instead of the Colebrook equation. In this case, $N_{Re} = fN_{Re}^2/fN_{Re} = fN_{Re}^2/16$. If this still gives $N_{Re} < 2000$, then $Q = \pi D\mu N_{Re}/4\rho$.

(6) If $2000 < N_{Re} < 4000$, the flow is in the transition zone, and the Churchill equation or the Moody diagram should be used to iterate for f and N_{Re}.

2. Power Law Fluid

The problem statement is

Given: DF, D, L, m, n, ρ *Find*: Q

The simplest approach for this problem is the iteration procedure, similar to the first procedure for the Newtonian fluid:

(1) Assume f = 0.005.

(2) Calculate Q from Eqn (6-65), i.e.

$$Q = \pi \left(\frac{D^5 DF}{32 fL} \right)^{1/2}$$ (6-72)

(3) Calculate the Reynolds number from Eqn (6-68), i.e.

$$N_{RePL} = \frac{2^{7-3n} \rho Q^{2-n}}{m \pi^{2-n} D^{4-3n}} \left(\frac{n}{3n+1} \right)^n$$ (6-73)

(4) Calculate f from Eqn (6-44).
(5) Compare the result of step 4 with step 1. If they do not agree, use the result of step 4 in step 2 and repeat steps 2–5 until agreement is reached. Convergence usually requires only two or three trials, at most, unless very unusual conditions are encountered.

3. Bingham Plastic

The procedure is very similar to that above:

Given: DF, D, L, μ_∞, τ_0, ρ *Find*: Q

(1) Assume f = 0.005.
(2) Calculate Q from Eqn (6-72).
(3) Calculate the Reynolds and Hedstrom numbers:

$$N_{Re} = \frac{4Q\rho}{\pi D \mu_\infty}, \qquad N_{He} = \frac{D^2 \rho \tau_0}{\mu_\infty^2}$$ (6-74)

(4) Calculate f from Eqn (6-62).
(5) Compare the result of step 4 with the assumed value in step 1. If they do not agree, use the f from step 4 in step 2 and repeat steps 2–5 until agreement. Note that an iteration is required to determine f_L in Eqn (6-62), but this procedure normally converges rapidly unless unusual conditions are encountered.

C. UNKNOWN DIAMETER

In this problem, it is desired to determine the size of the pipe (D) which will transport a given fluid (Newtonian or non-Newtonian) at a given flow rate Q over a given distance L with a given driving force DF. Since the unknown D appears in each of the dimensionless variables, it is appropriate to regroup these variables in a more convenient form for this problem.

1. Newtonian Fluid

The problem statement is

Given: DF, Q, L, ε, ρ, μ *Find*: D

We see that the three basic groups (N_{Re}, ε/D, and f) can be rearranged in the following way so that the unknown (D) appears in only one group (N_{Re}):

$$fN_{Re}^5 = \left(\frac{DF\pi^2D^5}{32LQ^2}\right)\left(\frac{4Q\rho}{\pi D\mu}\right) = \frac{32DF\rho^5Q^3}{\pi^3L\mu^5} \tag{6-75}$$

$$N_R = \frac{N_{Re}}{\varepsilon/D} = \frac{4Q\rho}{\pi\mu\varepsilon} \tag{6-76}$$

Thus, the three basic groups for this problem become fN_{Re}^5, N_R, and N_{Re}, with N_{Re} being the dimensionless "unknown" (since it is now the only group containing the unknown D). A suitable procedure for this problem is as follows:

(1)   Calculate fN_{Re}^5 from known quantities using Eqn (6-75).
(2)   Assume f = 0.005.
(3)   Calculate N_{Re} from

$$N_{Re} = \left(\frac{fN_{Re}^5}{0.005}\right)^{1/5} \tag{6-77}$$

(4)   Calculate D from N_{Re}:

$$D = \frac{4Q\rho}{\pi\mu N_{Re}} \tag{6-78}$$

(5)   Calculate ε/D.
(6)   Determine f from the Moody diagram or Churchill equation using the above values of N_{Re} and ε/D (if $N_{Re} < 2000$, use $f = 16/N_{Re}$).
(7)   Compare the result of step 6 and the value assumed in step 2. If they do not agree, use the result of step 6 for f in step 3 in place of 0.005, and repeat steps 3–7 until agreement.

2. Power Law Fluid

The problem statement is:

Given: DF, Q, m, n, ρ, L   *Find*: D

The procedure is analogous to that for the Newtonian fluid. In this case, the combined group $fN_{RePL}^{5/(4-3n)}$ (which we shall call K, for convenience) is independent of D:

$$fN_{RePL}^{5/(4-3n)} = \left(\frac{\pi^2 DF}{32LQ^2}\right)\left[\frac{2^{7-3n}Q^{2-n}}{m\pi^{2-n}}\left(\frac{n}{3n+1}\right)^n\right]^{5/(4-3n)} = K \qquad (6-79)$$

The following procedure may be used to find D:

(1) Calculate K from Eqn (6-79).
(2) Assume f = 0.005.
(3) Calculate N_{RePL} from

$$N_{RePL} = \left(\frac{K}{f}\right)^{(4n-3)/5} \qquad (6-80)$$

(4) Calculate f from Eqn (6-44), using the value of N_{RePL} from step 3.
(5) Compare the result of step 4 with the value of f in step 2. If they do not agree, use the value of f from step 4 in step 3, and repeat steps 3–5 until agreement. The diameter D is obtained from the last value of N_{RePL} from step 3:

$$D = \left[\frac{2^{7-3n}\rho Q^{2-n}}{m\pi^{2-n}N_{Re}}\left(\frac{n}{3n+1}\right)^n\right]^{1/(4-3n)} \qquad (6-81)$$

3. Bingham Plastic

The problem variables are

$$\textit{Given: } DF, Q, \mu_\infty, \tau_o, \rho, L \qquad \textit{Find: } D$$

The combined group that is independent of D is equivalent to Eqn (6-75), i.e.

$$fN_{Re}^5 = \left(\frac{\pi^2 DFD^5}{32LQ^2}\right)\left(\frac{4Q\rho}{\pi\mu_\infty D}\right)^5 = \frac{32DFQ^3\rho^5}{\pi^3L\mu_\infty^5} \qquad (6-82)$$

The procedure is:

(1) Calculate fN_{Re}^5 from Eqn (6-82).
(2) Assume f = 0.01.
(3) Calculate N_{Re} from

$$N_{Re} = \left(\frac{fN_{Re}^5}{f}\right)^{1/5} \qquad (6-83)$$

(4) Calculate D from

$$D = \frac{4Q\rho}{\pi\mu_\infty N_{Re}} \qquad (6-84)$$

(5) Calculate N_{He} from

$$N_{He} = \frac{D^2 \rho \tau_0}{\mu_\infty^2} \qquad (6\text{-}85)$$

(6) Calculate f from Eqn (6-62) using the values of N_{Re} and N_{He} from steps 3 and 5.

(7) Compare the result of step 6 with the value assumed in step 2. If they do not agree, use the result of step 6 into step 3, and repeat steps 3–7 until agreement.

The resulting value of D is given in step 4.

VII. TURBULENT DRAG REDUCTION— VISCOELASTIC FLUIDS

A very remarkable effect was observed by Toms during World War II when pumping napalm (a ''gel'' solution of a polymer in gasoline). He found that the polymer solution could be pumped through pipes in turbulent flow with a considerably lower friction loss than was exhibited by pure gasoline at the same flow rate in the same pipe without the polymer. This phenomenon, known as *turbulent drag reduction* (or the *Toms effect*), has been observed for solutions (mostly aqueous) of a variety of very high polymers (e.g. molecular weights in excess of 10^6) and has been the subject of a great amount of research. The effect is very significant, since as much as 85% *less* energy is required to pump some fluids through pipes, for solutions of certain high polymers at concentrations of 100 ppm or less, compared with the energy required to pump the solvent alone at the same flow rate through the same pipe. This is illustrated in Fig. 6-5, which shows some of Chang's data (Darby and Chang, 1984) for the Fanning friction factor versus (solvent) Reynolds number for fresh and ''degraded'' polyacrylamide solutions of concentrations of 100–500 ppm, in a 2 mm diameter tube. Note that the friction factor at low Reynolds numbers (laminar flow) is much larger than that for the (Newtonian) solvent, whereas it is much smaller at high (turbulent) Reynolds numbers. The non-Newtonian viscosity of these solutions is shown in Fig. 3-8, in Chapter 3.

Although the mechanism is still under debate, Darby and Chang (1984) and Darby and Pivsa-Art (1991) have presented a model for turbulent drag reduction based on the fact that solutions of very high polymers are viscoelastic and the concept that in any unsteady deformation (such as turbulence) elastic properties will store energy which would otherwise be dissipated by viscous properties. Since energy that is dissipated (i.e. the ''friction loss'') must be made up by adding energy (e.g. by a pump) to sustain the flow, that portion of the energy which is stored by elastic deformations remains in the flow and does not

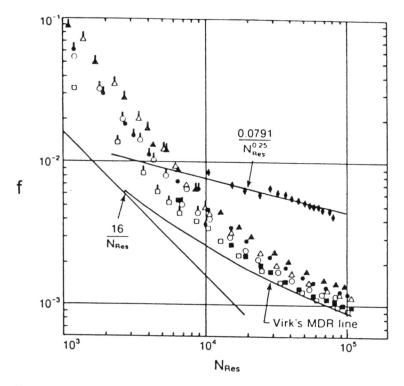

FIGURE 6-5 Drag reduction data for polyacrylamide solutions (Darby and Pivsa-Art, 1991). (N_{Res} is the Reynolds number based on solvent properties.)

have to be made up by external energy sources. Thus, less energy must be supplied externally to sustain the flow; i.e. the "drag" is reduced. This concept is analogous to bouncing an elastic ball. If there is no viscosity (i.e. "internal friction") to dissipate the energy, the ball will continue to bounce indefinitely with no external energy input needed. However, a viscous ball will not bounce at all, since all of the energy is dissipated by viscous deformation and is transferred to "heat". Thus, the greater the fluid elasticity in proportion to the viscosity, the less the energy that must be added to replace that which is dissipated by the turbulent motion of the flow.

The model for turbulent drag reduction developed by Darby and Chang (1984), and later modified by Darby and Pivsa-Art (1991), shows that for smooth tubes the friction factor vs Reynolds number relation for Newtonian fluids (e.g. the Colebrook or Churchill equation) may also be used for drag-reducing flows, provided: (1) the Reynolds number is defined using the properties (e.g. viscosity) of the Newtonian solvent and (2) the Fanning friction factor is modified as follows:

$$f_p = \frac{f_s}{\sqrt{1 + N_{De}^2}} \qquad (6\text{-}86)$$

Here, f_s is the Newtonian solvent Fanning friction factor, as predicted from the (Newtonian) Reynolds number, f_p is a "generalized" Fanning friction factor which applies to polymer (e.g. drag-reducing) solutions as well as Newtonian fluids, and N_{De} is the dimensionless Deborah number, which depends upon the fluid viscoelastic properties and accounts for the storage of energy by the elastic deformations (for Newtonian fluids, $N_{De} = 0$). Fig. 6-6 shows the data from Fig. 6-5 (and many other data sets, as well) replotted in terms of this generalized friction factor. The data are well represented by the classic Colebrook equation (for Newtonian fluids in smooth tubes) on this plot.

The complete expression for N_{De} is given by Darby and Pivsa-Art (1991), as a function of the viscoelastic fluid properties of the fluid (i.e. the Carreau parameters η_0, λ, and p). This expression is

$$N_{De} = \frac{0.0163 \, N_\zeta N_{Res}^{0.338}(\mu_s/\eta_0)^{0.5}}{[1/N_{Res}^{0.75} + 0.00476 \, N_\zeta^2 (\mu_s/\eta_0)^{0.75}]^{0.318}} \qquad (6\text{-}87)$$

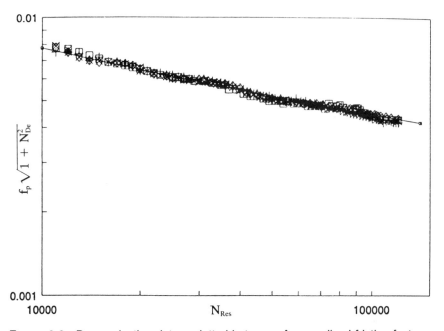

FIGURE 6-6 Drag reduction data replotted in terms of generalized friction factor.

where

$$N_\zeta = [(1 + N_\lambda^2)^p - 1]^{0.5} \quad (6\text{-}88)$$

and

$$N_\lambda = \frac{8V\lambda}{D} \quad (6\text{-}89)$$

N_{Res} is the Reynolds number based on the solvent properties, μ_s is the solvent viscosity, D is the pipe diameter, and V is the velocity in the pipe.

Inasmuch as the rheological properties are very difficult to measure for very dilute solutions (100 ppm or less), a simplified expression was developed by Darby and Pivsa-Art (1991), in which these rheological parameters are contained within two "constants", k_1 and k_2:

$$N_{De} = k_2 \left(\frac{8\mu_s N_{Res}}{\rho D^2} \right)^{k_1} N_{Res}^{0.34} \quad (6\text{-}90)$$

where k_1 and k_2 depend only on the specific polymer solution and its concentration. Darby and Pivsa-Art (1991) examined a variety of drag-reducing data sets from the literature for various polymer solutions in various sizes of pipes and determined the corresponding values of k_1 and k_2 that fit the model to the data. These values are given in Table 6-2. For any drag-reducing solution, k_1 and k_2 can be determined from two data points in the laboratory, at two different flow rates (Reynolds numbers) in turbulent flow, in any size pipe. The resulting values can be used with the model to predict friction loss for that solution at any Reynolds number in any size pipe. If the Colebrook equation for smooth tubes is used, the appropriate generalized expression for the friction factor is

$$f = \frac{0.41}{[\ln(N_{Res}/7)]^2} \left(\frac{1}{(1 + N_{De}^2)^{1/2}} \right) \quad (6\text{-}91)$$

Example 6-1: Determine the percentage reduction in the power required to pump water through a 3 in. I.D. smooth pipe at 300 gpm by adding 100 wppm of "degraded" Separan AP-30.

Solution—We first calculate the Reynolds number for the solvent (water) under the given flow conditions, using a viscosity of 0.01 poise and a density of 1 g/cm³:

$$N_{Res} = \frac{4Q\rho}{\pi D\mu} = \frac{4(300 \text{ gpm})[63.1 \text{ cm}^3/(\text{s gpm})](1 \text{ g/cm}^3)}{\pi(3 \text{ in.})(2.54 \text{ cm/in.})[0.01 \text{ g/(cm s)}]} = 3.15 \times 10^5$$

Then calculate the Deborah number from Eqn (6-90), using $k_1 = 0.088$ and $k_2 = 0.0431$ from Table 6-2:

TABLE 6-2 Parameters for Eqn (6-90) for Various Polymer Solutions

Polymer	Conc. (mg/kg)	Diam. (cm)	k_1	k_2 (s^{k_1})	Reference
Guar gum	20	1.27	0.05	0.009	Wang
(Jaguar A-20-D)	50		0.06	0.014	(1972)
	200		0.07	0.022	
	500		0.10	0.029	
	1000		0.16	0.028	
Guar gum	30		0.05	0.008	White
	60		0.06	0.010	(1966)
	240		0.08	0.016	
	480		0.11	0.018	
Polyacrylamide, fresh	100	0.176	0.093	0.0342	Darby
(Separan AP-30)	250	to	0.095	0.0293	(1991)
	500	1.021	0.105	0.0244	
Polyacrylamide,	100		0.088	0.0431	
degraded	250		0.095	0.0360	
	500		0.103	0.0280	
(AP-273)	10	1.090	0.12	0.0420	White
					(1975)
(PAM E198)	10	0.945	0.21	0.0074	Virk (1970)
	280			0.0078	
	300	2.0 & 3.0	0.40	0.0050	
(PAA)	700		0.53	0.0049	Hoffmann
	125	0.69	0.47	0.00037	(1975)
(ET-597)	250	1.1, &	0.39	0.0013	Astarita
	500	2.05	0.30	0.0061	(1969)
Hydroxyethyl	100	2.54	0.10	0.0074	Wang
cellulose	200		0.16	0.0072	(1972)
(OP-100M)	500		0.24	0.0068	
	1000		0.35	0.0063	
(HEC)	2860	4.8, 1.1 &	0.02	0.0310	Savins
		2.05			(1969)
Polyethylene oxide	10	5.08	0.22	0.017	Goren
(WSR 301)	20		0.21	0.016	(1967)
	50		0.19	0.014	
(W205)	10	0.945	0.31	0.0022	Virk (1970)
	105		0.26	0.0080	
Xanthan gum	1000	0.52	0.02	0.046	Bewersdorff
(Rhodopol 23)					(1988)

Source: Darby and Pivsa-Art(1991).

$$N_{De} = k_2 \left(\frac{8\mu_s N_{Res}}{\rho D^2} \right)^{k_1} N_{Res}^{0.34} = 5.45$$

These values can now be used to calculate the smooth pipe friction factor, from Eqn (6-91). Excluding the N_{De} term gives the friction factor for the Newtonian solvent (f_s), and including the N_{De} term gives the friction factor for the polymer solution (f_p), under the same flow conditions:

$$f_s = \frac{0.41}{[\ln (N_{Res}/7)]^2} = 0.00357$$

$$f_p = \frac{0.41}{[\ln(N_{Res}/7)]^2} \left(\frac{1}{(1 + N_{De}^2)^{1/2}} \right) = 0.000645$$

The power HP required to pump the fluid is given by $-\Delta P Q$. Since $-\Delta P$ is proportional to fQ^2, and Q is the same with and without the polymer, the fractional reduction in power is given by

$$DR = \frac{HP_s - HP_p}{HP_s} = \frac{f_s - f_p}{f_s} = 0.82$$

or 82% *reduction* in the power required to overcome "drag"!

REFERENCES

Astarita, G., G. Greco, Jr., and L. Nicodemo, *AIChE J.*, **15**, 564 (1969)
Bewersdorff, H. W., and N. S. Berman, *Rheol. Acta*, **27**, 130 (1988)
Churchill, S. W., *Chem. Eng.*, p. 91, Nov. 7 (1977)
Darby, R., and H. D. Chang, *AIChE J.*, **30**, 274 (1984)
Darby, R., and J. Melson, *Chem. Eng.*, p. 59, Dec. 28 (1981)
Darby, R., and J. Melson, *J. Pipelines*, **2**, 11 (1982)
Darby, R., and S. Pivsa-Art, *Canad. J. Chem. Eng.*, **69**, 1395 (1991)
Darby, R., R. Mun, and D. V. Boger, *Chem. Eng.*, p. 116, Sept (1992)
Dodge, D.W., and A. B. Metzner, *AIChE J.*, **5**, 189 (1959)
Goren, Y., and J. F. Norbury, *ASME J. Basic Eng.*, **89**, 816 (1967)
Hoffmann, L., and P. Schummer, *Rheol. Acta*, **17**, 98 (1978)
Savins, J. G., in *Viscous Drag Reduction*, C. S. Wells (Ed.), Plenum, 1969, p. 183
Virk, P. S., *AIChE J.*, **21**, 625 (1975)
Virk, P. S., and H. Baher, *Chem. Eng. Sci.*, **25**, 1183 (1970)
Wang, C.B., *Ind. Eng. Chem. Fund.*, **11**, 566 (1972)
White, A., *J. Mech. Eng. Sci.*, **8**, 452 (1966)
White, D., Jr. and R. J. Gordon, *AIChE J.*, **21**, 1027 (1975)

PROBLEMS

PIPE FLOWS

1. Show how the Hagen-Poiseuille equation, for the laminar flow of a Newtonian fluid in a tube, can be derived starting from the general microscopic equations of motion (e.g. the continuity and momentum equations).

2. Derive the relation between the friction factor and Reynolds number for smooth pipe, Eqn (6-32), starting with the von Karman equation for the velocity distribution in the turbulent boundary layer, Eqn (6-24).

3. Evaluate the kinetic energy correction factor α in Bernoulli's equation for turbulent flow, assuming that the "1/7" power law velocity profile (Eqn 6-34) is valid. Repeat this for laminar flow of a Newtonian fluid in a tube, for which the velocity profile is parabolic.

4. A Newtonian fluid with SG = 0.8 is forced through a capillary tube at a rate of 5 cm³/min. The tube has a downward slope of 30° to the horizontal, and the pressure drop is measured between two taps located 40 cm apart on the tube using a mercury manometer, which reads 3 cm. When water is forced through the tube at a rate of 10 cm³/min, the manometer reading is 2 cm.
    (a) What is the viscosity of the unknown Newtonian fluid?
    (b) What is the Reynolds number of the flow for each fluid?
    (c) If two separate pressure transducers, which read the total pressure directly in psig, were used to measure the pressure at each of the pressure taps directly instead of using the manometer, what would be the difference in the transducer readings?

5. You are given a liquid and are asked to find its viscosity. Its density is known to be 0.97 g/cm³. You place the fluid in an open vessel to the bottom of which a 20 cm long vertical tube, with an inside diameter of 2 mm, is attached. When the depth of the liquid in the container is 6 cm, you find that it drains out through the tube at a rate of 2.5 cm³/s. If the diameter of the open vessel is much larger than that of the tube, what is the fluid viscosity?

6. A tube is inclined downward at an angle of 30° to the horizontal, and a mercury manometer is attached to the tube at taps spaced 40 cm apart. When water is flowing through the tube at a rate of 10

cm^3/min, the manometer reading is 2 cm. When another liquid, with SG = 0.8, is flowing through the tube at a rate of 5 cm^3/min, the manometer reading is 3 cm. What is the viscosity of this fluid?

7. You must measure the viscosity of an oil that has a specific gravity of 0.92. To do this, you put the oil into a large container to the bottom of which is attached a small vertical tube, 25 cm long, through which the oil can drain by gravity. When the level of the oil in the container is 6 in. above the container bottom, you find that the flow rate through the tube is 50 cm^3/min. You run the same experiment with water instead of oil and find that under the same conditions the water drains out at a rate of 156 cm^3/min. If the loss coefficient for the energy dissipated in the contraction from the container to the tube is 0.5, what is the viscosity of the oil?

8. You want to transfer No. 3 fuel oil (30° API) from a storage tank to a power plant, at a rate of 2000 bbl/day. The diameter of the pipeline is 1 1/2 sch 40, with a length of 1200 ft. The discharge of the line is 20 ft higher than the suction end, and both ends are at 1 atm pressure. The inlet temperature of the oil is 60°F, and the transfer pump is 60% efficient. If the specific heat of the oil is 0.5 Btu/(lb$_m$°F), and the pipeline is perfectly insulated, determine:
   (a) The horsepower of the motor required to drive the pump.
   (b) The temperature of the oil leaving the pipeline.

9. You must specify a pump to deliver 800 bbl/day of a 35° API distillate at 90°F from a distillation column to a storage tank in a refinery. If the level in the tank is 20 ft above that in the column, the total equivalent length of pipe is 900 ft, and both the column and tank are at atmospheric pressure, what horsepower would be needed if you use 1 1/2 in. sch 40 pipe? What power would be needed if you use 1 in. sch 40 pipe?

10. Water is flowing at a rate of 700 gpm through a horizontal 6 in. sch 80 commercial steel pipe at 90°F. If the pressure drops by 2.23 psi over a 100 ft length of pipe:
   (a) What is the value of the Reynolds number?
   (b) What is the magnitude of the pipe wall roughness?
   (c) How much driving force (i.e. pressure difference) would be required to move the water at this flow rate through 10 miles of pipe if it were made of commercial steel?

(d) What size commercial steel pipe would be required to transport the water at the same flow rate over the same distance if the driving force is the static head in a water tower 175 ft above the pipe?

11. A 35° API distillate at 60°F is to be pumped a distance of 2000 ft through a 4 in. sch 40 horizontal pipeline at a flow rate of 500 gpm. What power must the pump deliver to the fluid if the pipeline is made of (a) drawn tubing; (b) commercial steel; (c) galvanized iron; (d) PVC plastic?

12. The Moody diagram illustrates the effect of roughness on the friction factor in turbulent flow but indicates no effect of roughness in laminar flow. Explain why this is so. Are there any restrictions or limitations that should be placed on this conclusion? Explain.

13. You have a large supply of very rusty 2 in. sch 40 steel pipe which you want to use for a pipeline. Since rusty metal is rougher than clean metal, you want to know its effective roughness before laying the pipeline. To do this, you pump water at a rate of 100 gpm through a 100 ft long section of the pipe, and find that the pressure drops by 15 psi over this length. What is the effective pipe roughness, in inches?

14. A 32 hp pump (100% efficient) is required to pump water through a 2 in. sch 40 pipeline, 6000 ft long, at a rate of 100 gpm.
    (a) What is the equivalent roughness of the pipe?
    (b) If the pipeline is replaced by new commercial steel 2 in. sch 40 pipe, what power would be required to pump water at a rate of 100 gpm through this pipe? What would be the percentage saving in power compared to the old pipe?

15. You have a piping system in your plant which has gotten old and rusty. The pipe is 2 in. sch 40 steel, 6000 ft long. You find that it takes 35 hp to pump water through the system at a rate of 100 gpm.
    (a) What is the equivalent roughness of the pipe?
    (b) If you replace the pipe with the same size new commercial steel pipe, what percentage savings in the required power would you expect at a flow rate of 100 gpm?

16. Water enters a horizontal tube through a flexible vertical rubber hose that can support no forces. If the tube is 1/8 in. sch. 40, 10 ft long,

and the water flow rate is 2 gpm, what force (magnitude and direction) must be applied to the tube to keep it stationary? Neglect the weight of the tube and the water in it. The hose I.D. is the same as that of the tube.

17. A water tower which is 90 ft high provides water to a residential subdivision. The water main from the tower to the subdivision is 6 in. sch 40 steel, 3 miles long. If each house uses a maximum of 50 gal/hr (at peak demand) and the pressure in the water main is not to be less than 30 psig at any point, how many homes can be served by the water main?

18. A heavy oil ($\mu = 100$ cP, SG = 0.85) is draining from a large tank, through a 1/8 in. sch 40 tube, into an open bucket. The level in the tank is 3 ft above the tube inlet, and the pressure in the tank is 10 psig. The tube is 30 ft long and is inclined downward at an angle of 45° to the horizontal. What is the flow rate of the oil, in gpm? What is the value of the Reynolds number in this problem?

19. SAE 10 lube oil (SG = 0.93) is being pumped upward through a straight 1/4 in. sch 80 pipe which is oriented at a 45° angle to the horizontal. The two legs of a manometer using water as the manometer fluid are attached to taps in the pipe wall which are 2 ft apart. If the manometer reads 15 in., what is the oil flow rate, in gal/hr?

20. Cooling water is fed by gravity from an open storage tank 20 ft above ground, through 100 ft of 1 1/2 in. I.D. steel pipe, to a heat exchanger at ground level. If the pressure entering the heat exchanger must be 5 psig for it to operate properly, what is the water flow rate through the pipe?

21. A water main is to be laid to supply water to a subdivision located 2 miles from a water tower. The water in the tower is 150 ft above ground, and the subdivision consumes a maximum of 10,000 gpm of water. What size pipe should be used for the water main? Assume sch 40 commercial steel pipe. The pressure above the water in the tank is 1 atm, and at the subdivision, 30 psig.

22. A water main is to be laid from a water tower to a subdivision which is 2 mi away. The water level in the tower is 150 ft above the ground. The main must supply a maximum of 1000 gpm with a minimum of 5 psig at the discharge end, at a temperature of 65°F. What size

commercial steel sch 40 pipe should be used for the water main? If plastic pipe (which is hydraulically smooth) were used instead, would this alter the result? If so, what diameter of plastic pipe should be used?

23. The water level in a water tower is 110 ft above ground level. The tower supplies water to a subdivision which is 3 miles away, through an 8 in. sch 40 steel water main. If the minimum water pressure entering the residential water lines at the houses must be 15 psig, what is the capacity of the water main in gpm? If there are 100 houses in the subdivision, and each consumes water at a peak rate of 20 gpm, how big should the water main be?

24. A hydraulic press is powered by a remote high-pressure pump. The gauge pressure at the pump is 20 MPa, and the pressure required to operate the press is 19 MPa (gauge), at a flow rate of 0.032 m^3/min. The press and pump are to be connected by 50 m of drawn stainless steel tubing. The fluid properties are those of SAE 10 lube oil at 40°C. What is the minimum tubing diameter that can be used?

25. Water is to be pumped at a rate of 100 gpm from a well which is 100 ft deep, through 2 miles of 4 in. sch 40 steel pipe, to a water tower which is 150 ft high.
   (a) Neglecting fitting losses, what horsepower will the pump require if it is 60% efficient?
   (b) If the elbow in the pipe at ground level below the tower breaks off, how fast will the water drain out of the tower?
   (c) How fast would it drain out if the elbow at the top of the well gave way instead?
   (d) What size pipe would you have to run from the water tower to the ground in order to drain it at a rate of 10 gpm?

26. A concrete pipe storm sewer, 4 ft in diameter, drops 3 ft in elevation per mile length. What is the maximum capacity of the sewer in gpm when it is flowing full?

27. A large tank contains SAE 10 lube oil at a temperature of 60°F and a pressure of 2 psig. The oil is 2 ft deep in the tank and drains out through a vertical tube in the bottom. The tube is 10 ft long and discharges the oil at atmospheric pressure. Assuming the oil to be Newtonian, how fast will it drain through the tube? If the oil is not

Newtonian, but instead can be described as a power law fluid with a flow index of 0.4 and an apparent viscosity of 80 cP at a shear rate of 1 s^{-1}, how would this affect your answer? Tube diameter = 1/2 in.

NON-NEWTONIAN PIPE FLOWS

28. A polymer solution is to be pumped at a rate of 3 gpm through a 1 in. diameter pipe. The solution behaves as a power law fluid with a flow index of 0.5, an apparent viscosity of 400 cP at a shear rate of 1 s^{-1}, and a density of 60 lb$_m$/ft^3.
    (a) What is the pressure gradient in psi/ft?
    (b) What is the shear rate at the pipe wall, and what is the apparent viscosity of the fluid at this shear rate?
    (c) If the fluid were Newtonian, with a viscosity equal to the apparent viscosity from (b) above, what would the pressure gradient be?
    (d) Calculate the Reynolds numbers for the polymer solution and for the above Newtonian fluid.

29. A coal slurry, which is characterized as a power law fluid, has a flow index of 0.4 and an apparent viscosity of 200 cP at a shear rate of 1 s^{-1}. If the coal has a specific gravity of 2.5 and the slurry is 50% coal by weight in water, what pump horsepower will be required to transport 25 million tons of coal per year through a 36 in. I.D., 1000 mile long pipeline? Assume that the entrance and exit of the pipeline are at the same pressure and elevation and that the pumps are 60% efficient.

30. A coal slurry is found to behave as a power law fluid, with a flow index of 0.3, a specific gravity of 1.5, and an apparent viscosity of 70 cP at a shear rate of 100 s^{-1}. What volumetric flow rate of this fluid would be required to reach turbulent flow in a 1/2 in. I.D. smooth pipe which is 15 ft long? What is the pressure drop in the pipe, in psi, under these conditions?

31. A coal slurry is to be transported by pipeline. It has been determined that the slurry may be described by the power law model, with a flow index of 0.4, an apparent viscosity of 50 cP at a shear rate of 100 s^{-1}, and a density of 90 lb$_m$/ft^3. What horsepower would be required to pump the slurry at a rate of 900 gpm through an 8 in. sch 40 pipe which is 50 miles long?

32. A sewage sludge is to be transported a distance of 3 miles through a 12 in. I.D. pipeline, at a rate of 2000 gpm. The sludge is a Bingham plastic, with a yield stress of 35 dyn/cm^2, a limiting viscosity of 80 cP, and a specific gravity of 1.2. What size motor (in horsepower) would be required to drive the pump if it is 50% efficient?

33. A coal suspension is found to behave as a power law fluid, with a flow index of 0.4, a specific gravity of 1.5, and an apparent viscosity of 90 cP at a shear rate of 100 s^{-1}. What would the volumetric flow rate of this suspension be in a 15 ft long, 5/8 in. I.D. smooth tube, with a driving force of 60 psi across the tube? What is the Reynolds number for the flow under these conditions?

34. A coal-water slurry containing 65% (by weight) coal is pumped at a rate of 15 gpm from a storage tank, through a 50 m long 1/2 in. sch 40 pipeline to a boiler, where it is burned. The storage tank is at 1 atm pressure at 80°F, and the slurry must be fed to the burner at 20 psig. The specific gravity of coal is 2.5, and it has a heat capacity of 0.5 Btu/(lb$_m$°F).
    (a) What power must the pump deliver to the slurry, if it is assumed to be Newtonian, with a viscosity of 200 cP?
    (b) In reality, the slurry is non-Newtonian and can best be described as a Bingham plastic, with a yield stress of 800 dyn/cm^2 and a limiting viscosity of 200 cP. Accounting for these properties, what would the required pumping power be?
    (c) If the pipeline is well insulated, what will the temperature of the slurry be when it enters the boiler, for both cases (a) and (b)?

35. A sludge is to be transported by pipeline. It has been determined that the slurry may be described by the power law model, with a flow index of 0.6, an apparent viscosity of 50 cP at a shear rate of 1 s^{-1}, and a density of 95 lb$_m$/ft^3. What hydraulic horsepower would be required to pump the slurry at a rate of 600 gpm through a 6 in. I.D. pipe which is 5 miles long?

36. (a) You must design a transfer system to feed a coal slurry to a boiler. However, you don't know the slurry properties, so you measure them in the lab using a cup-and-bob (Couette) viscometer. The cup has a diameter of 10 cm, the bob diameter is 9.8 cm, and the length of the bob is 8 cm. When the cup is rotated at a rate of 2 rpm, the torque measured on the bob is 2.4 x 10^4 dyn/cm^2, and at 20 rpm it is 6.5 x 10^4 dyn/cm^2.

(a) If you use the Bingham plastic model to describe the slurry properties, what are the values of the yield stress and the limiting viscosity?

(b) If the power law model is used instead, what would be the values of the flow index and consistency?

(c) Using the Bingham plastic model for the slurry, with a value of the yield stress of 35 dyn/cm^2, a limiting viscosity of 35 cP, and a density of 1.2 g/cm^3, what horsepower would be required to pump the slurry through a 1000 ft long, 3 in. I.D. sch 40 pipe, at a rate of 100 gpm?

37. A thick slurry is to be pumped through a 1 in. I.D. pipe which is 200 ft long. You don't know the properties of the slurry, so you test it in the lab by pumping it through a 4 mm I.D. tube which is 1 m long. At a flow rate of 0.5 cm^3/s, the pressure drop in this tube is 1 psi, and at a flow rate of 5 cm^3/s it is 1.5 psi. Estimate the pressure drop that would be required to pump the slurry through the 1 in. pipe at a rate of 2 gpm and also at 30 gpm. Clearly explain the procedure you use, and state any assumptions that you make. Comment in detail about the possible accuracy of your predictions.

38. Drilling mud has to be pumped down into an oil well that is 8000 ft deep. The mud is to be pumped at a rate of 50 gpm to the bottom of the well and back to the surface, through a pipe having an effective I.D. of 4 in. The pressure at the bottom of the well is 4500 psi. What pump head is required to do this? The drilling mud has properties of a Bingham plastic, with a yield stress of 100 dyn/cm^2, a limiting (plastic) viscosity of 35 cP, and a density of 1.2 g/cm^3.

39. A straight vertical tube, 100 cm long and 2 mm I.D. is attached to the bottom of a large vessel. The vessel is open to the atmosphere and contains a liquid with a density of 1 g/cm^3 to a depth of 20 cm above the bottom of the vessel.

(a) If the liquid drains through the tube at a rate of 3 cm^3/s, what is its viscosity?

(b) What is the largest tube diameter that can be used in this system to measure the viscosity of liquids which are at least as viscous as water, for the same liquid level in the vessel? Assume that the density is the same as water.

(c) A non-Newtonian fluid, represented by the power law model, is introduced into the vessel with the 2 mm diameter tube attached. If the fluid has a flow index of 0.65, an apparent viscosity of 5 cP at a shear rate of 10 s^{-1}, and a density of 1.2 g/cm^3, how fast will it drain through the tube if the level is 20 cm above the bottom of the vessel?

40. You want to siphon water from an open tank using a 1/4 in. diameter hose. The discharge end of the hose is 10 ft below the water level in the tank, and the siphon will not operate if the pressure falls below 1 psia anywhere in the hose. If you want to siphon the water at a rate of 2 gpm, what is the maximum height above the water level in the tank that the hose can extend and still operate?

41. A non-Newtonian fluid, described by the power law model, is flowing through a thin slit between two parallel planes of width W, separated by a distance H. The slit is inclined upward at an angle θ to the horizontal.
   (a) Derive an equation relating the volumetric flow rate of this fluid to the pressure gradient, slit dimensions, and fluid properties.
   (b) For a Newtonian fluid, the friction factor for this solution can be written in dimensionless form as

$$f = \frac{24}{N_{Reh}}$$

   where the Reynolds number, N_{Reh}, is based upon the hydraulic diameter of the channel. Arrange your solution for the power law fluid in dimensionless form, and solve for the friction factor, f.
   (c) Set your result in (b) equal to $24/N_{Reh}$ to determine an equivalent expression for the power law Reynolds number for slit flow.

42. You are drinking a milk shake through a straw which is 8 in. long and 0.3 in. in diameter. The milk shake has the properties of a Bingham plastic, with a yield stress of 300 dyn/cm^2, a limiting viscosity of 150 cP, and a density of 0.8 g/cm^3.
   (a) If the straw is inserted 5 in. below the surface of the milk shake, how hard must you suck to start the shake flowing through the straw (e.g. how much vacuum must you pull, in psi)?
   (b) If you pull a vacuum of 1 psi, how fast will the shake flow (in cm^3/s)?

43. Water is to be transferred at a rate of 500 gpm from a cooling lake through a 6 in. diameter sch 40 pipeline, to an open tank in a plant which is 30 miles from the lake.

    (a) If the transfer pump is 70% efficient, what horsepower motor is required to drive the pump?

    An injection station is installed at the lake, which injects a high polymer into the pipeline, to give a solution of 50 ppm concentration, with the following properties: a low shear limiting viscosity of 80 cP, a flow index of 0.5, and a transition point from low shear Newtonian to shear thinning behavior at a shear rate of $10 \ s^{-1}$.

    (b) What horsepower is now required to drive the same pump, to achieve the same flow rate?

44. You measure the viscosity of a sludge in the lab and conclude that it can be described as a power law fluid, with a flow index of 0.45, a viscosity of 7 poise at a shear rate of $1 \ s^{-1}$, and a density of 1.2 g/cm^3.

    (a) What horsepower would be required to pump the sludge through a 3 in. sch 40 pipeline, 1000 ft long, at a rate of 100 gpm?

    (b) The viscosity data show that the sludge could also be described by the Bingham plastic model, with a viscosity of 7 poise at a shear rate of $1 \ s^{-1}$ and a viscosity of 0.354 poise at a shear rate of $100 \ s^{-1}$. Using this model, what required horsepower would you predict for the above pipeline?

    (c) Which answer do you think would be the most reliable, and why?

45. An open drum, 3 ft in diameter, contains a mud which is known to be described by the Bingham plastic model, with a yield stress of 120 dyn/cm^2, a limiting viscosity of 85 cP, and a density of 98 lb_m/ft^3. A 1 in. I.D. hose, 10 ft long, is attached to a hole in the bottom of the drum, to drain the mud out. How far below the surface of the mud should the end of the hose be lowered in order to drain the mud at a rate of 5 gpm?

46. You would like to determine the pressure drop vs. flow rate characteristics of a slurry pipeline. In order to do this, you must determine the rheological properties of the slurry, so you test it in the lab by pumping it through a 1/8 in. I.D. pipe, 10 ft long. You find that it takes 5 psi pressure drop in the pipe to produce a flow rate of 100 cm^3/s and that 10 psi results in a flow rate of 300 cm^3/s.

(a) What can you deduce about the rheological characteristics of the slurry from these data?

(b) If it is assumed that the slurry can be adequately described by the power law model, what are the values of the fluid properties, as deduced by the data?

(c) If the Bingham plastic model is used instead of the power law model to describe the slurry, what are its properties?

47. Determine the power required to pump water at a rate of 300 gpm through a 3 in. I.D. pipeline, 50 miles long, if

(a) The pipe is new commercial steel.

(b) The pipe wall is hydraulically smooth.

(c) The pipe wall is smooth, and "degraded" Separan AP-30 polyacrylamide is added to the water at a concentration of 100 wppm.

7

Internal Flow Applications

I. NONCIRCULAR CONDUITS

All of the relations presented so far apply directly to pipe or conduits of circular cross section. However, many of these results may also be applied to conduits with noncircular cross sections, with suitable modification. It should be recalled that the derivation of the momentum equation for uniform flow in a tube (e.g. Eqn 5-44) involved no assumption as to the shape of the tube cross section. The result is that the friction loss is a function of a geometric parameter called the "hydraulic diameter":

$$D_h = 4\left(\frac{A}{W_p}\right) \tag{7-1}$$

where A is the cross section of the conduit and W_p is the wetted perimeter (i.e. the length of contact between the fluid and the solid boundary in the flow cross section). For a full circular pipe, $D_h = D$ (the pipe diameter). The hydraulic diameter is the key characteristic geometric parameter for a conduit with any cross-sectional shape.

A. LAMINAR FLOWS

The same approach that was taken to derive the relation between flow rate and driving force in a circular tube (e.g. momentum balance on a "slug" of fluid in the center of the conduit, or the integral continuity expression) can be applied

183

to analyze laminar flow in conduits with noncircular cross sections. A few examples will be given below. The results of these analyses are the equivalent of the Hagen-Poiseuille equation for a circular tube, expressed in both dimensional and dimensionless form.

1. Flow in a Slit

Flow between two flat, parallel plates which are closely spaced (h ≪ W) is shown in Fig. 7-1. From the definition, the hydraulic diameter for this geometry is $D_h = 2h$, and the laminar flow solution is

$$Q = -\frac{\Delta\Phi W h^3}{12\mu L}$$
(7-2)

The equivalent dimensionless form of this equation is

$$f N_{Reh} = 24$$
(7-3)

where

$$N_{Reh} = \frac{D_h V \rho}{\mu} = \frac{D_h Q \rho}{\mu A}$$
(7-4)

Here, $A = Wh$, and the Fanning friction factor is, by previous definition,

$$f = \frac{e_f}{\left(\dfrac{V^2}{2}\right)\left(\dfrac{4L}{D_h}\right)} = \frac{-\Delta\Phi}{\left(\dfrac{\rho V^2}{2}\right)\left(\dfrac{4L}{D_h}\right)}$$
(7-5)

2. Flow in a Film

The flow of a thin film down an inclined plane is shown in Fig. 7-2. The film thickness is h ≪ W, and the plate is inclined at an angle θ to the vertical direction.

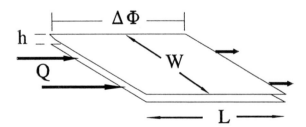

FIGURE 7-1   Flow in a slit.

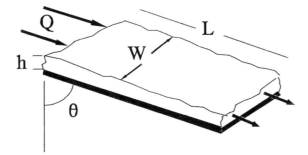

FIGURE 7-2 Flow in a film.

For this flow, the hydraulic diameter is $D_h = 4h$ (since only one boundary in the cross section is a wetted surface), and the laminar flow solution is

$$Q = -\frac{\Delta\Phi h^3 W}{3\mu L} = \frac{\rho g h^2 W \cos\Theta}{3\mu} \qquad (7-6)$$

The dimensionless form of this equation is

$$fN_{Reh} = 24 \qquad (7-7)$$

where the Reynolds number and friction factor are given by Eqns (7-4) and (7-5), respectively.

3. Annulular Flow

Flow in the annulus between two concentric cylinders, as illustrated in Fig. 7-3, is frequently encountered in heat exchangers. For this geometry, the hy-

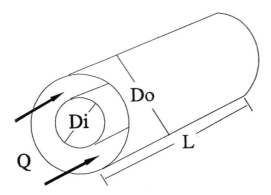

FIGURE 7-3 Flow in an annulus.

draulic diameter is $D_h = D_o - D_i$, and the Newtonian laminar flow solution is

$$Q = -\frac{\Delta\Phi\pi(D_o^2 - D_i^2)}{128\mu L}\left(D_o^2 + D_i^2 - \frac{D_o^2 - D_i^2}{\ln(D_o/D_i)}\right) \tag{7-8}$$

The dimensionless form of this expression is

$$fN_{Reh} = 16\alpha \tag{7-9}$$

where

$$\alpha = \frac{(D_o - D_i)^2}{D_o^2 + D_i^2 - \dfrac{D_o^2 - D_i^2}{\ln(D_o/D_i)}} \tag{7-10}$$

Note that as $D_i/D_o \to 0$, the flow approaches that for a circular tube, and $\alpha \to$ 1. Likewise, as $D_i/D_o \to 1$, the flow approaches that for a slit, and $\alpha \to 1.5$.

For a wide variety of geometries, the value of fN_{Reh} varies only by 50%, as illustrated by the annular flow result (e.g. from 16 to 24). The laminar flow solution for a Newtonian fluid in a variety of geometries has been determined, and the results are summarized in Table 7-1. This table gives the expressions for the cross-sectional area and hydraulic diameter for six different conduit geometries, and the corresponding values of fN_{Reh}, the dimensionless laminar flow solution. The maximum range of the values for fN_{Reh} for all of these geometries is seen to range from about 12 to 24. Thus, for any arbitrary geometry, the dimensionless expression $fN_{Reh} = 18$ would provide an approximate solution for fully developed flow, with an error of about 30% or less.

B. TURBULENT FLOWS

The effect of geometry on the flow field is much less pronounced for turbulent flows than for laminar flows. This is because the majority of the energy dissipation (e.g. flow resistance) occurs within the boundary layer. In turbulent flows, this region occupies a relatively small fraction of the total flow field, near the boundary. In contrast, for laminar flows the "boundary layer" occupies the entire flow field. Thus, while the total area of solid surface contacted by the fluid in turbulent flow is important to the flow resistance, the actual shape of the surface is not. Consequently, the hydraulic diameter provides an even better characterization of the effect of geometry for noncircular conduits for turbulent flows than for laminar flows. That is, the relations developed for turbulent flows in circular pipes can be applied directly to conduits of noncircular cross section simply by replacing the tube diameter by the hydraulic diameter in the relevant dimensionless groups. The accuracy of this procedure increases with increasing Reynolds number, since the higher the Reynolds number the greater the turbulence intensity and the smaller the boundary layer, hence, the less important is the *shape* of the cross section.

It is important to use the hydraulic diameter substitution ($D = D_h$) in the appropriate (original) form of the dimensionless group (e.g. $N_{Re} = DV\rho/\mu$), and not a form which has been adapted for circular tubes (e.g. $N_{Re} = 4Q\rho/\pi D\mu$). That is, the proper modification of the Reynolds number for a noncircular conduit is $D_h V\rho/\mu$, *not* $4Q\rho/\pi D_h\mu$. One clue that the dimensionless group is the wrong form for a noncircular conduit is the presence of π, since this normally arises only from circular geometries (remember: "pi are round, cornbread are square"). Hence, the appropriate dimensionless groups from the tube flow solutions may be modified for noncircular geometries as follows:

$$N_{Reh} = \frac{D_h V\rho}{\mu} = \frac{4Q\rho}{W_p\mu} \tag{7-11}$$

$$f = \frac{e_f D_h}{2LV^2} = \frac{2e_f}{LQ^2}\left(\frac{A^3}{W_p}\right) \tag{7-12}$$

$$N_R = \frac{N_{Reh}}{\varepsilon/D_h} = \frac{D_h^2 Q\rho}{\varepsilon A\mu} = \frac{16Q\rho}{\varepsilon\mu}\left(\frac{A}{W_p^2}\right) \tag{7-13}$$

$$fN_{Reh}^2 = \frac{32e_f\rho^2}{L\mu^2}\left(\frac{A}{W_p}\right)^3 = \frac{e_f\rho^2 D_h^3}{2L\mu^2} \tag{7-14}$$

$$fN_{Reh}^5 = \frac{2048e_f Q^3\rho^5}{L\mu^5}\left(\frac{A}{W_p^2}\right)^3 \tag{7-15}$$

The circular tube expressions can also be transformed to the equivalent expressions for a noncircular conduit by the substitution

$$\pi \Rightarrow \frac{W_p}{D_h} = 4\frac{A}{D_h^2} = \frac{1}{4}\left(\frac{W_p^2}{A}\right) \tag{7-16}$$

II. MOST ECONOMICAL DIAMETER

We have seen how to determine the pumping requirement (driving force) for a given pipe size and flow requirement and how to determine the proper pipe size for a given pump head (driving force) and flow requirement. However, when installing a pipeline or piping system we are free to select both the "best" pipe and the "best" pump. The term "best" in this case refers to that combination of pipe and pump which will minimize the total system cost.

The total cost of a pipeline or piping system includes both the capital cost of the pipe and pumps as well as operating costs, the primary operating cost

TABLE 7-1 Laminar Flow Factors for Noncircular Conduits

Circle

$A = \pi D^2/4$
$D_h = D$

$fN_{Reh} = 16$

Square

$A = D^2$
$D_h = D$

$fN_{Reh} = 14.2$

Isosceles Triangle

$A = (1/2)(d^2 \sin \theta)$
$D_h = (d \sin \theta)/[1 + \sin(\theta/2)]$

θ(deg)	fN_{Reh}
10	12.5
30	13.1
45	13.3
60	13.3
90	13.2
120	12.7
150	12.5

Rectangle

$A = Dd$

$D_h = 2dD/(D + d)$

D/d	fN_{Reh}
1	14.2
2	15.8
5	19.2
10	21.1

$fN_{Reh} = 16[2/3 + (11/24)(d/D)(2 - d/D)]$

Ellipse

$A = \pi dD$

$D_h = 4dD(64 - 16c^2)/[(d + D)(64 - 3c^4)]$

$c = (D - d)/(D + d)$ (for $0.1 < D/d < 10$)

$fN_{Reh} = 2D_h^2(D^2 + d^2)/(D^2 d^2)$

Right Triangle

$A = dD/2$

$D_h = 2dD/[d + D + (d^2 + D^2)^{1/2}]$

$\theta = \tan^{-1}(d/D)$

θ	fN_{Reh}
10	12.5
30	13.0
45	13.2
60	13.0
70	12.8
90	12.0

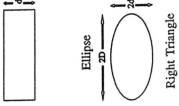

being that of the energy required to drive the pumps:

    Capital cost of pipe (CCP)

    Capital cost of pump stations (CCPS)

    Energy cost to power pumps (EC)

Although the energy cost is continuous and the capital costs are "one time", it is common to spread out (or amortize) the capital cost over a period of years (Y), i.e., the "economic lifetime" of the pipeline. The reciprocal of this (X = 1/Y) is the fraction of the total capital cost expended per year. Taking 1 year as the time basis, we can combine the capital cost per year and the energy cost per year to get the total cost (there are other costs, such as maintenance, but these are minor and will not materially influence the result).

Data on the cost of typical pipelines of various sizes were reported by Darby and Melson (1982). They showed that these data can be represented by the equation

$$CCP = aD_{ft}^p L \tag{7-17}$$

where D_{ft} is the pipe ID in feet, and the parameters a and p depend upon the grade of steel in the pipe, as shown in Table 7-2. Likewise, the capital cost of (installed) pump stations (for 500 hp and over) was shown to be a linear function of the pump power, as follows (see Fig. 7-4):

$$CCPS = A + \frac{B\ HP}{\eta_e} \tag{7-18}$$

where A = \$172,800, B = \$450.8/hp (in 1980 \$), and HP/η_e is the horsepower rating of the pump.

The energy cost is determined from the pumping energy requirement, as determined by the Bernoulli equation:

$$-w = \frac{\Delta\Phi}{\rho} + \frac{1}{2}\Delta V^2 + \sum e_f \tag{7-19}$$

TABLE 7-2   Cost of Pipe (1980 \$)[a]

	Pipe grade				
	ANSI 300#	ANSI 400#	ANSI 600#	ANSI 900#	ANSI 1500#
a	23.1	23.9	30.0	38.1	55.3
p	1.16	1.22	1.31	1.35	1.39

[a]Pipe cost (\$/ft) = $a(ID_{ft})^p$.

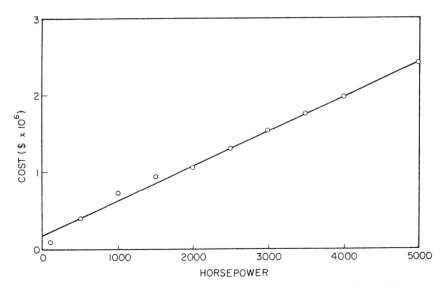

FIGURE 7-4   Cost of Pump stations (1980 $). Pump station cost ($) = CCPS = A + B HP, where A = $172,800 and B = $451/hp for stations of 500 hp or more.

where

$$\sum e_f = 4f \frac{V^2}{2} \sum \frac{L}{D} \tag{7-20}$$

and $\Sigma L/D$ is assumed to include the equivalent length of any fittings (which are usually negligible for a long pipeline). The required pumping power (HP) is

$$HP = -w\dot{m} = \dot{m} \left[\frac{2fLV^2}{D} + \frac{\Delta\Phi}{\rho} \right] = \frac{32fL\dot{m}^3}{\pi^2 \rho^2 D^5} + \dot{m} \frac{\Delta\Phi}{\rho} \tag{7-21}$$

The total energy cost per year is therefore

$$EC = \frac{C\ HP}{\eta_e} \tag{7-22}$$

where C is the unit energy cost ($/(hp yr), ¢/kWh, etc.) and η_e is the pump efficiency. Note that the capital cost increases almost linearly with the pipe diameter, whereas the energy cost decreases with about the fifth power of the diameter. The total annual cost of the pipeline is the sum of the capital and energy costs:

$$TC = X(CCP + CCPS) + EC \tag{7-23}$$

Substituting Eqns (7-17), (7-18), and (7-22) into Eqn (7-23), we get

$$TC = XaD^PL + XA + \frac{BX + C}{\eta_e}\left[\frac{32fL\dot{m}^3}{\pi^2\rho^2D^5} + \dot{m}\frac{\Delta\Phi}{\rho}\right] \tag{7-24}$$

Now we wish to find the pipe diameter that minimizes this total cost. To do this, we differentiate Eqn (7-24) with respect to D, set the derivative equal to zero, and solve for D (which is the most economical diameter, i.e. D_{ec}):

$$D_{ec} = \left[\left(\frac{B + CY}{ap\eta_e}\right)\left(\frac{160f\dot{m}^3}{\pi^2\rho^2}\right)\right]^{\frac{1}{p+5}} \tag{7-25}$$

where $Y = 1/X$.

A. NEWTONIAN FLUIDS

Equation (7-25) is implicit for D_{ec}, since the friction factor (f) depends upon D_{ec} through the Reynolds number and the relative roughness of the pipe. It can be solved by iteration in a straightforward manner, however, by first assuming a value for f (say, 0.005), calculating D_{ec} from Eqn (7-25), using this diameter to compute the Reynolds number and relative roughness, then using these values to find f (from the Moody diagram or Churchill equation). This value is used to replace the assumed value, and the process is repeated until the f values agree.

Another approach might be to regroup the characteristic dimensionless groups in the problem so that the unknown (D_{ec}) appears in only one group. After solving Eqn (7-25) for f, we see that the following group will be independent of D_{ec}:

$$fN_{Re}^{p+5} = \left(\frac{4}{\pi}\right)^{p+3}\frac{\rho^2ap\eta_e\dot{m}^{p+2}}{10(B + CY)\mu^{p+5}} = N_C \tag{7-26}$$

We can call this the "cost group" (N_C), since it contains all of the cost parameters. We can also define a roughness group that does not include the diameter, as follows:

$$N_R = \frac{\varepsilon/D_{ec}}{N_{Re}} = \frac{\pi\mu\varepsilon}{4\dot{m}} \tag{7-27}$$

The remaining group is the Reynolds number, which is the dependent group because it contains D_{ec}:

$$N_{Re} = \frac{4\dot{m}}{\pi D_{ec}\mu} \tag{7-28}$$

The Moody diagram can be used to construct a plot of N_{Re} versus $N_C = fN^{p+5}$ for various values of p and N_R (a double parametric plot), which permits a direct solution to this problem (see Darby and Melson, 1982). The above equations can also be used directly for an iterative solution. Since the value of N_C is known, assuming a value for f will give N_{Re} from Eqn (7-26). This, in turn, gives D_{ec} from Eqn (7-28), and hence ε/D_{ec}. These values of N_{Re} and ε/D_{ec} are used to find f (from the Moody diagram or Churchill equation), and the iteration is continued until successive values of f agree. The most difficult aspect of working with these groups is ensuring a consistent set of units for all of the variables (with appropriate use of the conversion factor g_c, if working in engineering units). For this reason, it is easier to work with consistent units in a scientific system (e.g. SI or cgs), which avoids the need for g_c.

Example 7-1: What is the most economical diameter for a pipeline that is required to transport crude oil, with a viscosity of 30 cP and a specific gravity of 0.95, at a rate of 1 million barrels per day using ANSI 1500# pipe, if the cost of energy is 5¢/kWh (in 1980 $)? Assume that the economical lifetime of the pipeline is 40 years and that the pumps are 50% efficient.
Solution—From Table 7-2, the pipe cost parameters are

$$p = 1.39, \qquad a = 55.3 \ \$/\text{ft}^{2.39} \times (3.28 \ \text{ft/m})^{2.39} = 945.5 \ \$/\text{m}^{2.39}$$

Using SI units will simplify the problem. After converting, we have

$$\dot{m} = \rho Q = 1748 \ \text{kg/s}, \qquad \mu = 0.03 \ \text{N s/m}^2, \qquad CY = \$17.52/\text{watt}$$

Also, from Fig. 7-4 we get the pump station cost factor B:

$$B = 451 \ \$/\text{hp} = 0.604 \ \$/\text{watt}$$

We can now calculate the "cost group", Eqn (7-26):

$$N_c = \left(\frac{4}{\pi}\right)^{p+3} \frac{\rho^2 a p \eta_e \dot{m}^{p+2}}{10(B + CY)\mu^{p+5}} = 5.07 \times 10^{27} = fN_{Re}^{6.39}$$

Assuming a roughness of 0.0018 in., we will solve for D_{ec} by iteration as follows. First, assume f = 0.005, and use this to get N_{Re} from $N_c = fN_{Re}^{6.39}$. From N_{Re}, we find D_{ec}, and thus ε/D_{ec}. Then, using the Churchill equation or Moody diagram, we find f, and compare with the assumed value. This is repeated until convergence:

Assumed f	N_{Re}	D_{ec} (m)	ε/D_{ec}	f(Churchill)
0.005	4.96×10^4	1.49	3.07×10^{-5}	0.00523
0.00523	4.93×10^4	1.50	3.05×10^{-5}	0.00524

This is close enough. The most economical diameter is 1.5 m, or 49.2 in. The "standard pipe size" closest to this value (or the closest size that can readily be manufactured) would be used.

B. NON-NEWTONIAN FLUIDS

A procedure analogous to that above may be used for non-Newtonian fluids that follow the power law or Bingham plastic models (Darby and Melson, 1982).

1. Power Law

For power law fluids, the basic dimensionless variables are the Reynolds number, the friction factor, and the flow index (n). The Reynolds number, in terms of the mass flow rate, is given by

$$N_{RePL} = \left(\frac{4}{\pi}\right)^{2-n} \left(\frac{4n}{3n+1}\right)^n \frac{\dot{m}^{2-n}\rho^{n-1}}{D_{ec}^{4-3n}8^{n-1}m} \tag{7-29}$$

Eliminating D_{ec} from Eqns (7-25) and (7-29), the equivalent cost group becomes

$$f^{4-3n}N_{RePL}^{5+p} \tag{7-30}$$

$$= \frac{(52.4)10^{3n}2^{[7p-3n(1+p)]}}{\pi^{(2+n)(1+p)}\dot{m}^{5+p}} \left(\frac{ap\eta_e}{B+CY}\right)^{4-3n} \left(\frac{\rho^{3-p+n(p-1)}\dot{m}^{2(p-1)+n(4-p)}}{[(3n+1)/n]^{n(5+p)}}\right)$$

Since all values on the right-hand side of Eqn (7-30) are known, assuming a value of f allows a corresponding value of N_{RePL} to be determined. This value can then be used to check the assumed value of f using the power law friction factor equations from the previous chapter (Eqn 6-44), and iterating until agreement is attained.

2. Bingham Plastic

The basic dimensionless variables for a Bingham plastic are the Reynolds number, the Hedstrom number, and the friction factor. By eliminating D_{ec} from the Reynolds number and Eqn (7-25) (as above) the cost group is found to be

$$fN_{Re}^{p+5} = \left(\frac{4}{\pi}\right)^{p+3} \frac{\rho^2 ap\eta_e\dot{m}^{p+2}}{10(B+CY)\mu_\infty^{p+5}} \tag{7-31}$$

D_{ec} can also be eliminated from the Hedstrom number by combining it with the Reynolds number:

$$N_{He}N_{Re}^2 = \left(\frac{4}{\pi}\right)^2 \frac{\tau_0\rho\dot{m}^2}{\mu_\infty^4} \tag{7-32}$$

These equations can readily be solved by iteration, as follows. Assuming a value of f allows N_{Re} to be determined from Eqn (7-31). This is then used with Eqn (7-32) to give N_{He}. The friction factor is then calculated using these values of N_{Re} and N_{He} and the Bingham plastic pipe friction factor equations from the previous chapter (Eqn 6-62). This is compared with the assumed value, and the process is repeated until agreement is attained.

Graphs have been presented by Darby and Melson (1982) that can be used to solve these problems directly without iteration. However, interpolation on doubleparametric log scales is required, so only approximate results can be expected from reading these plots. As mentioned before, the greatest difficulty in using these equations arises from ensuring consistent units. I have found it most convenient to use cgs units in most problems such as these, since fluid properties (density and viscosity) are most often presented in these units, and the scientific system (cgs) does not require the conversion factor g_c. In addition, the energy cost is frequently given in ¢/kWh, which is readily converted to cgs units (e.g. $/joule).

III. FRICTION LOSS IN VALVES AND FITTINGS

Evaluation of the friction loss in valves and fittings involves the determination of the appropriate loss coefficient (K_f), which is then used to calculate the energy loss per unit mass of fluid (e_f):

$$e_f = \frac{K_f V^2}{2} \qquad (7\text{-}33)$$

here V is (usually) the velocity in the pipe upstream of the fitting or valve (however, this is not always true, and care must be taken to ensure that the value of V that is used is the one for which the corresponding K_f is defined).

There are several methods for determining the value of K_f, which we will describe in the order of increasing accuracy. The 2-K (Hooper) method is recommended, because it accounts directly for the effect of both Reynolds number and fitting size on the loss coefficient. For highly turbulent flow, the Crane method agrees well with the 2-K method, but it is less accurate at low Reynolds numbers and is not recommended for laminar flow. The loss coefficient and (L/D)$_{eq}$ methods are more approximate, but they give acceptable results at high Reynolds numbers and when losses in valves and fittings are ''minor losses'' compared to the pipe friction. They are also appropriate for first estimates in problems requiring iterative solutions.

A. LOSS COEFFICIENT

Values of K_f for various types of valves, fittings, etc. can be found tabulated in various textbooks and handbooks. The assumption that these values are constant for a given type of valve or fitting is not accurate, however, because in reality the value of K_f does vary with both the size (scale) of the fitting and the level of turbulence (Reynolds number). One reason that K_f is not the same for all fittings of the same type (e.g. all 90° elbows) is that not all of the dimensions of a fitting, such as the diameter and radius of curvature, scale by the same factor when comparing large and small fittings. Most tabulated values for K_f are close to the values of K_∞ from the 2-K method.

B. EQUIVALENT (L/D) METHOD

The basis for this method is the presumption that there must be some length of pipe (L_{eq}) which has the same friction loss as that which occurs in the fitting, at a given (pipe) Reynolds number. Thus, the fittings are conceptually replaced by the equivalent additional length of pipe which has the same friction loss as the fitting, i.e.

$$e_f = \frac{4fV^2}{2} \sum \left(\frac{L}{D}\right)_{eq}$$ (7-34)

where f is the friction factor in the pipe at the given pipe Reynolds number and relative roughness. This is a convenient concept, because it allows the solution of pipe flow problems with fittings to be carried out in a manner identical to that without fittings if L_{eq} is known. Values of $(L/D)_{eq}$ are tabulated in various text-books and handbooks for a variety of fittings and valves. The method assumes that (1) all size fittings of a given type can be scaled by the corresponding pipe diameter (D), and (2) the influence of turbulence level (i.e. Reynolds number) on the friction loss in the fitting is the same as for the pipe (since the pipe f is used to determine the fitting loss). Neither of these assumptions is accurate (as pointed out above), although the approximation provided by this method gives reasonable results at high turbulence levels (high Reynolds numbers), especially if fitting losses are minor.

C. CRANE METHOD

This method is given in Crane Technical Paper 410 (Crane Co., 1978) and is a modification of the above methods. It is equivalent to the $(L/D)_{eq}$ method, except that it recognizes that there is generally a higher degree of turbulence in the fitting than in the pipe at a given (pipe) Reynolds number. This is accounted for by always using the "fully turbulent" value for f (e.g. f_T) in the expression for the friction loss in the fitting, regardless of the actual Reynolds number in the pipe, i.e.:

$$e_f = \frac{K_f V^2}{2} \quad \text{where} \quad K_f = 4f_T(L/D)_{eq}$$ (7-35)

The value of f_T can be calculated from the Colebrook equation, e.g.

$$f_T = \frac{0.0625}{\left[\log\left(\frac{3.7D}{\varepsilon}\right)\right]^2}$$ (7-36)

in which ε is the pipe roughness (0.0018 in. for commercial steel). This is a two constant model $[f_T$ and $(L/D)_{eq}]$, and values of these constants are tabulated in the Crane paper for a wide variety of fittings, valves, etc. This method gives satisfactory results for high turbulence levels (high Reynolds numbers) but is less accurate at low Reynolds numbers.

D. 2-K (HOOPER) METHOD

This method is based on experimental data in a variety of valves and fittings, over a wide range of Reynolds numbers. The effects of both the Reynolds number and scale (fitting size) are reflected in the expression for the loss coefficient:

$$e_f = \frac{K_f V^2}{2} \quad \text{where} \quad K_f = \frac{K_1}{N_{RE}} + K_\infty \left(1 + \frac{1}{ID_{in}}\right) \tag{7-37}$$

Here, ID_{in} is the internal diameter of the pipe that contains the fitting. This method is recommended as being the most accurate and is valid over the widest range of Reynolds numbers. Tables 7-3, 7-4, and 7-5 give values of K_1 and K_∞ for various fittings, valves, contractions and expansions, and entrance and exit losses from the articles by Hooper (1981, 1988).

The definition of K_f (i.e. $K_f = 2e_f/V^2$) includes the kinetic energy of the fluid, $V^2/2$. For sections that involve area changes (e.g. pipe entrance, exit, expansion, contraction, etc.) the entering and leaving velocity will be different. Since the value of the velocity used with the definition of K_f is arbitrary, it is very important to know which velocity is implied when values of the loss coefficient are used from various sources (handbooks, manuals, texts, etc.).

IV. FITTING LOSSES FOR NON-NEWTONIAN FLUIDS

There are insufficient data in the literature to enable reliable correlation or prediction of friction loss in valves and fittings for non-Newtonian fluids. As a first approximation, however, it may be assumed that a correlation similar to the 2-K method should apply to non-Newtonian fluids if the (Newtonian) Reynolds number in Eqn (7-37) could be replaced by a single corresponding dimensionless group that adequately characterizes the influence of the non-Newtonian properties. For the power law and Bingham plastic fluid models, two rheological parameters are required to describe the viscous properties, which generally results in two corresponding dimensionless groups (N_{Re} and n for the power law, and N_{Re} and N_{He} for the Bingham plastic). However, it is possible to define an "effective viscosity" for a non-Newtonian fluid model that has the same significance in the Reynolds number as the viscosity for a Newtonian fluid, incorporates

TABLE 7-3 2-K Constants for Loss Coefficients for Valves and Fittings

$$K_f = K_1/N_{Re} + K_\infty(1 + 1/ID_{in})$$

Fitting					K_1	K_∞
Elbows	90°		Std (r/D = 1), Threaded		800	0.40
			Std (r/D = 1), Flanged/Welded		800	0.25
			Long Radius (r/D = 1.5), All types		800	0.20
			Mitered	1 Weld (90°)	1000	1.15
			(r/D = 1.5)	2 Welds (45°)	800	0.35
				3 Welds (30°)	800	0.30
				4 Welds (22.5°)	800	0.27
				5 Welds (18°)	800	0.25
	45°		Std (r/D = 1), All Types		500	0.20
			Long Radius (r/D = 1.5)		500	0.15
			Mitered, 1 Weld (45°)		500	0.25
			Mitered, 2 Welds (22.5°)		500	0.15
	180°		Std (r/D = 1), Threaded		1000	0.70
			Std (r/D = 1), Flanged/Welded		1000	0.35
			Long Radius (r/D = 1.5), All types		1000	0.30
Tees	Used as		Std, Threaded		500	0.70
	elbow		Long Radius, Threaded		800	0.40
			Std, Flanged/Welded		800	0.80
			Stub-in Branch		1000	1.00
	Run-through		Threaded		200	1.10
			Flanged/Welded		150	0.05
			Stub-in Branch		100	0.00
Valves	Gate, ball,		Full Line Size, β = 1.0		300	0.10
	or plug		Reduced Trim, β = 0.9		500	0.15
			Reduced Trim, β = 0.8		1000	0.25
	Globe		Standard		1500	4.00
			Angle or Y-type		1000	2.00
	Diaphragm,		dam type		1000	2.00
	butterfly				800	0.25
	check		Lift		2000	10.00
			Swing		1500	1.50
			Tilting disk		1000	0.50

Source: William B. Hooper, Chemical Engineering, p. 97, 1981.

TABLE 7-4 Loss Coefficients for Expansions and Contractions

K_f to be used with upstream velocity head, $V_1^2/2$. $\beta = d/D$

Contraction

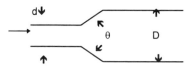

$\theta < 45°$

  $N_{Re1} < 2500$:

  $$K_f = 1.6\left[1.2 + \frac{160}{N_{Re1}}\right]\left[\frac{1}{\beta^4} - 1\right]\sin(\theta/2)$$

  $N_{Re1} > 2500$:

  $$K_f = 1.6[0.6 + 1.92f_1]\left[\frac{1 - \beta^2}{\beta^4}\right]\sin(\theta/2)$$

$\theta > 45°$

  $N_{Re1} < 2500$:

  $$K_f = \left[1.2 + \frac{160}{N_{Re1}}\right]\left[\frac{1}{\beta^4} - 1\right][\sin(\theta/2)]^{1/2}$$

  $N_{Re1} > 2500$:

  $$K_f = [0.6 + 0.48f_1]\left[\frac{1 - \beta^2}{\beta^4}\right][\sin(\theta/2)]^{1/2}$$

Expansion

$\theta < 45°$

  $N_{Re1} < 4000$:

  $$K_f = 5.2(1 - \beta^4)\sin(\theta/2)$$

  $N_{Re1} > 4000$:

  $$K_f = 2.6(1 + 3.2f_1)(1 - \beta^2)^2\sin(\theta/2)$$

$\theta > 45°$

  $N_{Re1} < 4000$:

  $$K_f = 2(1 - \beta^4)$$

  $N_{Re1} > 4000$:

  $$K_f = (1 + 3.2f_1)(1 - \beta^2)$$

N_{Re1} is the upstream Reynolds number, and f_1 is the pipe friction factor at this Reynolds number.
Source: William B. Hooper, Chemical Engineering, p. 89, Nov. 7, 1988.

TABLE 7-5 Loss Coefficients for Pipe Entrance and Exit

$$K_f = K_1/N_{Re} + K_\infty$$

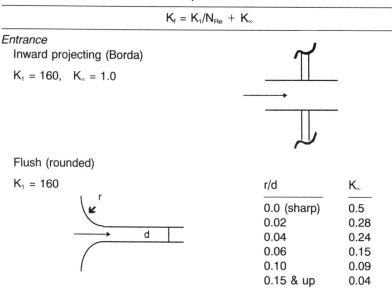

Entrance
Inward projecting (Borda)

$K_1 = 160,\quad K_\infty = 1.0$

Flush (rounded)

$K_1 = 160$

r/d	K_∞
0.0 (sharp)	0.5
0.02	0.28
0.04	0.24
0.06	0.15
0.10	0.09
0.15 & up	0.04

For pipe exit: $K_\infty = 1.0$ for all geometries
$K_1 = 0.0$

Orifice: $K_\infty = \dfrac{2.91}{\beta^4}\,(1 - \beta^2)(1 - \beta^4) = \dfrac{(1 - \beta^2)(1 - \beta^4)}{C_o^2\beta^4}$

$\beta = D_o/D_p$
$K_1 = 0.0$

Source: William B. Hooper, Chemical Engineering, p. 97, 1981.

all of the appropriate parameters for that model, and can be used to define an equivalent non-Newtonian Reynolds number (see Darby and Forsyth, 1992). For the power law model, the result is identical to that derived in Chapter 6:

$$N_{RePL} = \frac{2^{7-3n}\rho Q^{2-n}}{m\pi^{2-n}D^{4-3n}}\left[\frac{n}{3n + 1}\right]^n \tag{7-38}$$

For the Bingham plastic, the corresponding expression for the Reynolds number is

$$N_{ReBP} = \frac{4Q\rho}{\pi D\mu_\infty(1 + \pi D^3\tau_o/24Q\mu_\infty)} \tag{7-39}$$

This is determined by equating the exact laminar tube flow solution for a Bingham plastic (the Buckingham-Reiner equation) to the equivalent Newtonian solution (i.e. $f = 16/N_{ReBP}$), and using this to define N_{ReBP}. Darby and Forsyth (1992) have shown experimentally that mass transfer in Newtonian and non-Newtonian fluids can be correlated by this method; that is, the same correlation applies to both Newtonian and non-Newtonian fluids when the Newtonian Reynolds number is replaced by either Eqn (7-38) for the power law fluid model or Eqn (7-39) for the Bingham plastic fluid model. As a first approximation, therefore, we may assume that the same result would apply to friction loss in valves and fittings as described by the 2-K model (Eqn 7-37).

V. PIPE FLOW PROBLEMS WITH FITTINGS

The inclusion of significant fitting friction loss in piping systems requires a somewhat different procedure for the solution of flow problems than that which was used in the absence of fitting losses. We will consider the same classes of problems as before, i.e. unknown driving force, unknown flow rate, and unknown diameter for Newtonian, power law, and Bingham plastics. The governing equation, as before, is the Bernoulli equation, which can be written

$$DF = -\left(\frac{\Delta\Phi}{\rho} + w\right) = \sum e_f + \frac{1}{2}\Delta(\alpha V^2) \tag{7-40}$$

where

$$\sum e_f = \frac{1}{2}\sum (V^2 K_f) = \frac{8Q^2}{\pi^2}\sum \frac{K_f}{D^4} \tag{7-41}$$

and the summation is over each fitting and segment of pipe (of diameter D) in the system. The loss coefficients for the pipe and fittings are given by the Fanning friction factor and 2-K formula, as before:

$$K_{pipe} = \frac{4fL}{D}, \qquad K_{fit} = \frac{K_1}{N_{Re}} + K_{\infty}\left(1 + \frac{1}{ID_{in}}\right) \tag{7-42}$$

Substituting Eqn (7-41) into Eqn (7-40) gives the following for the Bernoulli equation:

$$DF = \frac{8Q^2}{\pi^2}\left(\sum_i \frac{K_i}{D_i^4} + \frac{\alpha_2}{D_2^4} - \frac{\alpha_1}{D_1^4}\right) \tag{7-43}$$

The α's are the kinetic energy correction factors at the upstream and downstream points (recall that $\alpha = 2$ in laminar flow and $\alpha = 1$ in turbulent flow for a Newtonian fluid).

A. UNKNOWN DRIVING FORCE

Here we wish to find the net driving force required to transport a given fluid at a given rate through a given pipeline containing a specified array of valves and fittings.

1. Newtonian Fluid

The "knowns" and "unknowns" for this case are

Given: Q, μ, ρ, D_i, L_i, ε_i, fittings   Find: DF

The driving force (DF) is given by Eqn (7-43), in which the K_i's are related to the other variables by the Moody diagram (or Churchill equation) for each pipe segment (K_{pipe}) and by the 2-K method for each valve and fitting (K_{fit}), as a function of the Reynolds number:

$$N_{Rei} = \frac{4Q\rho}{\pi D_i \mu} \tag{7-44}$$

The solution procedure is as follows:

(1) Calculate N_{Rei} from Eqn (7-44) for each pipe segment, valve, and fitting (i).

(2) For each pipe segment of diameter D_i, get f_i from the Churchill equation or Moody diagram using N_{Rei} and ε_i/D_i, and calculate $K_{pipe} = 4(fL/D)_i$.

(3) For each valve and fitting, calculate K_{fi} from N_{Rei} and D_i, using the 2-K method.

(4) Calculate the driving force, DF, from Eqn (7-43).

2. Power Law Fluid

The "knowns" and "unknowns" for this case are

Given: Q, D_i, L_i, ε_i, m, n, fittings   Find: DF

The appropriate expressions that apply are the Bernoulli equation, Eqn (7-43), the power law Reynolds number, Eqn (7-38), the pipe friction factor as a function of N_{RePL} and n, (Eqn 6-44), and the 2-K equation for fitting losses (Eqn 7-37) with the Reynolds number replaced by N_{RePL}. The procedure is:

(1) From given values, calculate N_{RePL} from Eqn (7-38).

(2) Using N_{RePL} and n, calculate f (and the corresponding K_{pipe}) for each pipe section from the power law friction factor equation (Eqn 6-44),

and calculate K_f for each valve and fitting using the 2-K method (Eqn 7-37).

(3) Calculate the driving force, DF, from the Bernoulli equation, Eqn (7-43).

3. Bingham Plastic

The procedure for the Bingham plastic is identical to that for the power law fluid, except that Eqn (7-39) is used for the Reynolds number in the 2-K equation for fittings instead of Eqn (7-38), and the expression for the Bingham pipe friction factor is given by Eqn (6-62).

B. UNKNOWN FLOW RATE

The Bernoulli equation (Eqn 7-43) can be rearranged for the flow rate, Q, as follows:

$$
Q = \frac{\pi}{2\sqrt{2}} \left[\frac{DF}{\sum_i \left(\dfrac{K_i}{D_i^4} \right) + \dfrac{\alpha_2}{D_2^4} - \dfrac{\alpha_1}{D_1^4}} \right]^{1/2}
\tag{7-45}
$$

Thus, the flow rate can be readily calculated if the loss coefficients can be determined. The procedure involves an iteration, starting with estimated values for the loss coefficients. These are used in Eqn (7-45) to find Q, which is used to calculate the Reynolds number(s), which are then used to determine revised values for the K_i's, as follows.

1. Newtonian Fluid

The "knowns" and "unknowns" are

Given: DF, D, L, ϵ, μ, ρ, fittings   *Find*: Q

(1) A first estimate for the pipe friction factor and the K_i's can be made by assuming the flow to be fully turbulent (and the α's = 1). Thus:

$$
f_1 = \frac{0.0625}{[\log(3.7D/\epsilon)]^2}
\tag{7-46}
$$

and

$$
K_{fit} = K_\infty \left(1 + \frac{1}{ID_{in}} \right)
\tag{7-47}
$$

(2) Using these values, calculate Q from Eqn (7-45), and then calculate the Reynolds number: $N_{Re} = 4Q\rho/\pi D\mu$.

(3) Using this Reynolds number, determine the revised pipe friction factor (and hence $K_{pipe} = 4fL/D$) from the Moody diagram (or Churchill equation) and the K_{fit}'s from the 2-K equation.

(4) Repeat steps 2 and 3 above until Q does not change.

The solution is the last value of Q calculated from step 2.

2. Power Law Fluid

The ''knowns'' and ''unknowns'' are

Given: DF, D, L, m, n, ρ, fittings   Find: Q

The procedure is essentially identical to that above for the Newtonian fluid, except that Eqn (7-38) is used for the Reynolds number in step 2, and Eqn (6-44) is used for the pipe friction factor in step 3.

3. Bingham Plastic

The ''knowns'' and ''unknowns'' are

Given: DF, D, L, μ_∞, τ_o, fittings   Find: Q

The procedure is, again, similar to that for the Newtonian fluid, except that the pipe friction factor in step 3 (thus K_{pipe}) is determined from Eqn (6-62) using $N_{Re} = 4Q\rho/\pi D\mu_\infty$ and $N_{He} = D^2\rho\tau_o/\mu_\infty^2$. The values of the K_{fit}'s are determined from the 2-K equation using Eqn (7-39) for the Reynolds number.

C. UNKNOWN DIAMETER

It is assumed that the system contains only one size (diameter) pipe. The Bernoulli equation can be rearranged to give D:

$$D = \left[\frac{8Q^2 \left(\sum_i K_i + \dfrac{\alpha_2 D^4}{D_2^4} - \dfrac{\alpha_1 D^4}{D_1^4} \right)}{\pi^2 DF} \right]^{1/4} \tag{7-48}$$

This is obviously implicit in D (the terms involving the α's may be neglected for the initial estimate). If the K_i's can be estimated, the diameter can be determined from Eqn (7-48). However, since D is unknown, so is ε/D, so a ''cruder'' first estimate for f and for the K_{fit}'s is required. Also, since $K_{pipe} = 4fL/D$, an estimated value for f still does not allow determination of K_{pipe}. Therefore, the initial estimate will be made by neglecting the fittings altogether, which is the procedure outlined in the previous chapter for pipe without fittings.

1. Newtonian Fluid

The "knowns" and "unknowns" are

$Given$: Q, DF, L, ϵ, μ, ρ, fittings $Find$: D

If fittings are neglected, the following group can be evaluated from known values:

$$fN_{Re}^5 = \frac{32DF\rho^5Q^3}{\pi^3L\mu^5} \tag{7-49}$$

The procedure is as follows:

(1) For a first estimate, assume f = 0.005.
(2) Use this in Eqn (7-49) to estimate the Reynolds number:

$$N_{Re} = \left(\frac{fN_{Re}^5}{0.005}\right)^{1/5} = \left(\frac{32DF\rho^5Q^3}{0.005\pi^3L\mu^5}\right)^{1/5} \tag{7-50}$$

(3) Get a first estimate for D from this Reynolds number:

$$D = \frac{4Q\rho}{\pi\mu N_{Re}} \tag{7-51}$$

Now the complete equations for f and K_{fit}'s can be used for further iteration.

(4) Using the first estimates of D and N_{Re} above, determine f (and K_{pipe}) from the Moody diagram (or Churchill equation) and the K_{fit}'s from the 2-K formula.
(5) Calculate D from Eqn (7-48), using the previous value of D (from step 3) in the α terms.
(6) If the values of D from steps 3 and 5 do not agree, calculate the value of N_{Re} using the D from step 5, and use these in step 4.
(7) Repeat steps 4–6 until D does not change.

2. Power Law Fluid

The "knowns" and "unknowns" are

$Given$: Q, DF, L, m, n, ρ, fittings $Find$: D

The basic procedure for the power law fluid is the same as above, i.e. we get a first estimate for the Reynolds number by ignoring fittings and assuming turbulent flow. We use this to estimate the value of f (hence K_{pipe}) using Eqn (6-44) and to estimate the K_{fit}'s from the equivalent 2-K equation. Inserting these into Eqn (7-48) then gives a first estimate for the diameter, which is then used to revise the Reynolds number. The iteration continues until successive values agree, as follows:

(1) Assume $f = 0.005$.

(2) Ignoring fittings, the first estimate for N_{RePL} is

$$N_{RePL} = \left(\frac{fN_{RePL}^{5/(4-3n)}}{0.005}\right)^{(4-3n)/5}$$

$$= \left(\frac{\pi^2 DF}{0.16LQ^2}\right)^{(4-3n)/5}\left[\frac{2^{7-3n}Q^{2-n}}{m\pi^{2-n}}\left(\frac{n}{3n+1}\right)^n\right] \qquad (7\text{-}52)$$

(3) Get a first estimate for D from this value and the definition of the Reynolds number:

$$D = \left[\frac{2^{7-3n}\rho Q^{2-n}}{m\pi^{2-n}N_{RePL}}\left(\frac{n}{3n+1}\right)^n\right]^{1/(4-3n)} \qquad (7\text{-}53)$$

(4) Using the values of N_{RePL} from step 2 and D from step 3, calculate the value of f (and K_{pipe}) from Eqn (6-44) and the K_{fit}'s from the 2-K equation.

(5) Insert the values of the K's into Eqn (7-48) to find a new value of D.

(6) If the value of D from step 5 does not agree with that from step 3, use the value from step 5 to calculate a revised N_{RePL}, and repeat steps 4–6 until agreement is attained.

3. Bingham Plastic

The "knowns" and "unknowns" are

Given: Q, DF, L, τ_o, μ_∞, ρ, fittings   *Find*: D

The procedure for a Bingham plastic is similar to that above, using Eqn (6-62) for the pipe friction factor:

(1) Assume $f = 0.02$.

(2) Calculate

$$fN_{Re}^5 = \frac{32DFQ^3\rho^5}{\pi^3L\mu_\infty^5} \qquad (7\text{-}54)$$

(3) Get a first estimate of Reynolds number from

$$N_{Re} = \left(\frac{fN_{re}^5}{0.02}\right)^{1/5} \qquad (7\text{-}55)$$

(4) Use this to estimate D:

$$D = \frac{4Q\rho}{\pi N_{Re}\mu_\infty} \qquad (7\text{-}56)$$

(5) Using this D and N_{Re}, calculate $N_{He} = D^2\rho\tau_o/\mu_\infty^2$, the pipe friction

factor from Eqn (6-62), $K_{pipe} = 4fL/D$, and the K_{fit}'s for the Bingham plastic.

(6) Insert the K_f's into Eqn (7-48) to get a revised value of D.

(7) Using this value of D, revise the values of N_{Re} and N_{He}, and repeat steps 5–7 until successive values agree.

VI. SLACK FLOW

A special condition can occur when pumping a liquid up and down over hilly terrain (for example), which is called "slack flow". Consider the situation shown in Fig. 7-5, in which the pump upstream provides the driving force to move the liquid up the hill, at a flow rate of Q. Since gravity works against the flow on the uphill side and aids the flow on the downhill side, the first job of the pump is to get the fluid to the top of the hill. Thus, the minimum pressure is at point 2 at the top of the hill, and the flow rate (Q) is determined by the balance between the pump head ($h_p = -w/g$) and the frictional and gravitational resistance to flow on the uphill side (i.e. the Bernoulli equation applied from point 1 to point 2):

$$h_p = h_{f(1-2)} + \frac{\Phi_2 - \Phi_1}{\rho g} \tag{7-57}$$

where

$$h_{f(1-2)} = \frac{4fL_{12}}{gD} \left(\frac{V^2}{2} \right) \tag{7-58}$$

Now the driving force in the pipe on the downhill side (from point 2 to point 3) is determined only by the potential (pressure and gravity) difference between

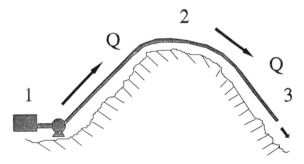

FIGURE 7-5 Condition for slack flow.

these two points, which is independent of the flow rate. However, this driving force is balanced by the friction loss (resistance) in the pipe:

$$\frac{\Phi_2 - \Phi_3}{\rho g} = h_{f(2-3)} \qquad (7\text{-}59)$$

The friction loss is determined by the fluid properties, the velocity, and the pipe size. If the pipe is full of liquid, the velocity is determined by the pipe diameter and flow rate (Q), both of which are the same on the downhill side as on the uphill side. Since the downhill driving force is primarily due to gravity, the higher the hill the greater will be the driving force relative to the "full pipe" flow resistance. Thus, it is quite feasible that, for a full pipe, the downhill conditions will be such that

$$\frac{\Phi_2 - \Phi_3}{\rho g} > (h_{f(2-3)})_{full} \qquad (7\text{-}60)$$

Since the balance equation (Eqn 7-59) must be satisfied, we see that the friction loss on the downhill side must increase to balance the driving force. The only way this can happen is for the velocity to increase, and the only way that this can occur is for the flow cross-sectional area to decrease (since Q is fixed). The only way that the flow area can change is for the liquid to fill only part of the pipe, i.e. it must flow partly full (with the remaining space filled with vapor). This condition, in which the pipe is full on the upstream side but only partly full on the downstream side, with a corresponding higher velocity in the downhill pipe such that the friction loss on the downhill side balances the driving force, is known as *slack flow*. Since the pressure in the vapor space is uniform, there will be no "pressure drop" in the slack flow downhill pipe, because the driving force in this section is due only to gravity.

Since the cross section of the fluid in the pipe, which is not full, will not be circular (see Fig. 7-6), the methods that we used for flow in a noncircular conduit are applicable, i.e. the hydraulic diameter applies. Thus, Eqn (7-59) becomes

$$z_2 - z_3 = h_{f(2-3)} = \frac{2fLQ^2}{gD_hA^2} \qquad (7\text{-}61)$$

where $D_h = 4A/W_p$. If the depth of the liquid in the pipe is h (see Fig. 7-6) (which can be either larger or smaller than R), the expressions for the flow cross section and wetted perimeter are

$$A = R^2\left[\cos^{-1}\left(1 - \frac{h}{R}\right) - \left(1 - \frac{h}{R}\right)\sqrt{1 - \left(1 - \frac{h}{R}\right)^2}\right] \qquad (7\text{-}62)$$

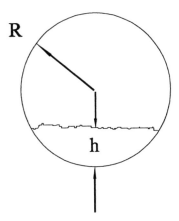

F_{IGURE} 7-6 Pipe flowing less than full.

and

$$W_p = 2R \cos^{-1}\left(1 - \frac{h}{R}\right) \tag{7-63}$$

In order to find h for a given pipe, fluid, and flow rate, a trial-and-error (iterative) procedure is required as follows:

(1) Assume a value of h/R, and calculate A, W_p, and D_h.
(2) Calculate $N_{Re} = (D_h Q\rho)/(A\mu)$, and determine f from the Moody diagram (or Churchill equation).
(3) Calculate the right-hand side (RHS) of Eqn (7-61). If $(z_2 - z_3)$ < RHS, then increase the assumed value of h/R, and repeat the process. If $(z_2 - z_3)$ > RHS, then decrease the assumed value of h/R, and repeat. The solution is obtained when $(z_2 - z_3)$ = RHS for Eqn (7-61).

Example 7-2: A commercial steel pipeline with a 10 in. I.D. carries water over a 300 ft high hill. The actual length of the pipe is 500 ft on the upstream side and 500 ft on the downstream side of the hill. Find: (a) The minimum flow rate at which slack flow will not occur in the pipe, and (b) the position of the interface in the pipe when the flow rate is 80% of this value.

Solution—Slack flow will occur when the driving force (due to gravity) on the downstream side of the hill (from 2 to 3 in Fig. 7-5) just balances the friction loss in this part of the pipeline. That is, Eqn (7-61) will be satisfied, with A = $\pi D^2/4$ and $D_h = D$.

(a) Since this an "unknown flow rate" problem, the flow rate can most easily be determined by first computing the value

$$fN_{Re} = \frac{g\Delta z\rho^2 D^3}{2L_{2-3}\mu^2} = 4.82 \times 10^{10}$$

where the fluid density has been assumed to be 1 g/cm^3 and the viscosity 1 cP. This is solved iteratively with the Churchill equation, by first assuming f = 0.005, then using this to get N_{Re} from the above equation. Using this N_{Re} value, with $\epsilon/D = 0.0018/10 = 0.00018$, we find f from the Churchill equation. This process is repeated until the values of f agree. This process gives f = 0.0035 and $N_{Re} = 3.73 \times 10^6$. The flow rate is then

$$Q = \frac{\pi D\mu N_{Re}}{4\rho} = 7.44 \times 10^5 \text{ cm}^3/\text{s} = 11,800 \text{ gpm}$$

(b) For a flow rate of 80% of this value, we will have slack flow, so Eqn (7-61) must be satisfied for the resulting noncircular flow section (partly full pipe). In this case, we cannot calculate either f, A, or $D_h = 4A/W_p$ a priori. Collecting these quantities on the left of the equation, we get

$$\frac{f}{D_h A^2} = \frac{g\Delta z}{2LQ^2} = 2.13 \times 10^{-9} \text{ cm}^{-3}$$

This value is used to determine f, D_h, A, W_p, and h/R by iteration using Eqns (7-62) and (7-63) and the Churchill equation, as follows. Assuming a value of h/R permits calculation of A and W_p from Eqns (7-62) and (7-63), which also gives $D_h = 4A/W_p$. The Reynolds number is then determined from $N_{Re} = D_h Q\rho/A\mu$, which is used to determine f from the Churchill equation. These values are combined to calculate the value of $f/D_h A^2$, and the process is repeated until this value equals 2.13×10^{-9} cm^{-3}. The results are

h/R = 1.37,   A = 57.2 in.^{2}, W_p = 19.5 in.,

$N_{Re} = 3.01 \times 10^6$,   f = 0.00337

That is, the water interface in the pipe is a little more than two-thirds of the pipe diameter above the bottom of the pipe.

VII. PIPE NETWORKS

Piping systems often involve interconnected segments, that may occur in various combinations of series and/or parallel arrangements. The principles required to analyze such systems are the same as we have used for other systems, e.g. the conservation of mass (continuity) and energy (Bernoulli) equations. For each pipe junction or "node" in the network, continuity tells us that the sum of all of the flow rates into the node must equal the sum of all of the flow rates out of the node. Also, the total driving force (pressure drop plus gravity head loss, plus pump head) between any two nodes is related to the flow rate and friction loss by the Bernoulli equation applied between these two points.

If we number each of the nodes in the network (including the entrance and exit points), then the continuity equation as applied at each node ("i") relates the flow rates into and out of the node:

$$\sum_{n=1}^{n} Q_{ni} = \sum_{m=1}^{m} Q_{im} \tag{7-64}$$

where Q_{ni} represents the flow rate from any upstream node "n" into node "i", and Q_{im} is the flow rate from node "i" out to any downstream node "m".

Also, the total driving force in a branch between any two nodes (e.g. "i" and "j") is determined by Bernoulli's equation (e.g. Eqn 7-43) as applied to this branch. If the driving force is expressed as the total head loss between nodes (e.g. $h_i = \Phi_i/\rho g$):

$$h_i - h_j - \frac{w_{ij}}{g} = \frac{8Q_{ij}^2}{g\pi^2 D_{ij}^4} \sum_{i}^{j} K_{fij} \tag{7-65}$$

where $-w_{ij}/g$ is the pump head (if any) in the branch between nodes "i" and "j", D_{ij} is the pipe diameter, Q_{ij} is the flow rate, and $\Sigma_i^j K_{fij}$ represents the sum of the loss coefficients for all of the fittings, valves, and pipe segments in the branch between nodes "i" and "j". The latter are determined by the 2-K equation for all valves and fittings, and the Churchill equation for all pipe segments, respectively, and are functions of the flow rates and pipe sizes (Q_{ij} and D_{ij}) in the branch between nodes "i" and "j". Thus, the total number of equations is equal to the number of branches plus the number of (internal) nodes, which then equals the number of unknowns that can be determined in the network.

These network equations can be solved for the unknown driving force (across each branch), or the unknown flow rate (in each branch of the network), or an unknown diameter for any one or more of the branches, subject to constraints on the pressure (driving force) and flow rates. Since the solution involves simultaneous, coupled, nonlinear equations, the process is best done by iteration on a computer and can usually be done by iteration using a spreadsheet. The simplest procedure is usually to assume values for the head at one or more intermediate nodes, since these values are bounded by upstream and downstream values which are usually known, and then iterate on these values.

A typical procedure for determining the flow rates in each branch of a network, given, say, the pipe sizes and pressures entering and leaving the network, can best be illustrated by an example.

Example 7-3:   **Flow in a Manifold**—A manifold, or "header", distributes fluid from a common source into various branch lines, as shown in the Figure. The manifold diameter is usually much larger than that of the branches, and the pressure drop in the manifold is much smaller than that in the branch lines, so that the pressure is essentially the same entering each branch. However, these conditions cannot always be satisfied in practice, especially if the total flow rate is large, and/or the manifold is not sufficiently larger than the branch lines, so the assumptions should be verified.

The header illustrated is 0.5 inch in diameter, and feeds three branch lines, each 0.25 inch in diameter. The five nodes are labeled in the diagram. The fluid exits each of the branches at atmospheric pressure and the same elevation, so each of the branch exit points is labeled "5", since the exit conditions are common to all three branches. The distance between the branches on the header is 60 feet, and each branch is 200 feet long. Water enters the header (node 1) at a pressure of 100 psig, and exits the branches (node 5) at atmospheric pressure. The entire network is assumed to be horizontal. Each branch contains two globe valves, in addition to the 200 feet of pipe and the entrance fitting from the header to the branch. We must determine the pressure at the entrance to each branch, the flow rate through each branch, and the total flow rate entering the system. The equations for this problem are set up in a spreadsheet (see figure). The procedure followed is outlined below.

The head at both the entrance to the header (h_1 = 230.8 ft) and the exit from the branches (h_5 = 0) is known. If the head at node 2 were known, Bernoulli's equation (Eqn 7-65) can be used to calculate the flow rate from 1 to 2 (Q_{12}) and the flow rate from 2 to 5 (Q_{25}). By continuity, the flow rate from 2 to 3 must be the difference between these ($Q_{23} = Q_{12} - Q_{25}$). This flow rate is then used in Eqn (7-65) to determine the head at node 3 (h_3). With h_3 known, Q_{35} can be determined from Eqn (7-65), as above. Q_{34} is then determined from

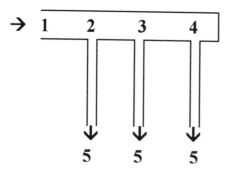

Example 7-3 Flow in a Header

continuity ($Q_{34} = Q_{23} - Q_{35}$). Knowing Q_{34} and h_3 allows h_4 to be determined from Eqn (7-65). Q_{45} can then be determined from the known values of h_4 and h_5 as well as by continuity, since $Q_{34} = Q_{45}$. Comparison of these two values of Q_{45} provides a check on the converged solution, as does the overall continuity equation: $Q_{12} = Q_{25} + Q_{35} + Q_{45}$.

The spreadsheet calculations are done by first assuming a value for h_2 and checking the continuity of the flow rates for agreement. The value of h_2 is adjusted until these checks are in reasonable agreement. The calculation of flow rate from Bernoulli's equation (Eqn 7-65) is also iterative, since the equation involves the loss coefficients that depend upon the flow rate through the Reynolds number. Thus, initial estimates for these loss coefficients must be made, as

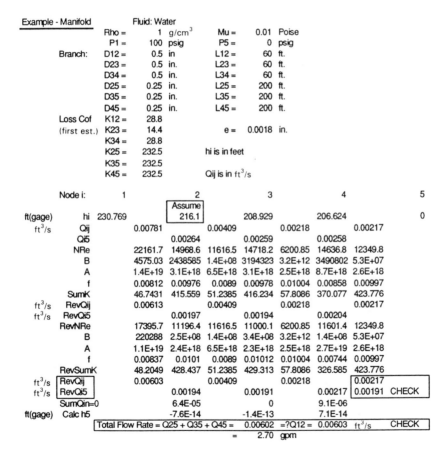

Spreadsheet Output for Example 7-3

214

shown on the spreadsheet output (these are based on $f = 0.005$ for the pipe segments, and $K_f = K_\infty$ for each fitting). The iteration revises these values using the Churchill equation for $K_{pipe} = 4fL/D$ and the 2-K method for the fittings. Only two iterations steps are needed for each unknown Q calculation, as shown on the output. The result shows that the head in the manifold drops about 10%, but this does not result in the flow rates in the branch lines being significantly different. The spreadsheet is also very convenient for ''what if?'' analyses, since it is easy to change any of the known conditions, pipe sizes, fluid properties, etc. and immediately observe the results.

REFERENCES

Crane Co., "Flow of Fluids Through Valves, Fittings, and Pipe", Technical Paper 410, Crane Co., 300 Park Ave., New York, NY 10022 (1978)
Darby, R., and J. Forsyth, *Canad. J. Chem. Eng.*, **70**, 97–103 (1992)
Darby, R., and J. D. Melson, *J. Pipelines*, **2**, 11–21 (1982)
Hooper, W. B., *Chem. Eng.*, p 97, August 24 (1981)
Hooper, W. B., *Chem. Eng.*, p 89, November (1988)

PROBLEMS

1. You must design a pipeline to carry crude oil at a rate of 1 million barrels per day. If the viscosity of the oil is 25 cP and its specific gravity is 0.9, what is the most economic diameter for the pipeline if the pipe cost is $3 per foot length per inch diameter, power cost $0.05/kWh, and the cost of the pipeline is to be written off over a 3-year period? The oil enters and leaves the pipeline at atmospheric pressure.

2. A crude oil pipeline is to be built to carry oil at the rate of 1 million barrels per day (1 bbl = 42 gal). If the pipe costs $12 per ft length per in. diameter, power to run the pumps costs $0.07/kWh, and the economic lifetime of the pipeline is 30 yr, what is the most economical diameter for the pipeline? What total pump horsepower would be required if the line is 800 mi long, assuming 100% efficient pumps? (Oil: $\mu = 35$ cP, $\rho = 0.85$ g/cm^3).

3. A coal slurry pipeline is to be built to transport 45 million tons/yr of coal slurry a distance of 1500 mi. The slurry can be approximately described as Newtonian, with a viscosity of 35 cP and SG = 1.25. The pipeline is to be built from ANSI 600# commercial steel pipe, and the costs of pipe and pump stations are given in the chapter in 1980 dollars. The pumps are 50% efficient, energy costs are $0.06/kWh, and the economic lifetime of the pipeline is 25 years. What would be the most economical pipe diameter, and what would the corresponding velocity be in the line?

4. The Alaskan pipeline was desired to carry crude oil at a rate of 1.2 million bbl/day (1 bbl = 42 gal). If the oil is assumed to be Newtonian, with a viscosity of 25 cP and SG = 0.85, the cost of energy is $0.1/kWh, and the pipe grade is ANSI 600#, what would be the most economical diameter for the pipeline?

5. What is the most economical diameter of a pipeline that is required to transport crude oil (μ = 30 cP, SG = 0.95) at a rate of 1 million barrels per day using ANSI 1500# pipe if the cost of energy is $0.05/kWh (in 1980 dollars), the economic lifetime of the pipeline is 40 years, and the pumps are 50% efficient.

6. Find the most economical diameter of Schedule 40 commercial steel pipe that would be needed to transport a petroleum fraction with a viscosity of 60 cP and SG = 1.3 at a rate of 1500 gpm. The economic life of the pipeline is 30 years, the cost of energy is $0.08/kWh, and the pump efficiency is 60%. The cost of pipe is $20/ft per inch I.D. What would be the most economical diameter to use if the pipe is stainless steel, at a cost of $85/ft per inch I.D., all other things being equal?

7. You must design and specify equipment for transporting 100% acetic acid (SG = 1.0) at a rate of 50 gpm from a large vessel at ground level into a storage tank that is 20 ft above the vessel. It is necessary to use stainless steel for the system (pipe is hydraulically smooth), and you must determine the most economical size pipe to use. You have 1.5 in. and 2 in. nominal sch 40 pipe available for the job. Cost may be estimated from the following approximate formulas:

Pump cost : Cost (\$) = 75.2 (gpm)$^{0.3}$(ft of head)$^{0.25}$

Motor cost : Cost (\$) = 75 (hp)$^{0.85}$

Pipe cost : Cost (\$)/ft = 2.5 (nom. diam., in.)$^{3/2}$

90° Elbow : Cost (\$) = 5 (nom. diam., in)$^{1.5}$

Power : Cost = 0.03 \$/kWh

(a)   Calculate the total pump head (i.e. pressure drop) required for each size pipe, in feet of head.

(b)   Calculate the motor hp required for each size pipe assuming 80% pump efficiency (motors available only in multiples of 1/4 hp).

(c)   Calculate the total capital cost for pump, motor pipe, and fittings for each size pipe.

(d)   Assuming the useful life of the installation is 5 yr, calculate the total operating cost over this period for each size pipe.

(e)   Which size pipe results in the lowest total cost over the 5 yr period?

8.   A large building has a roof with dimensions 50 ft × 200 ft, which drains into a gutter system. The gutter contains three smooth metal downspouts which have a square cross section, 3 in. on a side. The length of the downspouts from the roof to the ground is 20 ft. What is the heaviest rainfall (in in./hr) that the downspouts can handle before the gutter will overflow?

9.   An open concrete flume is to be constructed to carry water from a plant unit to a cooling lake by gravity flow. The flume has a square cross section and is 1500 ft long. The elevation at the upstream end is 10 ft higher than that of the lower end. If the flume is to be designed to carry 10,000 gpm of water when full, what should its size (i.e. width and depth) be? Assume rough cast concrete.

10. A concrete-lined drainage ditch has a triangular cross section that is an equilateral triangle 8 ft on each side. The ditch has a slope of 3 ft/mi. What is the flow capacity of the ditch when full, in gpm?

11. A concrete-lined open drainage canal is to be constructed to carry water from a storm sewer system to a nearby river. The elevation of the inlet to the canal is 10 ft higher than the discharge end, and it

is 3 mi long. The maximum design capacity of the canal is 1 million gpm. If the cross section of the canal is in the shape of a semicircle, how big should it be?

12. A drainage ditch is to be built to carry rainfall runoff from a subdivision. The maximum design capacity is to be 1 million gal/hr, and it will be concrete-lined. If the ditch has a cross section that is an equilateral triangle (open at the top) and if it has a slope of 2 ft/mi, what should the width at the top be?

13. A drainage canal is to be dug to keep a low-lying area from flooding during heavy rains. The canal would carry the water to a river, which is 1 mi away and 6 ft lower in elevation. The canal will be lined with cast concrete and will have a semicircular cross section. If it is sized to drain all of the water falling on a 1 mi^2 area during a rainfall of 4 in./hr, what should the diameter of the semicircle be?

14. An open drainage canal, having a rectangular cross section and a width of 20 ft, is lined with concrete. The canal has a slope of 1 ft/1000 yd. What is the depth of water in the canal when the water is flowing through it at a rate of 500,000 gpm?

15. An open concrete flume is to be constructed to carry water from a reservoir to a plant located 5 miles away. The plant elevation is 20 ft below that of the reservoir. The flume is to have a triangular cross section (equilateral) and is to be designed for a maximum capacity of 5000 gpm when running full. What should the length of each side of the triangle be?

16. An air ventilating system must be designed to deliver air at 20°F and atmospheric pressure at a rate of 150 ft^3/s, through 4000 ft of square duct. If the air blower is 60% efficient and is driven by a 30 hp motor, what size duct is required if it is made of sheet metal?

17. Oil with a viscosity of 25 cP and SG = 0.78 is contained in a large open tank. A vertical tube, made of commercial steel, with a 1 in. ID and a length of 6 ft, is attached to the bottom of the tank. You want the oil to drain through the tube at a rate of 30 gpm.
    (a) How deep should the oil in the tank be for it to drain at this rate?

(b) If a globe valve is installed in the tube, how deep must the oil be to drain at the same rate, with the valve wide open?

18. Heat is to be transferred from one process stream to another by means of a double-pipe heat exchanger. The hot fluid flows in a 1 in. sch 40 tube, which is inside (concentric with) a 2 in. sch 40 tube, with the cold fluid flowing in the annulus between the tubes. If both fluids are to flow at a velocity of 8 ft/s and the total equivalent length of the tubes is 1300 ft, what pump power is required to circulate the colder fluid? Properties at average temperature: $\rho = 55$ lb$_m$/ft^3, $\mu = 8$ cP.

19. Water at 68°F is flowing through a 45° pipe bend at a rate of 2000 gpm. The inlet to the bend is 3 in. ID and the outlet is 4 in. ID. The pressure at the inlet is 100 psig, and the pressure drop in the bend is equal to half of what it would be in a 3 in. 90° elbow. Calculate the net force (magnitude and direction) that the water exerts on the pipe bend.

20. Water flows through a 30° pipe bend at a rate of 200 gpm. The diameter of the entrance to the bend is 2.5 in., and that of the exit is 3 in. The pressure in the pipe is 30 psig, and the pressure drop in the bend is negligible. What is the total force (magnitude and direction) exerted by the fluid on the pipe bend?

21. What size pump (horsepower) is required to pump oil (SG = 0.85, $\mu = 60$ cP) from tank A to tank B at a rate of 2000 gpm through a 10 in. sch 40 pipeline, 500 ft long, containing twenty 90° elbows, one open globe valve, and two open gate valves? The oil level in tank A is 20 ft below that in tank B, and both are open to the atmosphere.

22. A plant piping system takes a process stream ($\mu = 15$ cP, $\rho = 0.9$ g/cm^3) from one vessel at 20 psig and delivers it to another vessel at 80 psig. The system contains 900 ft of 2 in. sch 40 pipe, 24 standard elbows, and 5 globe valves. If the downstream vessel is 10 ft higher than the upstream vessel, what horsepower pump would be required to transport the fluid at a rate of 100 gpm, assuming a pump efficiency of 100%?

23. Crude oil (μ = 40 cP, SG = 0.87) is to be pumped from a storage tank to a refinery through a 10 in. sch 20 commercial steel pipeline, at a flow rate of 2000 gpm. The pipeline is 50 miles long and contains 35 90° elbows and 10 open gate valves. The pipeline exit is 150 ft higher than the entrance, and the exit pressure is 25 psig. What horsepower is required to drive the pumps in the system, if they are 70% efficient?

24. The Alaskan pipeline is 48 in. I.D., is 800 miles long, and carries crude oil at a rate of 1.2 million bbl/day (1 bbl = 42 gal). Assuming the crude oil to be a Newtonian fluid with a viscosity of 25 cP and SG = 0.87, what is the total pumping horsepower required to operate the pipeline? The oil enters and leaves the pipeline at sea level, and the line contains the equivalent of 150 90° elbows and 100 open gate valves. Assume that inlet and discharge pressures are 1 atm.

25. A 6 in. sch 40 pipeline carries oil (μ = 15 cP, SG = 0.85) at a velocity of 7.5 ft/s from a storage tank at 1 atm pressure to a plant site. The line contains 1500 ft of straight pipe, 25 90° elbows, and four open globe valves. The oil level in the storage tank is 15 ft above ground, and the pipeline discharges at a point 10 ft above ground at a pressure of 10 psig. What is the required flow capacity in gpm and the pressure head to be specified for the pump needed for this job? If the pump is 65% efficient, what horsepower motor is required to drive the pump?

26. A pump takes water from a reservoir and delivers it to a water tower. The water in the tower is at atmospheric pressure and is 120 ft above the reservoir. The pipeline is composed of 1000 ft of straight 2 in. sch 40 pipe containing 32 gate valves, 2 globe valves, and 14 standard elbows. If the water is to be pumped at a rate of 100 gpm using a pump that is 70% efficient, what horsepower motor would be required to drive the pump?

27. You must determine the pump pressure and power required to transport a petroleum fraction (μ = 60 cP, ρ = 55 lb_m/ft^3) at a rate of 500 gpm from a storage tank to the feed plate of a distillation column. The pressure in the tank is 2 psig, and that in the column is 20 psig. The liquid level in the tank is 15 ft above ground, and the column inlet is 60 ft high. If the piping system contains 400 ft of 6 in. sch

80 steel pipe, 18 standard elbows, and 4 globe valves, calculate the required pump head (i.e. pressure rise) and the horsepower required if the pump is 70% efficient.

28. What horsepower pump would be required to transfer water at a flow rate of 100 gpm from tank A to tank B if the liquid surface in tank A is 8 ft above ground and that in tank B is 45 ft above ground? The piping between tanks consists of 150 ft of 1 1/2 in. sch 40 pipe, and 450 ft of 2 in. sch 40 pipe, including 16 90° standard elbows and 4 open globe valves.

29. An additive having a viscosity of 2 cP and a density of 50 lb_m/ft^3 is fed from a reservoir into a mixing tank. The pressure in the reservoir and in the tank is 1 atm, and the level in the reservoir is 2 ft above the end of the feed line in the tank. The feed line contains 10 ft of 1/4 in. sch 40 pipe, four elbows, two plug valves, and one globe valve. What will the flow rate of the additive be, in gpm, if the valves are fully open?

30. The pressure in the water main serving your house is 90 psig. The plumbing between the main and your outside faucet contains 250 ft of galvanized 3/4 in. sch 40 pipe, 16 elbows, and the faucet, which is an angle valve. When the faucet is wide open, what is the flow rate, in gpm?

31. You are filling your beer mug from a keg. The pressure in the keg is 5 psig, the filling tube from the keg is 3 ft long, 1/4 in. I.D., and the valve is a diaphragm dam type. The tube is attached to the keg by a (threaded) tee, used as an elbow. If the beer leaving the tube is 1 ft above the level of the beer inside the keg, and there is a 2 ft. long, 1/4 in. I.D., stainless steel tube inside the keg, how long will it take to fill your mug, if it holds 500 cm³? (Beer: μ = 8 cP, ρ = 64 lb_m/ft^3)

32. You must install a piping system to drain SAE 10 lube oil at 70°F (SG = 0.928) from tank A to tank B by gravity flow. The level in tank A is 10 ft above that in tank B, and the pressure in A is 5 psi greater than that in tank B. The system will contain 200 ft of sch 40 pipe, 8 standard elbows, two gate valves, and a globe valve. What size pipe should be used if the oil is to be drained at a rate of 100 gpm?

33. A new industrial plant requires a supply of water at a rate of 5.7 m^3/min. The gauge pressure in the water main, which is located 50 m from the plant, is 800 kPa. The supply line from the main to the plant will require a total length of 65 m of galvanized iron pipe, four standard elbows, and two gate valves. If the water pressure at the plant must be no less than 500 kPa, what size pipe should be used?

34. A pump is used to transport water at 72°F from tank A to tank B, at a rate of 200 gpm. Both tanks are vented to the atmosphere. Tank A is 6 ft above the ground with a water depth of 4 ft in the tank, and tank B is 40 ft above ground with a water depth of 5 ft. The water enters the top of tank B, at a point that is 10 ft above the bottom of the tank. The pipeline joining the tanks contains 185 ft of 2 in. sch 40 galvanized iron pipe, three standard elbows, and one gate valve. If the pump is 70% efficient, what horsepower motor would be required to drive the pump?

35. If the pump in Problem 34 is driven by a 5 hp motor, what is the maximum flow rate that can be achieved, in gpm?

36. A pipeline carrying gasoline (SG = 0.72, μ = 0.7 cP) is 5 miles long and is made of 6 in. sch 40 commercial steel pipe. The line contains 24 90° elbows, eight open gate valves, two open globe valves, and a pump capable of producing a maximum head of 400 ft. The line inlet pressure is 10 psig, and the exit pressure is 20 psig. The discharge end is 30 ft higher than the inlet end.
    (a) What is the maximum flow rate possible in the line, in gpm?
    (b) What is the horsepower of the motor required to drive the pump, if it is 60% efficient?

37. A 12 in. sch 40 pipe, 60 ft long, discharges water at atmospheric pressure from a reservoir. The pipe is horizontal, and the inlet is 12 ft below the surface of the water in the reservoir.
    (a) What is the flow rate in gpm?
    (b) In order to limit the flow rate to 3500 gpm, an orifice is installed at the end of the pipe. What should the orifice diameter be?
    (c) What size pump would have to be used to limit the flow rate to 3500 gpm without using an orifice?

38. Crude oil with a viscosity of 12.5 cP and SG = 0.88 is to be pumped through a 12 in. sch 30 commercial steel pipe at a rate of 1900 bbl/hr. The pipeline is 15 miles long, with a discharge that is 250 ft above the inlet, and contains 10 standard elbows and four gate valves.
    (a) What pump horsepower is required if the pump is 67% efficient?
    (b) If the cost of energy is $0.08/kWh and the pipe is ANSI 600# steel, is the 12 in. pipe the most economical one to use (assume a 30 year economical life of the pipeline)? If not, what is the most economical diameter?

39. A pipeline to carry crude oil at a rate of 1 million bbl/day is constructed with 50 in. I.D. pipe and is 700 miles long with the equivalent of 70 gate valves installed but no other fittings.
    (a) What is the total power required to drive the pumps if they are 70% efficient?
    (b) How many pump stations will be required if the pumps develop a discharge pressure of 100 psi?
    (c) If the pipeline must go over hilly terrain, what is the steepest downslope grade that can be tolerated without creating slag flow in the pipe line? (Crude oil viscosity is 25 cP, SG = 0.9.)

40. You are building a pipeline to transport crude oil (SG = 0.8, μ = 30 cP) from a seaport over a mountain to a tank farm. The top of the mountain is 3000 ft above the seaport and 1000 ft above the tank farm. The distance from the port to the mountain top is 200 miles, and from the mountain top to the tank farm is 75 miles. The oil enters the pumping station at the port at atmospheric pressure and is to be discharged at the tank farm at 20 psig. The pipeline is 20 in. sch 40, and the oil flow rate is 600 gpm.
    (a) Will stack flow occur in the line? If so, you must install a restriction (orifice) in the line to ensure that the pipe is always full. What should the pressure loss across the orifice be, in psi?
    (b) How much pumping power will be required if the pumps are 70% efficient? What pump head is required?
    (c) What are the answers to (a) and (b) if the flow rate is 2000 gpm instead of 6000 gpm?

41. You want to siphon water from an open tank using a hose. The discharge end of the hose is 10 ft below the water level in the tank.

The minimum allowable pressure in the hose for proper operation is 1 psia. If you wish the water velocity in the hose to be 10 ft/s, what is the maximum height that the siphon hose can extend above the water level in the tank for proper operation?

42. Water from a lake is flowing over a concrete spillway at a rate of 100,000 gpm. The spillway is 100 ft wide and is inclined at a 30° angle to the vertical. If the effective roughness of the concrete is 0.03 in., what is the depth of water in the stream flowing down the spillway?

43. A pipeline consisting of 1500 ft of 6 in. sch 40 pipe, containing 25 90° elbows and four open gate valves, carries oil at a velocity of 7.5 ft/s from a storage tank to a plant site. The storage tank is at atmospheric pressure, and the level in the tank is 15 ft above ground. The pipeline discharge is 10 ft above ground, and the discharge pressure is 10 psig.
    (a) What is the required pump capacity (in gpm) and pump head (in ft) needed in this pipeline?
    (b) If the pump has an efficiency of 65%, what horsepower motor would be required to drive it?

44. A centrifugal pump is located 4 ft above the surface of water in a tank. The suction line to the pump is 6 in. sch 40 pipe and extends 6 ft below the surface of the water. If the water temperature is 50°F, what is the pressure (in psia) at the pump inlet when the flow rate is 500 gpm?

45. Water is pumped at a rate of 500 gpm through a 10 in. I.D. pipeline, 50 ft long, containing two standard elbows and a swing check valve. The pressure is atmospheric entering and leaving the pipeline. Calculate the pressure drop (in psi) through the pipeline due to friction using (a) the 2-K method; (b) the $(L/D)_{eq}$ method.

46. A concrete-lined canal with a rectangular cross section is 10 ft wide. If the elevation of the canal drops 5 ft over a 1 mile length, what is the capacity of the canal (in ft^3/s) when the water in the canal is 10 ft deep?

47. Water, at 70°F, is flowing in a film down the outside of a 4 in. O.D. vertical tube, at a rate of 1 gpm. What is the thickness of the film?

48. What diameter of pipe would be required to transport a liquid with a viscosity of 1 cP and a density of 1 g/cm^3 at a rate of 1506 gpm if the length of the pipe is 213 ft, and the wall roughness is 0.006 in., and the total driving force is 100 ft lb_f/lb_m?

49. The ETSI pipeline was designed to carry a coal slurry from Wyoming to Texas at a rate of 30×10^6 tons/yr. The slurry behaves like a Bingham plastic, with a yield stress of 100 dyn/cm^2, a limiting viscosity of 40 cP, and a density of 1.4 g/cm^3. Using the cost of ANSI 1500# pipe and 7¢/kWh for electricity, determine the most economical diameter for the pipeline, if its economic lifetime is 25 years and the pumps are 50% efficient.

50. A roof with a projected horizontal area of 2500 ft^2, 10 ft above ground, is drained by a gutter that feeds into a vertical drain. The drain has a rectangular cross section, with an aspect ratio of 2:1, and must be sized to handle a maximum rainfall of 4 in./hr without overflowing. If the drain roughness is equivalent to that of commercial steel, how big should it be?

51. A mud slurry is drained from a tank through a 50 ft long plastic hose. The hose has an elliptical cross section, with a major axis of 4 in. and a minor axis of 2 in. The open end of the hose is 10 ft below the level in the tank. The mud is a Bingham plastic, with a yield stress of 100 dyn/cm^2, a limiting viscosity of 50 cP, and a density of 1.4 g/cm^3.
    (a) At what rate will the mud drain through the hose (in gpm)?
    (b) At what rate would water drain through the hose?

52. A rubber hose, 1/4 in. in diameter and 10 ft long, is bent through a gradual curve of 90°. The hose is connected to an air supply at 5 psig, and the air exits to the atmosphere. Assuming the hose to have the same roughness as a straight hydraulically smooth tube, determine the net force (magnitude and direction) exerted on the tube (T = 32°F).

53. A 90° threaded elbow is attached to the end of a 3 in. sch 40 pipe, and a reducer with an inside diameter of 1 in. is threaded into the elbow. If water is pumped through the pipe and out the reducer into the atmosphere at a rate of 500 gpm, calculate the forces exerted on the pipe at the point where the elbow is attached.

8

Pumps and Compressors

I. PUMPS

There exists a wide variety of pumps designed for a variety of applications. However, they can be broadly classified into two categories: positive displacement and centrifugal. The most significant characteristics of each of these are described below.

A. POSITIVE DISPLACEMENT PUMPS

The term *positive displacement pumps* is quite descriptive, since these pumps are designed to displace a more or less fixed volume of fluid during each cycle of operation. They include piston, diaphragm, screw, gear, progressing cavity, etc. pumps. The volumetric flow rate is determined by the displacement per cycle of the moving member (either rotating or reciprocating), times the cycle rate (e.g. rpm). The flow capacity is thus fixed by the design, size, and operating speed of the pump. The pressure (or head) that the pump develops depends upon the flow resistance of the system in which the pump is installed and is limited only by the size of the driving motor and the strength of the parts. Consequently, the discharge line from the pump should never be closed off without allowing for recycle around the pump, or damage to the pump could result. In general, positive displacement pumps have limited flow capacity but are capable of high pressures. These pumps operate at approximately constant flow rate, with variable head. They are appropriate for high-pressure requirements, very viscous fluids, and applications where a precisely controlled or metered flow rate is required.

B. CENTRIFUGAL PUMPS

This is also a very descriptive term, since centrifugal pumps operate by the transfer of energy (or angular momentum) from a spinning impeller to the fluid, normally inside a casing. A sectional view of a typical centrifugal pump is shown in Fig. 8-1. The fluid enters at the axis, or "eye", of the impeller (which may be open or closed and usually contains radial curved vanes) and is discharged from the impeller periphery. The kinetic energy of the fluid is increased by the momentum imparted by the high-speed impeller. This kinetic energy is then converted to pressure energy ("head") in a diverging area (the "volute") between the impeller discharge and the casing, before the fluid exits the pump. The head that these pumps can develop depends upon the pump design and the size, shape, and speed of the impeller. The flow capacity is determined by the flow resistance of the system in which the pump is installed. Thus, as will be shown, these pumps operate at approximately constant head, with variable flow rate (within limits, of course, determined by the size and design of the pump and the size of the driving motor).

Centrifugal pumps can be operated in a "closed-off" condition (i.e. closed discharge line), since the liquid will recirculate within the pump without causing

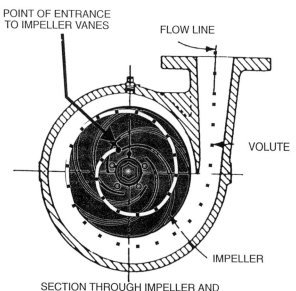

POINT OF ENTRANCE
TO IMPELLER VANES
FLOW LINE

VOLUTE

IMPELLER

SECTION THROUGH IMPELLER AND
VOLUTE ALONG MEAN FLOW SURFACE

FIGURE 8-1   Sectional view of a typical centrifugal pump.

damage. However, such conditions should be avoided, because energy dissipation within the pump could result in excessive heating of the fluid and/or the pump, or unstable operation, with adverse consequences. Centrifugal pumps are most appropriate for "ordinary" liquids (i.e. low to moderate viscosity) under a wide variety of flow conditions and are thus the most common type of pump. The following discussion will apply primarily to centrifugal pumps.

II. PUMP CHARACTERISTICS

Application of Bernoulli's equation from the suction to the discharge across a pump gives

$$-w = \frac{\Delta P}{\rho} = gH_p \qquad (8\text{-}1)$$

That is, the *net* energy or work put into the fluid by the pump goes to increasing the fluid pressure, which is equivalent to the pump head, H_p. However, since pumps are not 100% efficient, some of the energy delivered from the motor to the pump is dissipated or "lost" due to friction. It is very difficult to separately characterize this friction loss, so it is accounted for by a pump efficiency, η_e, which is the ratio of the useful work (or hydraulic work) done by the pump on the fluid ($-w$) to the work put into the pump by the motor ($-w_m$):

$$\eta_e = \frac{-w}{-w_m} \qquad (8\text{-}2)$$

The efficiency of a pump depends upon the pump and impeller design, the size and speed of the impeller, and the conditions under which it is operating, and is determined by testing by the pump manufacturer. This will be discussed in more detail later.

When selecting a pump for a particular application, it is first necessary to specify the flow capacity and head required of the pump. Although many pumps might be able to meet these specifications, the "best" pump is normally the one having the highest efficiency at the specified operating conditions. The required operating conditions, along with a knowledge of the pump efficiency, then allow us to determine the required size (e.g. "brake horsepower") of the driving motor for the pump:

$$HP = -w_m \dot{m} = \frac{\Delta P Q}{\eta_e} = \frac{\rho g H_p Q}{\eta_e} \qquad (8\text{-}3)$$

Now the power delivered from the motor to the pump is also the product of the torque on the shaft driving the pump (Γ) and the angular velocity of the shaft

(ω):

$$HP = \Gamma\omega = \frac{\rho g H_p Q}{\eta_e} \qquad (8\text{-}4)$$

If it is assumed that the fluid leaves the impeller tangentially at the speed at which the impeller is rotating (an approximation), then an angular momentum balance on the fluid in contact with the impeller gives

$$\Gamma = \dot{m}\omega R_i^2 = \rho Q\omega R_i^2 \qquad (8\text{-}5)$$

where R_i is the radius of the impeller and the angular momentum of the fluid entering the eye of the impeller has been neglected (a good assumption). By eliminating Γ from Eqns (8-4) and (8-5), we get

$$H_p \cong \frac{\eta_e \omega^2 R_i^2}{g} \qquad (8\text{-}6)$$

This shows that the pump head is determined primarily by the size and speed of the impeller and the pump efficiency and is independent of the flow rate of the fluid. This is approximately correct for most centrifugal pumps over a wide range of flow rates. However, there is a maximum flow that a given pump can handle, and as the flow rate approaches this limit the developed head will start to drop off. The maximum efficiency for most pumps occurs near this point where the head starts to drop.

Figure 8-2 shows typical pump characteristic curves, as determined by the pump manufacturer. "Size 2 × 3" means that the pump has a 2 in. discharge and a 3 in. suction port. "R&C" and "1 7/8 Pedestal" are the manufacturer's designations, and "3500 RPM" is the speed of the impeller. Performance curves for impellers with diameters from 6 1/4 to 8 3/4 in. are shown, and the efficiency is shown as contour lines of constant efficiency. The maximum efficiency for this pump is somewhat above 50%, although some pumps may operate as high as 80 or 90% efficiency. Operation at conditions on the right branch of the efficiency contours (the "maximum normal capacity" line on Fig. 8-2) should be avoided, since this could result in unstable operation. The pump whose characteristics are given in Fig. 8-2 is a slurry pump, with a semi-open impeller, designed to pump solid suspensions (this pump can pass solid particles as large as 1 1/4 in. in diameter). Such performance curves are normally determined by the manufacturer from operating data using water. Note that according to Eqn (8-6) the head is independent of fluid properties, although according to Eqn (8-4) the power is proportional to the fluid density. The horsepower curves on Fig. 8-2 are the motor horsepower required to pump water at 60°F and must be corrected for density when operating with other fluids and/or temperatures. Actually, it is better to use Eqn (8-4) to calculate the required motor horsepower

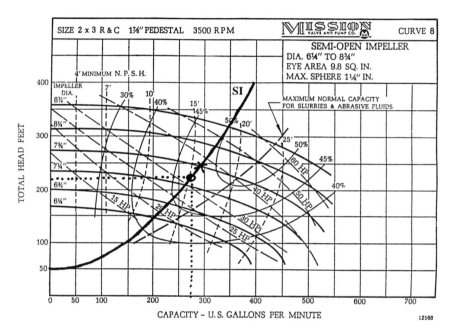

FIGURE 8-2 Typical pump characteristic curves. (From TRW Mission Pump Brochure.)

from the values of the head, flow rate, and efficiency at the operating point. The curves on Fig. 8-2 labeled "minimum N.P.S.H." refer to the cavitation characteristics of the pump, which will be discussed later.

III. PUMPING REQUIREMENTS AND PUMP SELECTION

When selecting a pump for a given application (e.g. a required flow capacity and head), we must specify the appropriate pump type, size (and type) of impeller, and size (horsepower) and speed (rpm) of the pump motor, which will do the "best" job. "Best" is normally the pump that will do the required job at the least cost (i.e. most efficient, or least power). We will concentrate on these factors and not get involved with the details of pump design (impeller vanes, casing dimensions, seals, etc.). More detail on these topics is given in Karassik et al. (1976).

A. REQUIRED HEAD

A typical piping application normally starts with a specified flow rate for a given fluid. The piping system is then designed with the necessary valves, fittings, etc., and should be sized for the most economical pipe size, as discussed in Chapter 7. Application of the energy balance (Bernoulli) equation to the entire system from the upstream end (point 1) to the downstream end (point 2) determines the overall net driving force (DF) in the system necessary to overcome the frictional resistance:

$$DF = \sum e_f \tag{8-7}$$

The total head-driving force is the net sum of the pump head, the total pressure drop, and the elevation drop:

$$\frac{DF}{g} = H_p + \frac{P_1 - P_2}{\rho g} + (z_1 - z_2) \tag{8-8}$$

The friction loss (Σe_f) is the sum of all of the losses from point 1 (upstream) to point 2 (downstream):

$$\sum e_f = \sum_i \left(\frac{V^2}{2} K_f\right)_i = \left(\frac{8Q^2}{\pi^2}\right) \sum_i \left(\frac{K_f}{D^4}\right)_i \tag{8-9}$$

where the loss coefficients (K_f's) include all pipes, valves, fittings, contractions, expansions, etc. in the system. Eliminating DF and Σe_f from Eqns (8-7), (8-8), and (8-9) and solving for the pump head, H_p, gives

$$H_p = \frac{P_2 - P_1}{\rho g} + (z_2 - z_1) + \frac{8Q^2}{g\pi^2} \sum_i \left(\frac{K_f}{D^4}\right)_i \tag{8-10}$$

This relates the system pump head requirement to the specified flow rate and the system loss parameters (e.g. the K_f's). Note that H_p is a quadratic function of Q, for highly turbulent flow (i.e. constant K_f's). A plot of H_p versus Q from Eqn (8-10), illustrated in Fig. 8-2 as line S1, is called the *operating line* for the system. For laminar flow, the K_f's are inversely proportional to the Reynolds number, which results in a linear relation between H_p and Q. The required pump head and flow capacity are determined by the system requirements, and we must select the best pump to meet this requirement.

B. COMPOSITE CURVES

Most pump manufacturers provide composite curves, such as those shown in Fig. 8-3, which can be used to determine the operating range of various pumps. For each pump that covers the specified conditions, the individual pump characteristics (Fig. 8-2) are then consulted. The intersection of the system curve with the pump characteristic curve for a given impeller determines the pump operating point. The impeller diameter is selected which will produce the required head (or greater). This is repeated for all possible pump, impeller, and speed combinations to determine the combination that will provide the highest

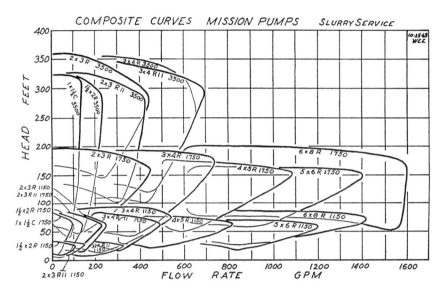

FIGURE 8-3 Typical pump composite curve. [From TRW Mission Pump Brochure (mfr.'s catalog)]

efficiency (i.e. least power requirement). Note that if the operating point (H_p, Q) does not fall exactly on one of the (impeller) curves, then the actual impeller diameter that produces the higher head at the required flow rate Q is chosen. However, when this pump is installed in the system, the actual operating point will correspond to the intersection of the system curve (Eqn 8-10) and the actual pump impeller curve at this point, as indicated by the "X" in Fig. 8-2.

Example 8-1: Consider a piping system that must deliver water at a rate of 275 gpm from one storage tank to another, both of which are at atmospheric pressure, with the level in the downstream tank being 50 ft higher than the level in the upstream tank. The piping system contains 65 ft of 2 in. sch 40 pipe, one globe valve, and six elbows. If the pump to be used has the characteristics shown in Fig. 8-2, what diameter impeller should be used with this pump, and what horsepower motor would be required?

Solution—The head requirement for the piping system is given by Eqn (8-10). Here, $z_2 - z_1 = 50$ ft and, since both upstream and downstream pressures are 1 atm, $\Delta P = 0$. The Reynolds number at 275 gpm for water at 60°F is 4.21 × 10^5, which gives a friction factor of 0.00497 in commercial steel pipe ($\varepsilon/D = 0.0018/2.067$). The corresponding loss coefficient for the pipe is $K_{pipe} = 4fL/D = 7.51$, and the loss coefficients for the fittings from Table 7-3 are (assuming flanged connections): elbow, $K_1 = 800$, $K_\infty = 0.25$; globe valve, $K_1 = 1500$, $K_\infty = 4.0$. At the pipe Reynolds number, this gives $\Sigma(K_f) = K_{pipe} + K_{Glbv} + 6K_{el} = 15.7$. The curve labeled S1 on Fig. 8-2 is H_p vs Q from Eqn (8-10), for this value of the loss coefficients. This neglects the variation of the K_f's over the range of flow rate indicated, which is a good assumption at this Reynolds number. At a flow rate of 275 gpm, the required head from Eqn (8-10) is 219 ft.

The point where the flow rate of 275 gpm intersects the system curve on Fig. 8-2 (at 219 ft of head) falls between impeller diameters of 7 1/4 and 7 3/4 in. as indicated by the "O" on the line. Thus, the 7 1/4 in. diameter would be too small, so we would need the 7 3/4 in. diameter. However, if the pump with this impeller is installed in the system, the operating point would move to the point indicated by the "X" on Fig. 8-2. This corresponds to a head of almost 250 ft and a flow rate of about 290 gpm (i.e. the excess head provided by the larger impeller results in a higher flow rate than desired, all other things being equal).

One way to achieve the desired flow rate of 275 gpm would obviously be to close down on the valve until this value is achieved. This is equivalent to increasing the resistance (i.e. the loss coefficient) for the system, which will shift the system curve upward (since it is steeper) until it intersects the 7 3/4 in. impeller curve at the desired flow rate of 275 gpm. The pump will still provide 250 ft of head, but about 30 ft of this head is ''lost'' in the additional resistance (e.g. dissipation) in the partly closed valve. The pump efficiency at this operating point is about 47%, and the motor power (HP) required to pump water at 60°F at this point is HP $= \rho g H_p Q / \eta_e = 37$ hp.

A control valve operates in this mode, but it is obviously not an efficient use of the available energy. A more efficient way of controlling the flow rate, instead of closing the valve, might be to adjust the speed of the impeller using a variable-speed drive. This would save energy, because it would not increase the friction loss by closing down on the valve but would require greater capital cost because variable-speed drives are more expensive than fixed-speed motors.

IV. CAVITATION AND NPSH

A. VAPOR LOCK AND CAVITATION

As previously mentioned, a centrifugal pump increases the fluid pressure by first imparting angular momentum (or kinetic energy) to the fluid, which is converted to pressure in the diffuser or volute section. Hence, the fluid velocity in and around the impeller is much higher than that either into or out of the pump, and the pressure is the lowest where the velocity is highest. Now the minimum pressure at which a pump will operate properly must be above the vapor pressure of the fluid; otherwise the fluid will vaporize (or "boil"). Obviously, the higher the temperature, the higher the vapor pressure, and the more likely that this condition will occur. Now when a centrifugal pump contains a gas or vapor it will still develop the same head, but since the pressure is proportional to the fluid density it will be several orders of magnitude lower than the pressure for a liquid at the same head. This condition, when the pump is filled with a gas or vapor, is known as *vapor lock*, and the pump will not function.

However, cavitation may result in an even more serious condition than vapor lock. When the pressure at any point within the pump drops below the vapor pressure of the liquid, vapor bubbles will form at that point (this generally occurs on or near the impeller). These bubbles will then be transported to another region in the fluid where the pressure is greater than the vapor pressure, at which point they will collapse. This formation and collapse of bubbles occurs very rapidly and can create local "shock waves", which can cause erosion and serious damage to the impeller or pump. (It is often obvious when a pump is cavitating, since it may sound as though there are rocks in the pump!)

B. NET POSITIVE SUCTION HEAD (NPSH)

To prevent cavitation, it is necessary that the pressure at the pump suction be sufficiently greater than the fluid vapor pressure that the minimum pressure anywhere in the pump will be above the vapor pressure. This required minimum suction pressure (in excess of the vapor pressure) depends upon the pump design, impeller size and speed, and flow rate and is called the *minimum required NPSH* (net positive suction head). Values of the minimum required NPSH for the pump in Fig. 8-2 are shown as dashed lines. The NPSH is almost independent of impeller diameter at low flow rates and increases with flow rate as well as with impeller diameter at higher flow rates. A distinction is sometimes made between the minimum NPSH "required" to prevent cavitation (sometimes termed the NPSHR) and the actual head (e.g. pressure) "available" at the pump suction (or NPSHA). A pump will not cavitate if NPSHA > NPSHR + vapor pressure head.

The NPSH at the operating point for the pump determines where the pump can be installed in a piping system to ensure that cavitation will not occur. The criterion is that the pressure head at the suction entrance to the pump (e.g. the NPSHA) must exceed the vapor pressure head by at least the value of the NPSH (or NPSHR) to avoid cavitation. Thus, if the pressure at the pump suction is P_s, and the fluid vapor pressure is P_v at the operating temperature, cavitation will be prevented if

$$\frac{P_s}{\rho g} \geq NPSH + \frac{P_v}{\rho g} \tag{8-11}$$

The suction pressure P_s is determined by applying the Bernoulli equation to the suction line upstream of the pump. For example, if the upstream suction line entrance pressure is P_1, the maximum distance above the suction line entrance that the pump can be located without cavitation (i.e. the maximum suction lift) is determined by Bernoulli's equation from P_1 to P_s:

$$h_{max} = \frac{P_1 - P_v}{\rho g} - NPSH + \frac{V_1^2 - V_s^2}{2g} - \frac{\sum (e_f)_s}{g} \tag{8-12}$$

where Eqn (8-11) has been used for P_s. V_1 is the velocity entering the suction line, V_s is the velocity at the pump inlet (suction), and $\Sigma(e_f)_s$ is the total friction loss in the suction line from the upstream entrance (point 1) to the pump inlet, including all pipe, fittings, etc. The diameter of the suction entering the pump is usually bigger than the discharge or exit diameter to minimize the kinetic energy head entering the pump, since this kinetic energy decreases the maximum suction lift and enhances cavitation. Note that if the maximum suction lift (h_{max}) is negative, this means that the pump must be located below the upstream entrance to the suction line to prevent cavitation.

It is best to be conservative when interpreting the NPSH requirements to prevent cavitation. The minimum required NPSH on the pump curves is normally determined using water at 60°F, with the discharge line fully open. However, even though a pump will run with a closed discharge line with no bypass, there will be much more recirculation within the pump if this occurs, which increases local turbulence and local velocities as well as dissipative heating, both of which increase the minimum required NPSH. This is especially true with high-efficiency pumps, which have close clearances between the impeller and pump casing.

Example 8-2: A centrifugal pump with the characteristics shown on Fig. 8-2 is to be used to pump an organic liquid from a reboiler to a storage tank, through a 2 in. sch 40 line, at a rate of 200 gpm. The pressure in the reboiler is 1 atm, and the liquid has a vapor pressure of 230 mm Hg, SG = 0.85, and a viscosity of 0.5 cP at the working temperature. If the suction upstream of the pump is also a 2 in. sch 40 line, containing two elbows and one globe valve, and the pump has a 7 3/4 in. impeller, what is the maximum height above the reboiler at which the pump can be located without cavitating?
Solution—The maximum suction lift is given by Eqn (8-12). From Fig. 8-2, the NPSH required of the pump at 200 gpm is about 11 ft. The velocity in the reboiler (V_1) can be neglected, and the velocity in the pipe (from Apx F) is $V_s = 200/10.45 = 19.1$ ft/s.
The friction loss is

$$e_f = \frac{V_s^2}{2} \sum (K_{pipe} + K_{GlbVal} + 2K_{el})$$

where $K_{pipe} = 4fh/D$ and the fitting losses are given by the 2-K formula and Table 7-3 (elbow, $K_1 = 800$, $K_\infty = 0.25$; globe valve, $K_1 = 1500$, $K_\infty = 4$). The value of the Reynolds number for this flow is 5.23×10^5, which, for commercial steel pipe ($\varepsilon/D = 0.0018/2.067$), gives $f = 0.00493$. Note that the pipe length is h in K_{pipe}, which is the same as the maximum suction length (h_{max}) on the left of Eqn (8-12) if the suction line is vertical. The unknown (h) thus appears on both sides of the equation. Thus, solving Eqn (8-12) gives $h = 14.2$ ft.

C. SPECIFIC SPEED

The flow rate, head, and impeller speed at the maximum or "best" efficiency point (BEP) of the pump characteristic can be used to define a dimensionless group called the *specific speed*:

$$N_s = \frac{N\sqrt{Q}}{(H)^{3/4}} \quad \text{or} \quad \frac{\text{rpm } \sqrt{\text{gpm}}}{\text{ft}^{3/4}} \tag{8-13}$$

Although this group is dimensionless (and hence unitless), it is common practice to use selected mixed (inconsistent) units when quoting the value of N_s, i.e. N in rpm, Q in gpm, and H in feet. The value of the specific speed represents the ratio of the pump flow rate to head at the speed corresponding to the maximum efficiency point (BEP), and depends primarily on the design of the pump and impeller. As previously stated, most centrifugal pumps operate at relatively low heads and high flow rates, e.g. high values of N_s. However, this value depends strongly on the impeller design, which can vary widely from almost pure radial flow to almost pure axial flow (like a fan). Examples of various types of impellers are shown in Fig. 8-4. Radial flow impellers have the highest head and lowest flow capacity (low N_s), whereas axial flow impellers have a high flow rate and low head characteristic (high N_s). Thus the magnitude of the specific speed is a direct indication of the impeller design and performance, as shown in Fig. 8-5. Figure 8-5 also indicates the range of flow rates and efficiencies of the various impeller designs, as a function of the specific speed. As indicated on Fig. 8-5, the maximum efficiency corresponds roughly to a specific speed of about 3000.

D. SUCTION SPECIFIC SPEED

Another "dimensionless" group, analogous to the specific speed, which relates directly to the cavitation characteristics of the pump is the *suction specific speed*, N_{ss},

$$N_{ss} = \frac{NQ^{1/2}}{(NPSH)^{3/4}} \tag{8-14}$$

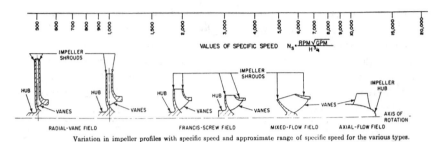

VALUES OF SPECIFIC SPEED $N_s = \frac{RPM\sqrt{GPM}}{H^{3/4}}$

RADIAL-VANE FIELD   FRANCIS-SCREW FIELD   MIXED-FLOW FIELD   AXIAL-FLOW FIELD

Variation in impeller profiles with specific speed and approximate range of specific speed for the various types.

Straight-vane, single-suction
closed impeller.
(Worthington Pump, Inc.)

Open mixed-flow impeller.
(Worthington Pump, Inc.)

Axial-flow impeller.
(Worthington Pump, Inc.)

Semiopen impeller.
(Worthington Pump, Inc.)

Open impellers. Notice that the impellers at left and
right are strengthened by a partial shroud.
(Worthington Pump, Inc.)

Open impeller with a partial
shroud.

Phantom view of a radial-
vane nonclogging impeller.
(Worthington Pump, Inc.)

Paper-pulp impeller.
(Worthington Pump, Inc.)

FIGURE 8-4   Impeller designs and specific speed characteristics (Karassik et al., 1976).

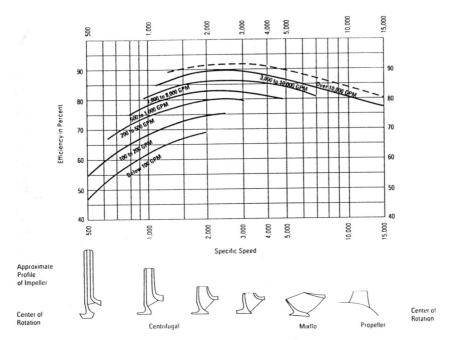

FIGURE 8-5 Correlation between impeller shape, specific speed, and efficiency (Raymer, 1993).

This identifies the *inlet conditions* that produce similar flow behavior in the inlet for geometrically similar pump inlet passages. Note that the suction specific speed (N_{ss}) relates only to the pump inlet, whereas the specific speed (N_s) relates to the entire pump. The suction specific speed can be used, for example, to characterize the conditions under which excessive recirculation may occur at the inlet to the impeller vanes. Recirculation involves flow reversal and re-entry resulting from undesirable pressure gradients at the inlet or discharge of the impeller vanes, and its occurrence generally defines the stable operating limits of the pump. For example, Fig. 8-6 shows the effect of the suction specific speed on the stable "recirculation-free" operating window, expressed as NPSH versus percent of capacity at BEP, for various values of N_s.

It should be noted that there are conflicting parameters in the proper design of a centrifugal pump. For example, Eqn (8-12) shows that the smaller the suction velocity (V_s), the less the tendency to cavitate, i.e. the less severe the NPSH requirement. This would dictate that the eye of the impeller should be as large as practical in order to minimize V_s. However, a large impeller eye means a large vane tip speed at the impeller inlet, which is destabilizing with

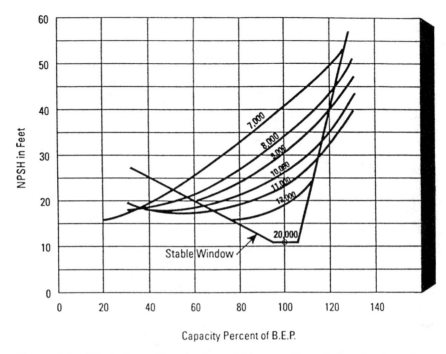

Capacity Percent of B.E.P.

FIGURE 8-6 Effect of specific speed on stable operating window due to recircu-lation. (Numbers on the curves are the values of the specific speed, N_{ss}.) (From Raymer, 1993.)

respect to recirculation. Hence, it is advisable to design the impeller with the smallest eye diameter that is practicable.

V. COMPRESSORS

A compressor may be thought of as a pump for a compressible fluid at high pressures. By ''high pressures'' is meant conditions under which the compress-ibility properties of the fluid (gas) must be considered, which normally occur when the pressure changes by as much as 40% or more. For ''low pressures'' (i.e. smaller pressure changes), a fan or blower may be an appropriate ''pump'' for a gas. Fan operation can be analyzed using the incompressible flow equa-tions, since the relative pressure difference is normally small. Just as for pumps, compressors may be either positive displacement or centrifugal, the former being suitable for relatively high pressures and low flow rates, while the latter are designed for higher flow rates but lower pressures. The major distinction in the governing equations, however, depends upon the conditions of operation, i.e.

whether the system is isothermal or adiabatic. The following analyses assume that the gas is adequately described by the ideal gas law. This assumption can be modified, however, by an appropriate compressibility correction factor, as necessary. For an ideal (frictionless) compression, the work of compression is given by the Bernoulli equation, which reduces to

$$-w = \int_{P_1}^{P_2} \frac{dP}{\rho} \tag{8-15}$$

Now the energy balance on the gas can be written:

$$\Delta h = q + e_f + \int_{P_1}^{P_2} \frac{dP}{\rho} \tag{8-16}$$

which says that the work of compression, plus the energy dissipated due to friction and any heat transferred into the gas during compression, all go to increasing the enthalpy of the gas. Assuming ideal gas properties, the density is

$$\rho = \frac{PM}{RT} \tag{8-17}$$

The compression work cannot be evaluated from Eqn (8-15) using Eqn (8-17) unless the operating condition or temperature is specified. We will consider two cases: isothermal compression and adiabatic compression.

A. ISOTHERMAL COMPRESSION

If the temperature is constant, eliminating ρ from Eqns (8-17) and (8-15) and evaluating the integral gives

$$-w = \frac{RT}{M} \ln \frac{P_2}{P_1} \tag{8-18}$$

where the ratio P_2/P_1 is the *compression ratio* r.

B. ISENTROPIC COMPRESSION

For an ideal gas under adiabatic frictionless (i.e. isentropic) conditions, the following expression holds:

$$\frac{P}{\rho^k} = \text{const.}, \quad \text{where } k = \frac{c_p}{c_v} \quad \text{and} \quad c_p = c_v + R \tag{8-19}$$

The specific heat ratio k is approximately 1.4 for diatomic gases (O_2, N_2. etc.) and 1.3 for triatomic and higher gases (NH_3, H_2O, CO_2, etc.). The expression for isothermal conditions that corresponds to Eqn (8-19) follows from Eqn (8-17):

$$\frac{P}{\rho} = \text{const.} \tag{8-20}$$

Note that the isothermal condition can be considered a special case of the isentropic condition if $k = 1$. The "constant" in Eqn (8-19) or (8-20) can be evaluated from known conditions at some point in the system [e.g., (P_1, T_1)]. Using Eqn (8-19) to eliminate the density from Eqn (8-15) and evaluating the integral leads to

$$-w = \frac{RT_1 k}{M(k-1)} \left[\left(\frac{P_2}{P_1} \right)^{(k-1)/k} - 1 \right] \tag{8-21}$$

Although it is not obvious by inspection, setting $k = 1$ in Eqn (8-21) reduces that equation to Eqn (8-18) (this follows by application of l'Hospital's rule).

If we compare the work required to compress a given gas to a given compression ratio by isothermal and isentropic processes, we see that the isothermal work is always less than the isentropic work. That is, less energy would be required if compressors could be made to operate under isothermal conditions. However, in most cases a compressor operates under more nearly adiabatic (isentropic, if frictionless) conditions because of the relatively short residence time of the gas in the compressor, which allows very little time for heat generated by compression to be transferred away. The temperature rise during an isentropic compression is determined by eliminating ρ from Eqns (8-17) and (8-19):

$$\frac{T_2}{T_1} = \left(\frac{P_2}{P_1} \right)^{(k-1)/k} = r^{(k-1)/k} \tag{8-22}$$

In reality, most compressor conditions are neither purely isothermal nor isentropic but somewhere in between. This is often accounted for in calculating the compression work by using the isentropic equation (Eqn 8-21) but replacing the specific heat ratio k by another constant, γ, where $1 < \gamma < k$. The constant γ is called the "polytropic" constant and is a function of the compressor design as well as the properties of the gas.

C. STAGED OPERATION

It is often not possible to reach a desired compression ratio using a single compressor, especially with centrifugal compressors. In such cases multiple compressor "stages" can be arranged in series to increase the overall compression ratio. Furthermore, in order to increase the overall efficiency it is common to cool the gas between stages by using "interstage coolers". It can be shown that as the number of stages increases, the total compression work for isentropic compression with interstage cooling to the initial temperature (T_1) approaches that of isothermal compression at T_1.

For multiple-stage operation, there will be an optimum compression ratio for each stage that will minimize the total compression work. This can be easily seen by considering a two-stage compressor with interstage cooling. The gas enters stage 1 at (P_1, T_1), leaves stage 1 at (P_2, T_2) and is then cooled to T_1. It then enters stage 2 at (P_2, T_1) and leaves at P_3. By computing the total isentropic work for both stages (using Eqn 8-21), and setting the derivative of this with respect to the interstage pressure (P_2) equal to zero, the value of P_2 that results in the least total work can be found. The result is that the optimum interstage pressure that minimizes the total work for a two-stage compression with intercooling to T_1 is

$$P_2 = (P_1P_3)^{1/2}, \quad \text{or} \quad \frac{P_2}{P_1} = \frac{P_3}{P_2} = r = \left(\frac{P_3}{P_1}\right)^{1/2} \tag{8-23}$$

That is, the total work is minimized if the compression ratio for each stage is the same. This result can easily be generalized to any number (n) of stages (with interstage cooling to the initial temperature), as follows:

$$r = \frac{P_2}{P_1} = \frac{P_3}{P_2} = \cdots = \frac{P_{n+1}}{P_n} = \left(\frac{P_{n+1}}{P_1}\right)^{1/n} \tag{8-24}$$

If there is no interstage cooling, or interstage cooling to a temperature other than T_1, it can be shown that the optimum compression ratio for each stage (i) is related to the temperature entering that stage (T_i) by

$$T_i\left(\frac{P_{i+1}}{P_i}\right)^{(k-1)/k} = T_i r_i^{(k-1)/k} = \text{const.} \tag{8-25}$$

D. EFFICIENCY

The above equations apply to ideal (frictionless) compressors. To account for friction losses, the ideal computed work is divided by the compressor efficiency, η_e, to get the total work that must be supplied to the compressor:

$$(-w)_{total} = \frac{(-w)_{ideal}}{\eta_e} \tag{8-26}$$

The energy "lost" due to friction is actually dissipated into thermal energy, which raises the temperature of the gas. This temperature rise is in addition to that due to the isentropic compression, so that the total temperature rise across an adiabatic compressor stage is given by

$$T_2 = T_1^{(k-1)/k} + \left(\frac{1 - \eta_e}{\eta_e}\right)\left(\frac{-W_{ideal}}{c_v}\right) \qquad (8\text{-}27)$$

REFERENCES

1. Karassik, I. J., W. C. Krutzsch, W. H. Fraser, and J. P. Messina, *Pump Handbook*, McGraw-Hill (1976)
2. Raymer, R. E., "Watch Suction Specific Speed", *Chemical Engineering Progress*, v. 89 no. 3, pp 79–84, March (1993)

PROBLEMS

PUMPS

1. The pressure developed by a centrifugal pump for Newtonian liquids that are not highly viscous depends upon the liquid density, the impeller diameter, the rotational speed, and the volumetric flow rate.

   (a) Determine a suitable set of dimensionless groups that should be adequate to relate all of these variables.

   You want to know what pressure a pump will develop with a liquid having SG = 1.4 at a flow rate of 300 gpm using an impeller with a diameter of 12 in., driven by a motor running at 1100 rpm. You have a similar test pump in the lab with a 6 in. impeller driven by a 1800 rpm motor. You want to run a test with the lab pump under conditions which will allow you to determine the pressure developed by the larger pump.

   (b) Should you use the same liquid in the lab as in the larger pump, or can you use a different liquid? Why?

   (c) If you use the same liquid, at what flow rate will the operation of the lab pump simulate that of the larger pump?

   (d) If the lab pump develops a pressure of 150 psi at the proper flow rate, what pressure will the field pump develop at 300 gpm?

   (e) What pressure will the field pump develop with water at 300 gpm?

2. The propeller of a speed boat is 1 ft in diameter and is 1 ft below the surface of the water. At what speed (rpm) will cavitation occur at the propeller? Water density = 64 lb_m/ft^3, P_v of water = 18.65 mm Hg.

3. You must specify a pump to be used to transport water at a rate of 5000 gpm through 10 miles of 18 in. sch 40 pipe. The friction loss in valves and fittings is equivalent to 10% of the pipe length, and the pump is 70% efficient. If a 1200 rpm electric motor is used to drive the pump, determine:
   (a) The required horsepower and torque rating of the motor.
   (b) The diameter of the impeller which should be used in the pump.

4. You must select a centrifugal pump which will develop a pressure of 40 psi when pumping a liquid with a specific gravity of 0.88 at a rate of 300 gpm. From all the pump characteristic curves in Appendix G, select the best pump for this job. Specify pump head, impeller diameter, motor speed, efficiency, and motor horsepower.

5. An oil with a 32.6° API gravity at 60°F is to be transferred from a storage tank to a process unit, which is 10 ft above the tank, at a rate of 200 gpm. The piping system contains 200 ft of 3 in. sch 40 pipe, 25 90° screwed elbows, six stub-in tees used as elbows, two lift check valves, and four standard globe valves. From the pump performance curves in Appendix H, select the best pump to do this job. Specify the pump size, motor speed, impeller diameter, operating head, efficiency, and the horsepower of the motor required to drive the pump.

6. You must purchase a centrifugal pump to circulate cooling water which will deliver 5000 gpm at a pressure of 150 psi. If the pump is driven by an 1800 rpm motor, what should the horsepower and torque rating of the motor be, and how large (diameter) should the pump impeller be, assuming an efficiency of 60%?

7. In order to pump a fluid of SG = 0.9 at a rate of 1000 gpm through a piping system, a hydraulic power of 60 hp is required. Determine the required pump head, the torque of the driving motor, and the estimated impeller diameter if an 1800 rpm motor is used.

8. From your prior analysis of pumping requirement for a water circulating system, you have determined that a pump capable of delivering 500 gpm at a pressure of 60 psi is required. If a motor operating at 1800 rpm is chosen to drive the pump, which is 70% efficient, determine:
   (a) The required horsepower rating of the motor.
   (b) The required torque rating of the motor.

(c) The diameter of the impeller that should be used in the pump.

(d) What color the pump should be painted.

9. You want to pump water at 70°F from an open well, 200 ft deep, at a rate of 30 gpm through a 1 in. sch 40 pipe, using a centrifugal pump having an NPSH of 8 ft. What is the maximum distance above the water level in the well that the pump can be located without cavitating? (Vapor pressure of water at 60°F = 18.7 mm Hg)

10. Steam condensate at 1 atm and 95°C (P_v = 526 mm Hg) is returned to a boiler from the condenser by a centrifugal boiler feed pump. The flow rate is 100 gpm through a 2.5 in. sch 40 pipe. If the equivalent length of the pipe between the condenser and pump is 50 ft, and the pump has an NPSH of 6 ft, what is the maximum height above the condenser that the pump can be located?

11. Water at 160°F is to be pumped at a rate of 100 gpm through a 2 in. sch 80 steel pipe from one tank to another located 100 ft. directly above the first. The pressure in the lower tank is 1 atm. If the pump to be used has a required NPSH of 6 ft of head, what is the maximum distance above the lower tank that the pump may be located?

12. A pump with a 1 in. diameter suction line is used to pump water from an open hot water well at a rate of 15 gpm. The water temperature is 90°C, with a vapor pressure of 526 mm Hg and density of 60 lb_m/ft^3. If the pump NPSH is 4 ft, what is the maximum distance above the level of the water in the well that the pump can be located and still operate properly?

13. Hot water is to be pumped out of an underground geothermally heated aquifer located 500 ft below ground level. The temperature and pressure in the aquifer are 325°F and 150 psig. The water is to be pumped out at a rate of 100 gpm through 2.5 in. pipe using a pump which has a required NPSH of 6 ft. The suction line to the pump contains four 90° elbows and one gate valve. How far below ground level must the pump be located in order to operate properly?

14. You must install a centrifugal pump to transfer a volatile liquid from a remote tank to a point in the plant 500 ft from the tank. To minimize the distance that the power line to the pump must be strung, it is desirable to locate the pump as close to the plant as possible. If the liquid has a vapor pressure of 20 psia, the pressure in the tank

is 30 psia, the level in the tank is 30 ft above the pump inlet, and the required pump NPSH is 15 ft, what is the closest that the pump can be located to the plant without the possibility of cavitation? The line is 2 in. sch 40, the flow rate is 100 gpm, and the fluid properties are $\rho = 45$ lb$_m$/ft^3 and $\mu = 5$ cP.

15. It is necessary to pump water at 70°F ($P_v = 0.35$ psia) from a well that is 150 ft deep, at a flow rate of 25 gpm. You do not have a submersible pump, but you do have a centrifugal pump with the required capacity which cannot be submerged. If a 1 in. sch 40 pipe is used, and the NPSH of the pump is 15 ft, how close to the surface of the water must the pump be lowered for it to operate properly?

16. You must select a pump to transfer an organic liquid with a viscosity of 5 cP and SG = 0.87 at a rate of 1000 gpm through a piping system which contains 1000 ft of 8 in. sch 40 pipe, 4 globe valves, 16 gate valves, and 43 standard 90° elbows. The discharge end of the piping system is 30 ft above the entrance, and the pressure at both ends is 10 psia.
    (a) What pump head is required?
    (b) What is the hydraulic horsepower to be delivered to the fluid?
    (c) Which combination of pump size, motor speed, and impeller diameter from the pump charts in Appendix H would you choose for this application?
    (d) For the pump selected, what size motor would you specify to drive it?
    (e) If the vapor pressure of the liquid is 5 psia, how far directly above the liquid level in the upstream tank could the pump be located without cavitating?

17. You need a pump which will develop at least 40 psi at a flow rate of 300 gpm of water. What combination of pump size, motor speed, and impeller diameter from the pump characteristics in Appendix H would be the best for this application? State your reasons for the choice you make. What is the pump efficiency, motor horsepower and torque requirement, and NPSH for the pump you choose at these operating conditions?

18. A centrifugal pump takes water from a well at 120°F ($P_v = 87.8$ mm Hg) and delivers it at a rate of 50 gpm to a storage tank through a piping system. The pressure in the storage tank is 20 psig, and the water level is 40 ft above that in the well. The piping system contains

300 ft of 1.5 in. sch 40 pipe, 10 standard 90° elbows, 6 gate valves, and an orifice meter with a diameter of 1 in.

(a) What are the specifications required for the pump?

(b) Would any of the pumps represented by the characteristic curves in the Appendix be satisfactory for this application? If more than one of them would work, which would be the best? What would be the pump head, impeller diameter, efficiency, NPSH, and required horsepower for this pump at the operating point?

(c) If the pump you select is driven by an 1800 rpm motor, what impeller diameter should be used?

(d) What should be the minimum torque and horsepower rating of the motor if the pump is 50% efficient?

(e) If the NPSH rating of the pump is 6 ft. at the operating conditions, where should it be located in order to prevent cavitation?

19. Water at 20°C is pumped at a rate of 300 gpm from an open well in which the water level is 100 ft below ground level into a storage tank that is 80 ft above ground. The piping system contains 700 ft of 3 1/2 in. sch 40 pipe, 8 threaded elbows, 2 globe valves, and 2 gate valves. The vapor pressure of water is 17.5 mm Hg.

(a) What pump head and hydraulic horsepower are required?

(b) Would a pump whose characteristics are similar to those shown in Fig. 8-2 be suitable for this job? If so, what impeller diameter, motor speed, and motor horsepower should be used?

(c) What is the maximum distance above the surface of the water in the well at which the pump can be located and still operate properly?

20. An organic fluid is to be pumped at a rate of 300 gpm from a distillation column reboiler to a storage tank. The liquid in the reboiler is 3 ft above ground level, the storage tank is 20 ft above ground, and the pump will be at ground level. The piping system contains 14 standard elbows, 4 gate valves, and 500 ft of 3 in. sch 40 pipe. The liquid has a specific gravity of 0.85, a viscosity of 8 cP, and a vapor pressure of 600 mm Hg. If the pump to be used has characteristics similar to those given in Appendix H, and the pressure in the reboiler is 5 psig, determine:

(a) The motor speed to be used.

(b) The impeller diameter.

(c) The motor horsepower and required torque.

(d) Where the pump must be located to prevent cavitation.

21. A liquid with a viscosity of 5 cP, density of 45 lb_m/ft^3, and vapor pressure of 20 psia is transported from a storage tank in which the pressure is 30 psia to an open tank 500 ft downstream, at a rate of 100 gpm. The liquid level in the storage tank is 30 ft above the pump, and the pipeline is 2 in. sch 40 commercial steel. If the transfer pump has a required NPSH of 15 ft, how far downstream from the storage tank can the pump be located without danger of cavitation?

22. You must choose a centrifugal pump to pump a coal slurry. You have determined that the pump must deliver 200 gpm at a pressure of at least 35 psi. Given the pump characteristic curves in Appendix H, tell which pump you would specify (give pump size, speed, and impeller diameter), and why. What is the efficiency of this pump at its operating point, what horsepower motor would be required to drive the pump, and what is the required NPSH of the pump? The specific gravity of the slurry is 1.25.

23. You must specify a pump to take an organic stream from a distillation reboiler to a storage tank. The liquid has a viscosity of 5 cP, SG = 0.78, and a vapor pressure of 150 mm Hg. The pressure in the storage tank is 35 psig, and the inlet to the tank is located 75 ft above the reboiler, which is at a pressure of 25 psig. The pipeline in which the pump is to be located is $2\frac{1}{2}$ in. sch 40, 175 ft long, and there will be two flanged elbows and a globe valve in each of the pump suction and discharge lines. The pump must deliver a flow rate of 200 gpm. If the pump you use has the same characteristics as that illustrated in Fig. 8-2, determine:
    (a) The proper impeller diameter to use with this pump.
    (b) The required head that the pump must deliver.
    (c) The actual head that the pump will develop.
    (d) The horsepower rating of the motor required to drive the pump.
    (e) The maximum distance above the reboiler that the pump can be located without cavitating.

24. A slurry pump must be selected to transport a coal slurry from an open storage tank to a rotary drum filter, at a rate of 250 gpm. The slurry is 40% solids by volume, and has a specific gravity of 1.2. The level in the filter is 10 ft above that in the tank, and the line contains 400 ft of 3 in. sch 40 pipe, two gate valves, and six 90° elbows. A lab test shows that the slurry can be described as a Bing-

ham plastic with $\mu_\infty = 50$ cP and $\tau_o = 80$ dyn/cm^2. Assume that the equivalent L/D method for friction loss in the fittings is adequate.
(a) What pump head is required?
(b) Using the pump curves in Appendix H, choose the pump that would be the best for this job. Specify the pump size, motor speed, impeller diameter, efficiency, and NPSH. Tell what criteria you used to make your decision.
(c) What horsepower motor would you need to drive the pump?
(d) Assuming the pump you choose has an NPSH of 6 ft at the operating conditions, what is the maximum elevation above the tank that the pump could be located, if the maximum temperature is 80°C? (P_v of water is 0.4736 bar at this temperature)

25. A red mud slurry residue from a bauxite processing plant is to be pumped from the plant to a disposal pond at a rate of 1000 gpm, through a 6 in. sch 40 pipeline, 1200 ft long. The line is horizontal, and the inlet and exit pressures are both atmospheric. The mud can be described as a Bingham plastic, with a yield stress of 250 dyn/cm^2, a limiting viscosity of 50 cP, and a density of 1.4 g/cm^3. The vapor pressure of the slurry at the operating temperature is 50 mm Hg. You are to use a 6×8 R pump, with a 1750 rpm motor, which has the characteristics given in Appendix H.
    (a) What impeller diameter and motor horsepower should you use with this pump for this job?
    (b) How close to the disposal pond could the pump be located without cavitating?
    (c) Although this pump is designed to pump suspensions, it probably would not operate properly with a mud as "thick" as this one. What type of pump do you think might be better suited for this job?

26. A red mud slurry residue from a bauxite processing plant is to be pumped from the plant to a disposal pond at a rate of 800 gpm, through a 10 in. I.D. pipeline that is 2.5 mi long. The pipeline is horizontal, and the inlet and discharge of the line are both at atmospheric pressure. The mud has properties of a Bingham plastic, with a yield stress of 250 dyn/cm^2, a limiting viscosity of 50 cP, and a density of 1.4 g/cm^3. The vapor pressure of the slurry at the operating temperature is 50 mm Hg. You have available several 6×8 R slurry pumps (see Appendix H), with 1750 rpm motors also available.
    (a) Could you use these pumps, in any configuration, to pump this slurry? If so, what impeller diameter would you choose for the pumps, assuming all impellers are to be the same size?

(b) What horsepower motor would you use with each pump?

(c) How close to the disposal pond could the pumps be located without cavitating?

COMPRESSORS

27. Calculate the work per pound of gas required to compress air from 70°F and 1 atm to 2000 psi with a 100% efficient compressor under the following conditions:

    (a) Single-stage isothermal compression.

    (b) Single-stage adiabatic compression.

    (c) Five-stage adiabatic compression with intercooling to 70°F and optimum interstage pressures.

    (d) Three-stage adiabatic compression with interstage cooling to 100°F and optimum interstage pressures.

    (e) Calculate the outlet temperature of the air for cases (b), (c), and (d). For air, $c_p = 0.24$ Btu/(lb$_m$ °F), $k = 1.4$.

28. It is desired to compress ethylene gas [MW = 28, $k = 1.3$, $c_p = 0.357$ Btu/(lb$_m$ °F)] from 1 atm and 80°F to 10,000 psia. Assuming ideal gas behavior, calculate the compression work required per pound of ethylene under the following conditions:

    (a) A single-stage isothermal compressor.

    (b) A four-stage adiabatic compressor, with interstage cooling to 80°F and optimum interstage pressures.

    (c) A four-stage adiabatic compressor with no intercooling, assuming the same interstage pressures as in (b) and 100% efficiency.

29. You have a requirement to compress natural gas ($k = 1.3$, M = 18) from 1 atm and 70°F to 5000 psig. Calculate the work required to do this per pound of gas in a 100% efficient compressor under the following conditions:

    (a) Isothermal single-stage compressor.

    (b) Adiabatic three-stage compressor with interstage cooling to 70°F.

    (c) Adiabatic two-stage compressor with interstage cooling to 100°F

30. Air is to be compressed from 1 atm and 70°F to 2000 psia. Calculate the work required to do this per pound of air using the following methods:

    (a) A single-stage 80% efficient isothermal compressor.

    (b) A single-stage 80% efficient adiabatic compressor.

(c) A five-stage 80% efficient adiabatic compressor with interstage cooling to 70°F.

(d) A three-stage 80% efficient adiabatic compressor with interstage cooling to 120°F. Determine the expression relating the pressure ratio and inlet temperature for each stage for this case by induction from the corresponding expression for optimum operation of the corresponding two-stage case.

(e) Calculate the final temperature of the gas for cases (b), (c), and (d).

31. It is desired to compress 1000 scfm of air from 1 atm and 70°F to 10 atm. Calculate the total horsepower required if the compressor efficiency is 80% for

(a) Isothermal compression.

(b) Adiabatic single-stage compression.

(c) Adiabatic three-stage compression with interstage cooling to 70°F and optimum interstage pressures.

(d) Calculate the gas exit temperature for cases (b) and (c). Note: $c_p = 7$ Btu/(lb mol °F), and assume ideal gas.

32. You want to compress air from 1 atm, 70°F, to 2000 psig, using a staged compressor with interstage cooling to 70°F. The maximum compression ratio per stage you can use is about 6, and the compressor efficiency is 70%.

(a) How many stages should you use?

(b) Determine the corresponding interstage pressures.

(c) What power would be required to compress the air at a rate of 10^5 scfm?

(d) Determine the temperature leaving the last stage.

(e) How much heat (in Btu/hr) must be removed by the interstage coolers?

9

Compressible Flows

I. GAS PROPERTIES

The main difference between the flow behavior of incompressible and compressible fluids and the equations that govern them is the effect of variable density, e.g. the dependence of density upon pressure and temperature. At low velocities (relative to the speed of sound), relative changes in pressure and associated effects are often small, and the assumption of incompressible flow with a constant (average) density is usually reasonable. It is when the gas velocity approaches the speed of sound that the effects of compressibility become the most significant. It is this condition of high-speed gas flow (e.g. ''fast gas'') which is of greatest concern to us here.

A. IDEAL GAS

All gases are ''non-ideal'', in that there are conditions under which the density of the gas may not be accurately represented by the ideal gas law, i.e.

$$\rho = \frac{PM}{RT} \tag{9-1}$$

However, there are also conditions under which this law provides a very good representation of the density for virtually any gas. In general, the further the temperature and pressure are from the critical temperature and pressure of the gas (i.e. the higher the temperature and lower the pressure), the better the ideal

253

gas law is for representing gas properties. For example, the critical conditions for CO_2 are 304 K, 72.9 atm, whereas for N_2 they are 126 K, 33.5 atm. Thus, at normal atmospheric conditions (300 K, 1 atm) N_2 can be described very accurately by the ideal gas law, whereas CO_2 deviates significantly from this law under such conditions. For the most common gases (e.g. air) at conditions which are not extreme, the ideal gas law provides a quite acceptable representation for most engineering purposes.

We will consider the ideal gas under two possible conditions: isothermal and isentropic (or adiabatic). The *isothermal*, or constant-temperature, condition may be approximated, for example, in a long pipeline in which the residence time of the gas is long enough that it has plenty of time to reach thermal equilibrium with its surroundings. Under these conditions,

$$\frac{P}{\rho} = \text{constant} = \frac{P_1}{\rho_1} = \frac{P_2}{\rho_2}, \text{ etc.} \tag{9-2}$$

The *adiabatic* condition occurs, for example, when the residence time of the fluid is short, as for flow through a short pipe, valve, orifice, etc., and/or for well-insulated boundaries. When friction loss is small, the system can be described as *locally isentropic*. It can readily be shown that an ideal gas under isentropic conditions obeys the following relation:

$$\frac{P}{\rho^k} = \text{constant } \frac{P_1}{\rho_1^k} = \frac{P_2}{\rho_2^k}, \text{ etc.} \tag{9-3}$$

where $k = c_p/c_v$ is the "isentropic exponent" and, for an ideal gas, $c_p = c_v + R$. For diatomic gases, $k \approx 1.4$, whereas for triatomic and higher gases, $k \approx 1.3$. A table of properties of various gases, including the isentropic exponent, and a chart of k values for steam as a function of temperature and pressure, are given in Appendix B.

B. SPEED OF SOUND

Sound is nothing more than a small-amplitude pressure wave, and the speed of sound is the velocity at which this wave will travel through a medium. An expression for the speed of sound can be derived as follows. With reference to Fig. 9-1, we consider a sound wave moving from left to right with velocity c. If we take the wave as our reference, this is equivalent to considering a standing wave with the medium moving from right to left with velocity c. Since the conditions are different upstream and downstream of the wave, we represent these differences by ΔV, ΔT, ΔP, and $\Delta \rho$. The conservation of mass principle applied to the flow through the wave reduces to

$$\dot{m} = \rho Ac = (\rho + \Delta\rho)(A)(c - \Delta V) \tag{9-4}$$

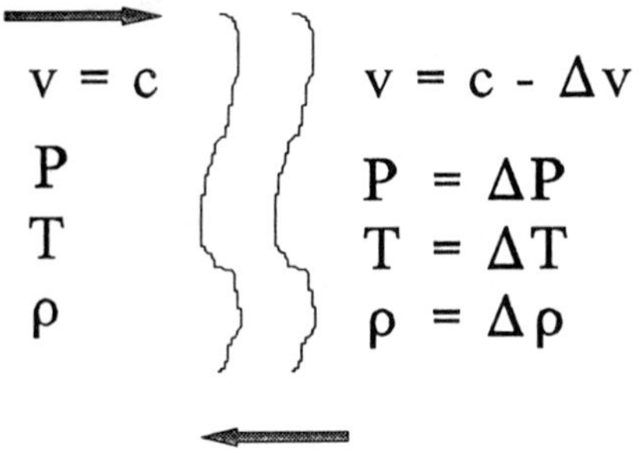

$$v = c \qquad v = c - \Delta v$$
$$P \qquad P = \Delta P$$
$$T \qquad T = \Delta T$$
$$\rho \qquad \rho = \Delta \rho$$

FIGURE 9-1 Sound wave moving at velocity c.

or

$$\Delta V = c \frac{\Delta \rho}{\rho + \Delta \rho} \tag{9-5}$$

Likewise, a momentum balance on the fluid "passing through" the wave gives

$$\sum F = \dot{m}(V_2 - V_1) \tag{9-6}$$

or

$$PA - (P + \Delta P)A = \rho Ac(c - \Delta V - c) \tag{9-7}$$

$$\Delta P = \rho c \Delta V \tag{9-8}$$

Eliminating ΔV from Eqns (9-5) and (9-8) and solving for c^2 gives

$$c^2 = \frac{\Delta P}{\Delta \rho}\left(1 + \frac{\Delta \rho}{\rho}\right) \tag{9-9}$$

For an infinitesimal wave under isentropic conditions, this becomes

$$c = \sqrt{\left(\frac{\partial P}{\partial \rho}\right)_s} = \sqrt{k\left(\frac{\partial P}{\partial \rho}\right)_T} \tag{9-10}$$

where the equivalence of the terms in the radicals follows from Eqns (9-2) and (9-3).

For an ideal gas, Eqn (9-10) becomes

$$c = \sqrt{\frac{kP}{\rho}} = \sqrt{\frac{kRT}{M}} \tag{9-11}$$

For solids and liquids,

$$\left(\frac{\partial P}{\partial \rho}\right)_s = \frac{K}{\rho} \tag{9-12}$$

where K is the bulk modulus (or "compressive stiffness") of the material. It is evident that the speed of sound in a completely incompressible medium would be infinite. From Eqn (9-11), we see that the speed of sound in an ideal gas is determined entirely by the nature of the gas (M and k) and the temperature (T).

II. PIPE FLOW

Consider a gas flowing in a uniform (constant cross section) pipe. The mass flow rate, and hence the mass velocity, $G = \dot{m}/A$, is the same at all locations along the pipe:

$$G = \dot{m}/A = \rho V = \text{constant} \tag{9-13}$$

Now the pressure will drop along the pipe because of energy dissipation (e.g. friction), just as for an incompressible fluid. However, since the density decreases with decreasing pressure, and the product of the density and velocity must be constant, the velocity must increase as the gas moves through the pipe. There is a limit as to how high this velocity can get in a straight pipe, however, which we will discuss shortly.

Since the fluid velocity and properties are changing from point to point along the pipe, in order to analyze the flow we apply the Bernoulli equation in differential form to a differential length of the pipe (dL):

$$\frac{dP}{\rho} + gdz + d\left(\frac{V^2}{2}\right) + \delta e_f = -\delta w = 0 \tag{9-14}$$

There is no shaft work done on the fluid in this system, and the elevation (potential energy) change can be neglected, so Eqn (9-14) can be rewritten using Eqn (9-13) as follows:

$$\frac{dP}{\rho} + \frac{G^2}{\rho} d\left(\frac{1}{\rho}\right) = -\delta e_f = -\frac{2fV^2 dL}{D} = -\frac{2f}{D}\left(\frac{G}{\rho}\right)^2 dL \tag{9-15}$$

Now the friction factor f is a function of the Reynolds number:

$$f = fn\left(N_{re} = \frac{DG}{\mu} \doteq \text{const}\right) \tag{9-16}$$

Since the gas viscosity is not highly sensitive to pressure changes, the Reynolds number, and hence the friction factor, will be very nearly constant along the pipe. Equation (9-15) is valid for any prescribed conditions, and we will apply it to an ideal gas in both isothermal and isentropic flow.

A. ISOTHERMAL FLOW

Substituting Eqn (9-1) for the density into Eqn (9-15), rearranging, integrating from the inlet of the pipe (point 1) to the outlet (point 2), and solving for G gives

$$G = \left[\frac{M(P_1^2 - P_2^2)/2RT}{2fL/D + \ln(P_1/P_2)} \right]^{1/2} \tag{9-17}$$

If the logarithmic term in the denominator (which comes from the change in kinetic energy of the gas) is neglected, the resulting equation is called the *Weymouth equation*. Furthermore, if the average density of the gas is used in the Weymouth equation, i.e.

$$\bar{\rho} = \frac{(P_1 + P_2)M}{2RT} \quad \text{or} \quad \frac{M}{2RT} = \frac{\bar{\rho}}{P_1 + P_2} \tag{9-18}$$

Eqn (9-17) reduces identically to the Bernoulli equation for an incompressible fluid in a straight, uniform pipe, which is

$$G = \left[\frac{\bar{\rho}(P_1 - P_2)}{2fL/D} \right]^{1/2} \tag{9-19}$$

Inspection of Eqn (9-17) shows that as P_2 decreases, both the numerator and denominator increase, i.e. they have an opposing effect on G. By setting the derivative of Eqn (9-17) with respect to P_2 equal to zero, the value of P_2 that maximizes G, and the corresponding expression for the maximum G, can be found. If the conditions at this state of maximum flow rate are denoted by an asterisk, e.g. P_2^*, G^*, the result is

$$G^* = P_2^* \sqrt{\frac{M}{RT}} \tag{9-20}$$

or

$$V_2^* = \sqrt{\frac{RT}{M}} = c \tag{9-21}$$

That is, as P_2 decreases, the mass velocity will increase up to a maximum value of G^*, at which point the velocity at the end of the pipe reaches the speed of sound. Any further reduction in the downstream pressure can have no effect on the flow in the pipe, since the speed at which pressure information can be transmitted is the speed of sound. That is, since pressure changes are transmitted at the speed of sound, they cannot propagate upstream in a gas that is already traveling at the speed of sound. Therefore, the pressure *inside* the downstream end of the pipe will remain at P_2^*, regardless of how low the pressure outside the end of the pipe (P_2) may fall. This condition is called *choked flow* and is a very important concept, since it establishes the conditions under which maximum gas flow can occur in a conduit. When the flow becomes choked, the mass flow rate in the pipe will be insensitive to the exit pressure but will still increase with an increase in the upstream pressure.

Although Eqn (9-17) appears to be explicit for G, it is actually implicit because the friction factor depends upon the Reynolds number, which depends on G. However, the Reynolds number under choked flow conditions is typically high enough that fully turbulent flow occurs, in which case the friction factor depends only on the relative pipe roughness, i.e.

$$\frac{1}{\sqrt{f}} = -4 \log\left[\frac{\varepsilon/D}{3.7}\right] \tag{9-22}$$

If the upstream pressure and flow rate are known, the downstream pressure (P_2) can be found by rearranging Eqn (9-17), as follows:

$$P_2 = \left[P_1^2 - \frac{2RTG^2}{M}\left(\frac{2fL}{D} + \ln\left(\frac{P_1}{P_2}\right)\right)\right]^{1/2} \tag{9-23}$$

This is obviously implicit in P_2. A first estimate for P_2 can be obtained by neglecting the last term on the right (e.g. the Weymouth approximation). This value may then be inserted into the last term in Eqn (9-23) to provide a second estimate for P_2, and the process can be repeated as necessary.

B. ADIABATIC FLOW

In this case, we use Eqns (9-1) and (9-3) to eliminate density and temperature from Eqn (9-15). This can be called the *locally isentropic* approach, since the friction loss is still included in the energy balance. Actual flow conditions are often somewhere between isothermal and adiabatic. In this case the flow behavior can be described by the isentropic equations, with the isentropic constant k replaced by the "polytropic" constant γ, where $1 < \gamma < k$, as was done for compressors. (The isothermal condition corresponds to $\gamma = 1$, whereas isentropic flow corresponds to $\gamma = k$.)

Combining Eqns (9-1) and (9-3) leads to the following expressions for density and temperature:

$$\rho = \left[\frac{MP_1^{\frac{k-1}{k}}}{RT_1}\right]P^{1/k}, \quad T = T_1\left(\frac{P}{P_1}\right)^{\frac{k-1}{k}} \tag{9-24}$$

Using these expressions to eliminate ρ and T from Eqn (9-15) and solving for G gives

$$G = \left[\frac{\left(\dfrac{k}{k+1}\right)\left(\dfrac{P_1^2 M}{RT_1}\right)\left(1 - \left(\dfrac{P_2}{P_1}\right)^{\frac{k+1}{k}}\right)}{\dfrac{2fL}{D} + \dfrac{\ln(P_1/P_2)}{k}} \right]^{1/2} \tag{9-25}$$

If the system contains fittings as well as straight pipe, the term $2fL/D$ is replaced by $K_f/2$, where K_f represents the sum of all loss coefficients in the system.

C. CHOKED FLOW

In isentropic flow (just as in isothermal flow), the mass velocity reaches a maximum when the downstream pressure drops to the point where the velocity becomes sonic at the end of the pipe (e.g the flow is choked). This can be shown by differentiating Eqn (9-25) with respect to P_2 (as before) or, alternatively, as follows:

$$G = \frac{\dot{m}}{A} = \rho V$$

$$\frac{\partial G}{\partial P} = \frac{\partial(\rho V)}{\partial P} = \rho \frac{\partial V}{\partial P} + V \frac{\partial \rho}{\partial P} = 0 \quad \text{(for max G)} \tag{9-26}$$

Now for isentropic conditions, the differential form of the Bernoulli equation is

$$\frac{dP}{\rho} + VdV = 0 \quad \text{or} \quad \frac{\partial V}{\partial P} = -\frac{1}{\rho V} \tag{9-27}$$

Substituting this into Eqn (9-26) gives

$$-\frac{1}{V} + V \frac{\partial \rho}{\partial P} = 0 \tag{9-28}$$

However, since

$$c^2 = \left(\frac{\partial P}{\partial \rho}\right)_s \tag{9-29}$$

Eqn (9-28) can be written

$$-\frac{1}{V} + \frac{V}{c^2} = 0 \quad \text{or} \quad V = c \tag{9-30}$$

This shows that when the mass velocity reaches a maximum (e.g. the flow is choked), the velocity is sonic.

1. Isothermal

Under isothermal conditions, choked flow occurs when

$$V_2 = c = V_2^* = \sqrt{\frac{RT}{M}} \tag{9-31}$$

where the asterisk denotes the sonic state. Thus,

$$G^* = \rho_2 V_2^* = \frac{P_2^* M}{RT} \sqrt{\frac{RT}{M}} = P_1 \left(\frac{P_2^*}{P_1}\right) \sqrt{\frac{M}{RT}} \tag{9-32}$$

If G^* is eliminated from Eqns (9-17) and (9-32), and the result is solved for K_f, the result is

$$K_f = \left(\frac{P_1}{P_2^*}\right)^2 - 2 \ln\left(\frac{P_1}{P_2^*}\right) - 1 \tag{9-33}$$

where $2fL/D$ in Eqn (9-17) has been replaced by $K_f/2$. Equation (9-33) shows that the pressure at the (inside of the) end of the pipe at which the flow becomes sonic (P_2^*) is a unique function of the upstream pressure (P_1) and the sum of the loss coefficients in the system (K_f). Since Eqn (9-33) is implicit in P_2^*, it can be solved by iteration for given values of K_f and P_1.

2. Adiabatic

For adiabatic (or locally isentropic) conditions, the corresponding expressions are

$$V_2 = c = V_2^* = \sqrt{\frac{kRT_2}{M}}, \qquad \frac{T_2}{T_1} = \left(\frac{P_2}{P_1}\right)^{\frac{k-1}{k}} \tag{9-34}$$

and

$$G^* = \frac{P_2^* M}{RT_2^*} \sqrt{\frac{kRT_2^*}{M}} = P_1 \left[\frac{kM}{RT_1}\left(\frac{P_2^*}{P_1}\right)^{\frac{k+1}{k}}\right]^{1/2} \tag{9-35}$$

Eliminating G^* from Eqns (9-25) and (9-35) and solving for K_f gives

$$K_f = \frac{2}{k+1}\left[\left(\frac{P_1}{P_2^*}\right)^{\frac{k+1}{k}} - 1\right] - \frac{2}{k} \ln\left(\frac{P_1}{P_2^*}\right) \tag{9-36}$$

As for isothermal flow, this is an implicit expression for the downstream pressure at the sonic state (P_2^*) in terms of the upstream pressure (P_1), the loss coefficients (K_f), and the specific heat ratio (k), which is most easily solved by iteration. It is very important to realize that once the pressure at the end of the pipe falls to P_2^* and choked flow occurs, all of the conditions within the pipe ($G = G^*$, $P_2 = P_2^*$, etc.) will remain the same *regardless* of how low the pressure outside the end of the pipe falls. The pressure drop *within the pipe* (which determines the flow rate) is always $P_1 - P_2^*$ when the flow is choked.

D. THE EXPANSION FACTOR

The adiabatic flow equation (Eqn 9-25) can be represented in a more convenient form as follows:

$$G = Y \sqrt{\frac{-2\rho_1 \Delta P}{K_f}} \tag{9-37}$$

where $\rho_1 = P_1 M/RT_1$, $\Delta P = P_2 - P_1$, and Y is called the *expansion factor*. Note that Eqn (9-37) without the Y term is the solution of the Bernoulli equation for an incompressible fluid of density ρ_1. Thus, the expansion factor $Y = (G_{adiabatic}/G_{incompressible})$ is simply the ratio of the adiabatic solution (Eqn 9-25) to the corresponding incompressible solution and is a unique function of P_2/P_1, k, and K_f. For convenience, values of Y are shown in Fig. 9-2a for k = 1.3 and Fig. 9-2b for k = 1.4 (from Crane, 1978), as a function of $\Delta P/P_1$ and K_f. The conditions corresponding to the end of the lines on the plots (i.e. the "button") represent the sonic (choked flow) state, where $P_2 = P_2^*$. These same conditions are given in the tables accompanying the plots, which enables the relationships for choked flow to be determined more precisely than is possible from reading the plots. Note that it is not possible to extrapolate beyond the "button" at the end of the lines in Figs. 9-2a and 9-2b, since this represents the choked flow state, in which $P_2 = P_2^*$ (inside the pipe) and is independent of the external exit pressure.

Figures 9-2a and 9-2b provide a convenient way of solving compressible adiabatic flow problems for piping systems. Some iteration is normally required, because the value of K_f depends upon the Reynolds number, which cannot be determined until G is found. An example of the procedure for solving a typical problem follows.

Given: P_1, D, L, ε, k, M   Find: P_2^* and G^*

(1) Estimate K_f by assuming fully turbulent flow. This requires a knowledge of ε/D to get $K_f = 4fL/D$ for the pipe and K_∞ for each fitting.

(2) On Fig. 9-2(a) (for k = 1.3) or Fig. 9-2(b) (for k = 1.4), at the end of the line corresponding to the value of K_f (or from the table beside the plot), read the values of Y and $\Delta P/P_1$.

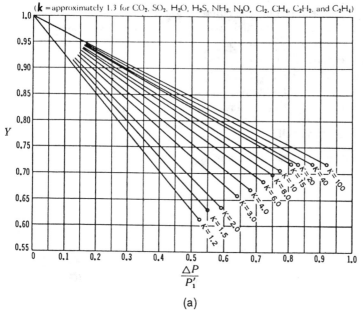

(a)

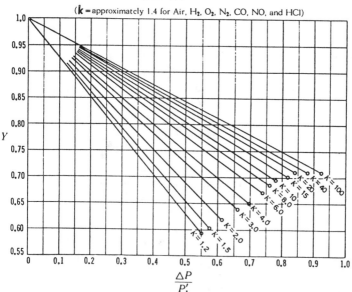

(b)

(3) Calculate $G = G^*$ from Eqn (9-37).
(4) Calculate $N_{Re} = DG/\mu$, and use this to revise the K_f's for the pipe ($K_f = 4fL/D$) and fittings (2-K's) accordingly.
(5) Repeat steps 2–4 until there is no change in G.

The value of downstream pressure (P_2) at which the flow becomes sonic ($P_2 = P_2^*$) is given by $P_2^* = P_1(1 - \Delta P/P_1)$. If the exit pressure is equal to or less than this value, the flow will be choked. Otherwise, the flow will be subsonic, and the flow rate will be determined by the pressure difference $P_1 - P_2$.

E. IDEAL ADIABATIC FLOW

The adiabatic flow of an ideal gas flowing through a frictionless conduit or a constriction (such as an orifice, valve, etc.) can be analyzed as follows. The total energy balance is

$$\Delta h + g\Delta z + \frac{1}{2} \Delta V^2 = q + w \tag{9-38}$$

For horizontal, adiabatic flow with no external work, this becomes

$$\Delta h + \frac{1}{2} \Delta V^2 = 0 \tag{9-39}$$

where

$$\Delta h = \Delta(c_p T) = \frac{k}{k-1} \Delta\left(\frac{P}{\rho}\right) \tag{9-40}$$

which follows from the ideal gas relation $c_p - c_v = R/M$, and the definition of k (i.e. $k = c_p/c_v$). Thus, Eqn (9-39) becomes

$$\left(\frac{k}{k-1}\right)\left(\frac{P_2}{\rho_2} - \frac{P_1}{\rho_1}\right) + \frac{V_2^2 - V_1^2}{2} = 0 \tag{9-41}$$

Using the isentropic condition ($P/\rho^k = \text{const}$) to eliminate ρ_2, this can be written

$$V_2^2 - V_1^2 = \left(\frac{2k}{k-1}\right)\left(\frac{P_1}{\rho_1}\right)\left[1 - \left(\frac{P_2}{P_1}\right)^{\frac{k-1}{k}}\right] \tag{9-42}$$

FIGURE 9-2 (a) Expansion factor for adiabatic flow in piping systems, k = 1.3 (Crane, 1978). (k = approximately 1.3 for CO_2, SO_2, H_2O, H_2S, NH_3, N_2O, Cl_2, CH_4, C_2H_2, and C_2H_4). (b) Expansion factor for adiabatic flow in piping systems, k = 1.4 (Crane, 1978). (k = approximately 1.4 for Air, H_2, O_2, N_2, CO, NO, and HCl).

If V_1 is eliminated using the continuity equation, $(\rho VA)_1 = (\rho VA)_2$, this becomes

$$V_2 = \left\{ \left(\frac{2k}{k-1}\right)\left(\frac{P_1}{\rho_1}\right) \frac{1 - \left(\frac{P_2}{P_1}\right)^{\frac{k-1}{k}}}{1 - \left(\frac{A_2}{A_1}\right)^2 \left(\frac{P_2}{P_1}\right)^{\frac{2}{k}}} \right\}^{\frac{1}{2}}$$

(9-43)

Since

$$G = V_2\rho_2 = V_2\rho_1 \left(\frac{P_2}{P_1}\right)^{\frac{1}{k}}$$

(9-44)

and assuming that the flow is from a larger conduit through a small constriction such that $A_1 \gg A_2$ (i.e. $V_1 \ll V_2$), Eqn (9-44) becomes

$$G = \left\{ \left(\frac{2k}{k-1}\right)\left(\frac{P_1^2 M}{RT_1}\right)\left[\left(\frac{P_2}{P_1}\right)^{\frac{2}{k}} - \left(\frac{P_2}{P_1}\right)^{\frac{k+1}{k}}\right] \right\}^{\frac{1}{2}}$$

(9-45)

Equation (9-45) represents flow through an "ideal nozzle", i.e. an isentropic constriction. From the derivative of Eqn (9-45), setting $\partial G/\partial r = 0$ where $r = P_2/P_1$, it can be shown that the mass flow is a maximum when

$$\frac{P_2^*}{P_1} = \left(\frac{2}{k+1}\right)^{\frac{k}{k-1}}$$

(9-46)

which, for $k = 1.4$ (e.g. air), has a value of 0.528. That is, if the downstream pressure is approximately one half or less of the upstream pressure, the flow will be choked. In such a case, the mass velocity can be determined by Eqn (9-35), with P_2^* from Eqn (9-46):

$$G^* = P_1\left(\frac{2}{k+1}\right)^{\frac{k+1}{2(k-1)}} \sqrt{\frac{kM}{RT_1}}$$

(9-47)

For $k = 1.4$, this reduces to

$$G^* = 0.684 \, P_1 \sqrt{\frac{M}{RT_1}}$$

(9-48)

The mass flow rate under adiabatic conditions is always somewhat greater than the rate under isothermal conditions, but the difference is normally <20%.

In fact, for long piping systems (L/D > 1000), the difference is usually less than 5% (e.g. Holland, 1973).

The flow of compressible (as well as incompressible) fluids through nozzles and orifices will be considered in the following chapter on flow-measuring devices.

III. GENERALIZED EXPRESSIONS

For adiabatic flow in a constant-area duct, the governing equations may be formulated in a more generalized dimensionless form, which is useful for solution of both subsonic and supersonic flows. We will present the resulting expressions and illustrate how to apply them here, but we will not show the derivation of all of the equations. For this, the reader is referred to the references by Shapiro (1953) and Hall (1951).

A. GOVERNING EQUATIONS

For steady flow of a gas (at a constant mass flow rate) in a uniform pipe, the pressure, temperature, velocity, density, etc. all vary from point to point along the pipe. The governing equations are the conservation of mass (continuity), conservation of energy, and conservation of momentum, all applied to a differential length of the pipe, as follows.

Continuity:

$$\frac{\dot{m}}{A} = G = \rho V = \text{constant} \tag{9-49}$$

or

$$\frac{d\rho}{\rho} + \frac{dV}{V} = 0$$

Energy:

$$h + \frac{1}{2} V^2 = \text{constant} = h_0 = c_p T_0 = c_p T + \frac{1}{2} V^2 \tag{9-50}$$

or

$$dh + V dV = 0$$

Since the fluid properties are determined by entropy and enthalpy, Eqn (9-50) represents a curve on an h-s diagram, which is called a *Fanno line*.

Momentum:

$$\frac{dP}{\rho} + VdV = -\frac{4\tau_w}{\rho D_h}\,dL = -\frac{2fV^2}{D}\,dL \tag{9-51}$$

By making use of the isentropic condition, the following relations can be shown:

$$h = c_p T = \frac{kRT}{(k-1)M} \tag{9-52}$$

$$\frac{P}{\rho} = \frac{RT}{M} = \frac{c^2}{k} \tag{9-53}$$

$$N_{Ma} = \frac{V}{c} \tag{9-54}$$

An "impulse function" (F) is also useful in some problems where the force exerted on bounding surfaces is desired:

$$F = PA + \rho AV^2 = PA(1 + kN_{Ma}^2) \tag{9-55}$$

These equations can be combined to yield the following dimensionless forms:

$$\frac{dP}{P} = -\frac{kN_{Ma}^2[1 + (k-1)N_{Ma}^2]}{2(1 - N_{Ma}^2)}\,4f\,\frac{dL}{D} \tag{9-56}$$

$$\frac{dN_{Ma}^2}{N_{Ma}^2} = \frac{kN_{Ma}^2[1 + (k-1)N_{Ma}^2/2]}{1 - N_{Ma}^2}\,4f\,\frac{dL}{D} \tag{9-57}$$

$$\frac{dV}{V} = \frac{kN_{Ma}^2}{2(1 - N_{Ma}^2)}\,4f\,\frac{dL}{D} = -\frac{d\rho}{\rho} \tag{9-58}$$

$$\frac{dT}{T} = -\frac{k(k-1)N_{Ma}^2}{2(1 - N_{Ma}^2)}\,4f\,\frac{dL}{D} \tag{9-59}$$

$$\frac{dP_0}{P_0} = -\frac{kN_{Ma}^2}{2}\,4f\,\frac{dL}{D} \tag{9-60}$$

$$\frac{dF}{F} = \frac{dP}{P} + \frac{kN_{Ma}^2}{1 + kN_{Ma}^2}\left(\frac{dN_{Ma}^2}{N_{Ma}^2}\right) \tag{9-61}$$

The subscript 0 represents the "stagnation" state, i.e. the conditions that would prevail if the gas were to be slowed to a stop and all kinetic energy converted reversibly to internal energy. For a given gas, these equations show that all conditions in the pipe depend uniquely on the Mach number and dimensionless pipe length. In fact, if $N_{Ma} < 1$, an inspection of these equations shows that as the distance down the pipe (dL) increases, V will increase but P, ρ, and T will decrease. However, if $N_{Ma} > 1$, just the opposite is true, i.e. V decreases while P, ρ, and T will increase with distance down the pipe. That is, a flow that is initially subsonic will approach (as a limit) sonic flow as L increases, while an initially supersonic flow will also approach sonic flow. Thus all flows, regardless of their starting conditions, will tend toward the speed of sound as the gas progresses down a uniform pipe. The only way that a subsonic flow can be transformed into a supersonic flow is through a converging-diverging nozzle, where the speed of sound is reached at the nozzle throat. We will not be concerned here with supersonic flows, but the interested reader can find this subject treated in many fluid mechanics books (e.g. Hall, 1951; Shapiro, 1953).

B. APPLICATIONS

It is convenient to take the sonic state ($N_{Ma} = 1$) as the "reference" state for application of these equations. Thus, if the upstream Mach number is N_{Ma}, the length of pipe through which this gas must flow to reach the speed of sound ($N_{Ma} = 1$) will be L*. This can be found by integrating Eqn (9-57) from (L = 0, N_{Ma}) to (L = L*, $N_{Ma} = 1$). The result is

$$\frac{4\bar{f}L^*}{D} = \frac{1 - N_{Ma}^2}{kN_{Ma}^2} + \frac{k + 1}{2k} \ln\left[\frac{(k + 1)N_{Ma}^2}{2 + (k - 1)N_{Ma}^2}\right] \tag{9-62}$$

where $\bar{f}$ is the average friction factor over the pipe length L*. Since the mass velocity is constant along the pipe, the Reynolds number (and hence f) will vary only as a result of variation in the viscosity, which is usually not large. If $\Delta L = L = L_1^* - L_2^*$ is the pipe length over which the Mach number changes from N_{Ma1} to N_{Ma2}, then

$$\frac{4\bar{f}\Delta L}{D} = \left(\frac{4\bar{f}L^*}{D}\right)_1 - \left(\frac{4\bar{f}L^*}{D}\right)_2 \tag{9-63}$$

Likewise, the following relations between the problem variables and their values at the sonic (reference) state may be obtained by integrating Eqns (9-56) to (9-60):

$$\frac{P}{P^*} = \frac{1}{N_{Ma}}\left[\frac{k + 1}{2 + (k - 1)N_{Ma}^2}\right]^{1/2} \tag{9-64}$$

$$\frac{T}{T^*} = \left(\frac{c}{c^*}\right)^2 = \frac{k + 1}{2 + (k - 1)N_{Ma}^2} \tag{9-65}$$

$$\frac{\rho}{\rho^*} = \frac{V^*}{V} = \frac{1}{N_{Ma}}\left[\frac{2 + (k - 1)N_{Ma}^2}{k + 1}\right]^{1/2} \tag{9-66}$$

$$\frac{P_0}{P_0^*} = \frac{1}{N_{Ma}} \left[\frac{2 + (k - 1)N_{Ma}^2}{k + 1} \right]^{\frac{k+1}{2(k-1)}} \tag{9-67}$$

With these relations in mind, the conditions at any two points (1 and 2) in the pipe are related by

$$\frac{T_2}{T_1} = \frac{T_2/T^*}{T_1/T^*}, \qquad \frac{P_2}{P_1} = \frac{P_2/P^*}{P_1/P^*} \tag{9-68}$$

$$\frac{4\bar{f}\Delta L}{D} = \frac{4\bar{f}}{D} (L_1^* - L_2^*) = \left(\frac{4\bar{f}L^*}{D} \right)_1 - \left(\frac{4\bar{f}L^*}{D} \right)_2 \tag{9-69}$$

Also, the mass velocity at N_{Ma} and at the sonic state is given by

$$G = N_{Ma}P \sqrt{\frac{kM}{RT}}, \qquad G^* = P^* \sqrt{\frac{kM}{RT^*}} \tag{9-70}$$

For pipe containing fittings, the term 4fL/D would be replaced by the sum of the loss coefficients (K_f) for all pipe sections and fittings. These relations apply to adiabatic flow in a constant-area duct, for which the sum of the enthalpy and kinetic energy is constant (e.g. Eqn 9-50), which is also known as a Fanno line. It is evident that each of the dependent variables, at any point in the system, is a unique function of the nature of the gas (k) and the Mach number of the flow (N_{Ma}) at that point. Since the Mach number is often the unknown quantity, an iterative or trial-and-error procedure for solving the above set of equations is required. However, these relations may be presented in tabular (see Appendix H) or in graphical form (see Fig. 9-3), which can be used directly for solving various types of problems, as shown below.

C. SOLUTION OF HIGH-SPEED GAS PROBLEMS

We will illustrate the procedure for solving the three types of pipe flow problems for high-speed gas flows: unknown driving force, unknown flow rate, and unknown diameter.

1. Unknown Driving Force

The unknown driving force could be either the upstream pressure, P_1, or the downstream pressure, P_2. However, one of these must be known, and the other can be determined as follows:

Given: P_1, T_1, G, D, L   *Find*: P_2

(1) Calculate $N_{Re} = DG/\mu_1$ and use this to find f_1 from Moody diagram or Churchill equation.

(2) Calculate $N_{Ma1} = (G/P_1)(RT_1/kM)^{1/2}$. Use this with Eqns (9-62), (9-64), and (9-65), or Fig. 9-3, or Appendix H, to find $(4fL_1^*/D)_1$,

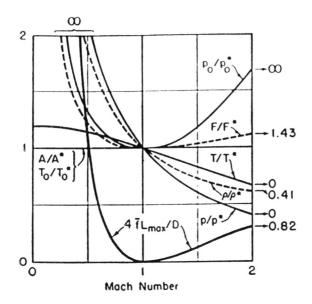

(a)

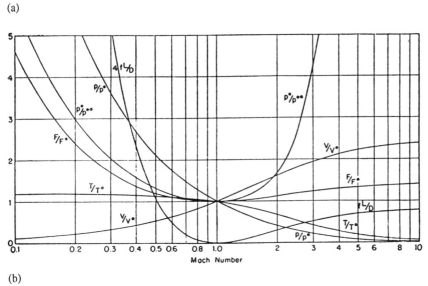

(b)

FIGURE 9-3 Fanno line functions for k = 1.4 (Hall, 1951; Shapiro, 1953).

P_1/P^*, and T_1/T^*. From these values and the given quantities, calculate L_1^*, P^*, and T^*.

(3) Calculate $L_2^* = L_1^* - L$, and use this to calculate $(4f_1L_2^*/D)_2$. Use this with Fig. 9-3, or Appendix H, or Eqns (9-62), (9-64), and (9-65) to get N_{Ma2}, P_2/P^*, and T_2/T^* [Note that Eqn (9-62) is implicit for N_{Ma2}.] From these values, determine P_2 and T_2.

(4) Revise μ by evaluating it at an average temperature $(T_1 + T_2)/2$ and pressure $(P_1 + P_2)/2$. Use this to revise N_{Re}, and thus f, and repeat steps 3 and 4 until no change occurs.

2. Unknown Flow Rate

The mass velocity (G) is the unknown that determines the mass flow rate since the pipe diameter is known. This requires a trial-and-error procedure, since neither the Reynolds nor the Mach numbers can be calculated a priori.

Given: P_1, P_2, T_1, L, and D   *Find*: G

(1) Assume a value for N_{Ma1}. Use Eqns (9-62), (9-64), and (9-65), or Fig. 9-3, or Appendix H, with this value to find P_1/P^*, T_1/T^*, and $(4fL_1^*/D)_1$. From these and known quantities, determine P^* and T^*.

(2) Calculate $G_1 = N_{Ma1}P_1(kM/RT_1)^{1/2}$ and $N_{Re1} = DG/\mu$. From the latter, find f_1 from the Moody diagram or Churchill equation.

(3) Calculate $(4fL_2^*/D)_2 = (4fL_1^*/D)_1 - (4f_1L/D)$. Use this with Eqn (9-62) (implicitly) and Eqns (9-64) and (9-65), or Fig. 9-3, or Appendix H, to find N_{Ma2}, P_2/P^*, and T_2/T^* at point 2.

(4) Calculate $P_2 = (P_2/P^*)P^*$, $T_2 = (T_2/T^*)T^*$, $G_2 = N_{Ma2}P_2(kM/RT_2)^{1/2}$, $N_{Re2} = DG_2/\mu$. Use the latter to determine a revised value of f (e.g. f_2).

(5) Using $f = (f_1 + f_2)/2$ for the revised friction factor, repeat steps 3 and 4 until there is no change.

(6) Compare the given value of P_2 with the calculated value from step 4. If they agree, the answer is the calculated value of G_2 from step 4. If they do not agree, return to step 1 with a new guess for N_{Ma1}, and repeat the procedure until agreement is achieved.

3. Unknown Diameter

The procedure involves a trial-and-error procedure similar to that for the unknown flow rate.

Given: P_1, T_1, L, P_2, $\dot{m}$ *Find*: D

(1) Assume a value for N_{Ma1}, and use Eqns (9-62), (9-64), and (9-65) or Fig. 9-3 or Appendix H to find P_1/P^*, T_1/T^*, and $(4fL_1^*/D)_1$. Also,

calculate $G = N_{Ma1}P_1(kM/RT_1)^{1/2}$, $D = (4\dot{m}/\pi G)^{1/2}$, and $N_{Re1} = DG/\mu$.
Use N_{Re1} to find f_1 from the Moody diagram or Churchill equation.

(2) Calculate $P_2/P^* = (P_1/P^*)(P_2/P_1)$, and use this with Fig. 9-3 or Appendix H or Eqns (9-64) (implicitly), (9-62), and (9-65) to find N_{Ma2}, $(4fL_2^*/D)_2$, and T_2/T^*. Calculate $T_2 = (T_2/T^*)(T^*/T_1)T_1$, and use P_2 and T_2 to determine μ_2. Then use μ_2 to determine $N_{Re2} = DG/\mu_2$, which determines f_2 from the Moody diagram or Churchill equation.

(3) Using $f = (f_1 + f_2)/2$, calculate $L = L_1^* - L_2^* = [(4fL_1^*/D)_1 - (4fL_2^*/D)_2](D/4f)$.

(4) Compare the value of L calculated in step 3 with the given value. If they agree, the value of D determined in step 1 is correct. If they do not agree, return to step 1, revise the assumed value of N_{Ma1}, and repeat the entire procedure until agreement is achieved.

REFERENCES

Crane Company, "Flow of Fluids Through Valves, Fittings, and Pipe", Technical Manual 410 (1978)

Hall, Newman A., *Thermodynamics of Fluid Flow*, Prentice-Hall, Englewood Cliffs, NJ (1951)

Holland, F. A., *Fluid Flow for Chemical Engineers*, Chemical Publishing Co., New York (1973)

Shapiro, Asher H., *The Dynamics and Thermodynamics of Compressible Fluid Flow*, Vol. I, Ronald Press, New York (1953)

PROBLEMS

COMPRESSIBLE FLOW

1. A 12 in. I.D. gas pipeline carries methane (MW = 16) at a rate of 20,000 scfm. The gas enters the line at a pressure of 500 psia, and a compressor station is located every 100 miles to boost the pressure back up to 500 psia. The pipeline is isothermal at 70°F, and the compressors are adiabatic with an efficiency of 65%. What is the required horsepower for each compressor? Assume ideal gas.

2. Natural gas (CH_4) is transported through a 6 in. I.D. pipeline, at a rate of 10,000 scfm. The compressor stations are 150 miles apart, and the compressor suction pressure is to be maintained at 10 psig above that at which choked flow would occur in the pipeline at the specified flow rate. The compressors are each two-stage, operate adiabatically with interstage cooling to 70°F, and have an efficiency of 60%. If the pipeline temperature is 70°F, calculate:

(a) The discharge pressure, interstage pressure, and compression ratio for the compressor stations.

(b) The horsepower required at each compressor station.

3. Natural gas (methane) is transported through a 20 in. sch 40 pipeline at a rate of 30,000 scfm. The gas enters the line from a compressor at 100 psi and 70°F. Identical compressor stations are located every 10 miles along the line, and at each station the gas is recompressed to 100 psia and cooled to 70°F.

(a) Determine the suction pressure at each compressor station.

(b) Determine the horsepower required at each station if the compressors are 80% efficient.

(c) How far apart could the compressor stations be located before the flow in the pipeline becomes choked?

4. Natural gas (methane) is transported through an uninsulated 6 in. I.D. smooth pipeline 1 mile long. The inlet pressure is 100 psi, and the outlet pressure is 1 atm. What is the mass flow rate of the gas and the compressor power required to pump it? $T = 70°F$, $\mu_{gas} = 0.02$ cP.

5. It is desired to transfer natural gas (CH_4) at a pressure of 200 psia and a flow rate of 1000 scf/s through a 1 mile long uninsulated smooth pipeline, into a storage tank at 20 psia. Can this be done using either a 6 in. or 12 in. I.D. pipe? What diameter pipe would you recommend? $T = 70°F$, $\mu = 0.02$ cP.

6. A natural gas (methane) pipeline is to be designed to transport gas at a rate of 5000 scfm. The pipe is to be 6 in. I.D., and the maximum pressure that the compressors can develop is 1500 psig. The compressor stations are to be located in the pipeline at the point at which the pressure drops to 100 psi above that at which choked flow would occur (i.e. the suction pressure for the compressor stations). If the design temperature for the pipeline is 60°F, the compressors are 60% efficient, and the compressor stations each operate with three stages and interstage cooling to 60°F, determine:

(a) The proper distance between compressor stations, in miles.

(b) The optimum interstage pressure and compression ratio for each compressor stage.

(c) The total horsepower required for each compressor station.

7. Ethylene leaves a compressor at 3500 psig and is carried in a 2 in. sch 40 pipeline, 100 ft long, to a unit where the pressure is 500 psig.

The line contains two plug valves, one swing check valve, and eight flanged elbows. If the temperature is 100°F, what is the flow rate (in scfm)?

8. A 12 in. I.D. natural gas (e.g. methane) pipeline carries gas at a rate of 20,000 scfm. The compressor stations are 100 mi apart, and the discharge pressure of the compressors is 500 psia. If the temperature of the surroundings is 70°F, what is the required horsepower of each compressor station, assuming 65% efficiency? If the pipeline breaks 10 mi downstream from a compressor station, what would be the flow rate through the break?

9. The pressure in a reactor fluctuates between 10 and 30 psig. It is necessary to feed air to the reactor at a constant rate of 20 lb_m/hr from an air supply at 100 psig, and the temperature is 70°F. To do this, you insert an orifice into the air line which will provide the required constant flow rate. What should the diameter of the orifice be?

10. Oxygen is to be fed to a reactor at a constant rate of 10 lb_m/s from a storage tank in which the pressure is constant at a 100 psig and the temperature is 70°. The pressure in the reactor fluctuates between 2 and 10 psig, so you want to insert a choke in the line to maintain the flow rate constant. If the choke is a 2 ft length of tubing, what should the diameter of the tubing be?

11. Methane is to be fed to a reactor at a rate of 10 lb_m/min. The methane is available in a pipeline at a pressure of 20 psia and a temperature of 70°, but the pressure in the reactor fluctuates between 2 and 10 psia. To control the flow rate, you want to install an orifice plate that will choke the flow at the desired flow rate. What should the diameter of the orifice be?

12. Ethylene gas (MW = 28, k = 1.3, μ = 0.1 cP) at 100°F is fed to a reaction vessel from a compressor through 100 ft of 2 in. sch 40 pipe containing two plug valves, one swing check valve, and eight flanged elbows. If the compressor discharge pressure is 3500 psig and the pressure in the vessel is 500 psig, what is the flow rate of the gas in scfm (1 atm, 60°F)?

13. Methane must be fed to a reactor at a constant rate of 10 lb_m/min. The source of the gas is at a pressure of 20 psia, and the pressure

in the reactor fluctuates between 2 and 10 psia. You want to install a choke in the gas feed line to produce choked flow that will permit the gas to flow at the desired rate, independent of the reactor pressure. What should the diameter of the choke be?

14. Nitrogen is fed from a high-pressure cylinder, through 1/4 in. I.D. stainless steel tubing, to an experimental unit. The line ruptures at a point 10 ft from the cylinder. If the pressure of the nitrogen in the cylinder is 3000 psig, and the temperature is 70°F, what are the mass flow rate of the gas through the line and the pressure in the tubing at the point of the break?

15. A storage tank contains ethylene at 200 psig and 70°F. If a 1 in. line, 6 ft long with a globe valve on the end, is attached to the tank, what would be the rate of leakage of the ethylene (in scfm) if:
(a) The valve is fully open?
(b) The line breaks off right at the tank?

16. Saturated steam at 200 psig (388°F, 2.13 ft^3/lb$_m$, μ = 0.015 cP) is fed from a header to a direct contact evaporator which operates at 10 psig. If the steam line is 2 in. sch 40 pipe, 50 ft long, and includes four flanged elbows and one globe valve, what is the steam flow rate in lb$_m$/hr?

17. Air is flowing from a tank at a pressure of 200 psia and T = 70°F through a Venturi meter into another tank at a pressure of 50 psia. The meter is mounted in a 6 in. I.D. pipe section (which is quite short) and has a throat diameter of 3 in. What is the mass flow rate of air?

18. A tank containing air at 100 psia and 70°F is punctured with a hole 1/4 in. in diameter. What is the mass flow rate of the air out of the tank?

19. A pressurized tank containing nitrogen at 800 psig is fitted with a globe valve to which is attached a line with 10 ft of 1/4 in. I.D. stainless steel tubing and three standard elbows. The temperature of the system is 70°F. If the valve is left wide open, what is the flow rate of nitrogen, in lb$_m$/s and also in scfm?

20. Gaseous chlorine (MW = 71) is transferred from a high-pressure storage tank at 500 psia and 60°F, through an insulated 2 in. sch 40

pipe 200 ft long, into another vessel where the pressure is 200 psia. What is the mass flow rate of the gas and its temperature at the point where it leaves the pipe?

21. A storage tank containing ethylene at a pressure of 200 psig and a temperature of 70°F springs a leak. If the hole through which the gas is leaking is 1/2 in. in diameter, what is the leakage rate of the ethylene, in scfm?

22. A high-pressure cylinder containing N_2 at 200 psig and 70°F is connected to a reactor in which the pressure is 15 psig by 1/4 in. I.D. stainless steel tubing, 20 ft long. A pressure regulator at the upstream end of the tubing is used to control the pressure, and hence the flow rate, of the N_2 in the tubing.
    (a) If the regulator controls the pressure entering the tubing at 25 psig, what is the flow rate of the N_2, in scfm?
    (b) If the regulator fails, so that the full cylinder pressure is applied at the tubing entrance, what will the flow rate of the N_2 into the reactor be, in scfm?

23. Oxygen is supplied to an astronaut through an umbilical hose that is 7 m long. The pressure in the oxygen tank is 200 kPa, at a temperature of 10°C, and the pressure in the space suit is 20 kPa. If the umbilical hose has a roughness equivalent to 0.01 mm, what should the hose diameter be to supply oxygen at a rate of 0.05 kg/s? If the suit springs a leak and the pressure drops to zero, at what rate would the oxygen escape?

24. Ethylene (MW = 28) is transported from a storage tank, at 250 psig and 70°F, to a compressor station where the suction pressure is 100 psig. The transfer line is 1 in. sch 80, 500 ft long, and contains two ball valves and eight threaded elbows. An orifice meter, with a diameter of 0.75 in., using radius taps, is installed near the entrance to the pipeline.
    (a) What is the flow rate of the ethylene through the pipeline, in scfh?
    (b) If the pipeline breaks at a point 200 ft from the storage tank, and there are four elbows and one valve in the line between the tank and the break, what is the flow rate of the ethylene, in scfh?
    (c) What is the differential pressure across the orifice, for both (a) and (b) above, in inches of water?

25. Air passes from a large reservoir at 70°F through an isentropic converging-diverging nozzle into the atmosphere. The area of the throat is 1 cm^2 and that of the exit is 2 cm^2. What is the reservoir pressure at which the flow in the nozzle just reaches sonic velocity, and what are the mass flow rate and exit Mach number under these conditions?

26. Air is fed from a reservoir through a converging nozzle into a 1/2 in. I.D. drawn steel tube which is 15 ft long. The flow in the tube is adiabatic and the reservoir temperature and pressure are 70°F and 100 psia.
    (a) What is the maximum flow rate (in lb$_m$/s) that can be achieved in the tube?
    (b) What is the maximum pressure at the tube exit at which this flow rate will be reached?
    (c) What is the temperature at this point under these conditions?

10

Flow Measurement and Control

I. SCOPE

In this chapter, we will illustrate and analyze some of the more common methods for measuring flow rate in conduits, namely, the pitot tube, venturi, nozzle, and orifice meters. This is by no means intended to be a comprehensive or exhaustive treatment, however, as there are a great many other devices in use for measuring flow rate—turbine, vane, Coriolis, ultrasonic, magnetic, etc. flowmeters, just to name a few. The examples considered here demonstrate the application of the fundamental conservation principles to the analysis of several of the most common devices. Also included is a consideration of control valves, since these are frequently employed in conjunction with the measurement of flow rates to provide a means of controlling flow.

II. THE PITOT TUBE

As previously discussed, the volumetric flow rate of a fluid through a conduit may be determined by integrating the local ("point") velocity over the cross section of the conduit:

$$Q = \int_A v\, dA \tag{10-1}$$

If the conduit cross section is circular, this becomes

$$Q = \int_0^{\pi R^2} v(r) \, d(\pi r^2) = 2\pi \int_0^R v(r) r \, dr \qquad (10\text{-}2)$$

The pitot tube is a device for measuring $v(r)$, the local velocity at a given position in the conduit, as illustrated in Fig. 10-1. The measured velocity is then used in Eqn (10-2) to determine the flow rate. It consists of a differential pressure-measuring device (e.g. a manometer, transducer, or "DP" cell) which measures the pressure difference between two tubes. One is attached to a hollow probe that can be positioned at any radial location in the conduit and the other attached to the wall of the conduit in the same axial plane as the end of the probe. The local velocity where the stream impinges on the end of the probe is $v(r)$. The fluid element that impacts the open end of the probe must come to rest at that point, since there is no flow through the probe or the tube. This point is known as the *stagnation point*. The Bernoulli equation may be applied to the fluid streamline which impacts the probe tip, where point 1 is in the free stream just upstream of the probe and point 2 is just inside the open end of the probe (the stagnation point), as follows:

$$\frac{P_2 - P_1}{\rho} + \frac{1}{2} (v_2^2 - v_1^2) = 0 \qquad (10\text{-}3)$$

The friction loss is negligible in the free stream from 1 to 2, and $v_2 = 0$ because the fluid in the probe is stagnant, so Eqn (10-3) can be solved for v_1 to give

$$v_1 = \sqrt{\frac{-2(P_2 - P_1)}{\rho}} \qquad (10\text{-}4)$$

The measured pressure difference ΔP is the difference between the "stagnation" pressure in the velocity probe, at the point where it connects to the differential

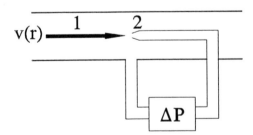

FIGURE 10-1   Pitot tube.

pressure measuring device, and the "static" pressure at the corresponding point in the tube connected to the wall. Since there is no flow in the vertical direction, the difference in pressure between any two vertical elevations is strictly hydrostatic. Thus, the pressure difference measured at the pressure cell is the same as that at the elevation of the probe, since the static head between point 1 and the pressure device is the same as that between point 2 and the pressure device, e.g. $\Delta P = P_2 - P_1$.

Now we usually want to determine the total flow rate (Q) through the conduit, rather than the velocity at a point. This can be done using Eqn (10-1) or Eqn (10-2), if the local velocity is measured at a sufficient number of radial points across the conduit to enable accurate evaluation of the integral. For example, the integral in Eqn (10-2) could be evaluated by plotting the measured $v(r)$ values as $v(r)$ vs r^2, or as $rv(r)$ vs r [in accordance with either the first or second form of Eqn (10-2), respectively], and the area under the curve from $r = 0$ to $r = R$ determined numerically.

The pitot tube is a relatively complex device and requires considerable effort and time to obtain an adequate number of velocity data points and to integrate these over the cross section to determine the total flow rate. On the other hand, the probe offers minimal resistance to the flow and hence is very efficient from the standpoint that it contributes a negligible amount to the friction loss in the system. It is also the only practical means for determining the flow rate in very large conduits (e.g. smokestacks) and/or very low pressure flows. There are standardized methods for applying this method to determine the total amount of material emitted through a stack, for example.

III. THE VENTURI AND NOZZLE

There are other devices, however, which can be used to determine the flow rate from a single measurement. These are sometimes referred to as *obstruction meters*, since the basic principle involves introducing of an "obstruction" (e.g. a constriction) into the flow channel and then measuring the pressure drop across this obstruction, which depends upon the flow rate. Two such devices are the venturi meter and the nozzle, illustrated in Figs. 10-2 and 10-3, respectively. In both cases, the fluid flows through a reduced area, which causes the velocity to increase at that point. The corresponding change in pressure between point 1 upstream of the constriction and point 2 at the position of the minimum area is measured and can be related to the flow rate through the energy balance. The velocities are related by the continuity equation, and the Bernoulli equation relates the velocity change to the pressure change, as follows:

$$\rho_1 V_1 A_1 = \rho_2 V_2 A_2 \qquad (10\text{-}5)$$

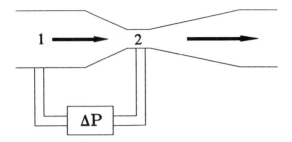

FIGURE 10-2 Venturi meter.

Or, assuming constant density,

$$V_2 = V_1 \frac{A_1}{A_2} \tag{10-6}$$

The Bernoulli equation is

$$\frac{P_2 - P_1}{\rho} + \frac{1}{2}(V_2^2 - V_1^2) + e_f = 0 \tag{10-7}$$

where plug flow has been assumed. Using Eqn (10-6) to eliminate V_1, and neglecting the friction loss, Eqn (10-7) can be solved for V_2:

$$V_2 = \sqrt{\frac{-2\Delta P}{\rho(1-\beta^4)}} \tag{10-8}$$

where $\Delta P = P_2 - P_1$ and $\beta = d_2/D_1$. The value d_2 is the minimum diameter, at the throat of the venturi or nozzle. In order to account for the inaccuracies

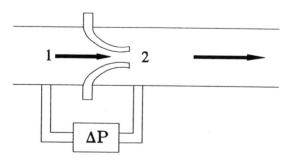

FIGURE 10-3 Nozzle.

introduced by assuming plug flow and neglecting friction, Eqn (10-8) is written

$$V_2 = C_v \sqrt{\frac{-2\Delta P}{\rho(1-\beta^4)}} \qquad (10\text{-}9)$$

where C_v is the "discharge" or venturi coefficient and may be determined empirically as a function of the Reynolds number in the conduit. Typical values are shown in Fig. 10-4. Since the discharge coefficient accounts for the non-idealities in the system, such as the friction loss, one would expect it to decrease with increasing Reynolds number, which is contrary to the trend in Fig. 10-4. However, the coefficient also accounts for deviation from plug flow, and this is greater at lower Reynolds numbers. In any event, the coefficient is not greatly different from 1.0, having a value of about 0.985 for (pipe) Reynolds numbers above about 2×10^5, which indicates that these non-idealities are small.

According to Miller (1983), for $N_{ReD} > 4000$ the discharge coefficient for the venturi, as well as for the nozzle and orifice, can be described as a function of N_{ReD} and β by the general equation

$$C = C_\infty + \frac{b}{N_{ReD}^n} \qquad (10\text{-}10)$$

where the parameters C_∞, b, and n are given in Table 10-1 as a function of β. The range over which Eqn (10-10) applies and its approximate accuracy are given in Table 10-2 (Miller, 1983). Because of the gradual expansion designed

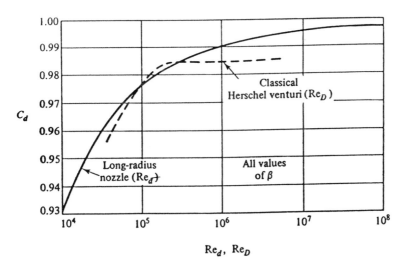

FIGURE 10-4 Venturi and nozzle discharge coefficient vs Reynolds number (White, 1994).

TABLE 10-1 Values for Discharge Coefficient Parameters in Eqn (10-10)

Primary device	Discharge coefficient C_∞ at infinite Reynolds number	Reynolds number term	
		Coefficient b	Exponent n
Venturi			
Machined inlet	0.995	0	0
Rough cast inlet	0.984	0	0
Rough welded sheet-iron inlet	0.985	0	0
Universal venturi tube[b]	0.9797	0	0
Lo-Loss tube[c]	$1.05 - 0.471\beta + 0.564\beta^2 - 0.514\beta^3$	0	0
Nozzle:			
ASME long radius	0.9975	$-6.53\beta^{0.5}$	0.5
ISA	$0.9900 - 0.2262\beta^{4.1}$	$1708 - 8936\beta + 19{,}779\beta^{4.7}$	1.15
Venturi nozzle (ISA inlet)	$0.9858 - 0.195\beta^{4.5}$	0	0

Orifice:			
Corner taps	$0.5959 + 0.0312\beta^{2.1} - 0.184\beta^8$	$91.71\beta^{2.5}$	0.75
Flange taps (D in inches)			
$D \geq 2.3$	$0.5959 + 0.0312\beta^{2.1} - 0.184\beta^8 + 0.09\dfrac{\beta^4}{D(1-\beta^4)} - 0.0337\dfrac{\beta^3}{D}$	$91.71\beta^{2.5}$	0.75
$2 \leq D \leq 2.3^{d}$	$0.5959 + 0.0312\beta^{2.1} - 0.184\beta^8 + 0.039\dfrac{\beta^4}{1-\beta^4} - 0.0337\dfrac{\beta^3}{D}$	$9171\beta^{2.5}$	0.75
Flange taps (D^* in millimeters)			
$D^* \geq 58.4$	$0.5959 + 0.0312\beta^{2.1} - 0.184\beta^8 + 2.286\dfrac{\beta^4}{D^*(1-\beta^4)} - 0.856\dfrac{\beta^3}{D^*}$	$91.71\beta^{2.5}$	0.75
$50.8 \leq D^* \leq 58.4$	$0.5959 + 0.0312\beta^{2.1} - 0.184\beta^8 + 0.039\dfrac{\beta^4}{1-\beta^4} - 0.856\dfrac{\beta^3}{D^*}$	$91.71\beta^{2.5}$	0.75
D and $D/2$ taps	$0.5959 + 0.0312\beta^{2.1} - 0.184\beta^8 + 0.039\dfrac{\beta^4}{1-\beta^4} - 0.0158\beta^3$	$91.71\beta^{2.5}$	0.75
$2\tfrac{1}{2}D$ and $8D$ taps^{d}	$0.5959 + 0.461\beta^{2.1} + 0.48\beta^8 + 0.039\dfrac{\beta^4}{1-\beta^4}$	$91.71\beta^{2.5}$	0.75

[a]Detailed Reynolds number, line size, beta ratio, and other limitations are given in Table 10-2.
[b]From BIF CALC-440/441; the manufacturer should be consulted for exact coefficient information.
[c]Derived from the Badger Meter, Inc. Lo-Loss tube coefficient curve; the manufacturer should be consulted for exact coefficient information.
[d]From Stolz (1978).
Source: Miller (1983)

TABLE 10-2 Applicable Range and Accuracy of Eqn (10-10), with Parameters from Table 10-1

Primary device	Nominal pipe diameter D, in (mm)	Beta ratio β	Pipe Reynolds number N_{Re_D} range	Coefficient accuracy, % †
Venturi				
Machine inlet	2–10 (50–250)	0.4–0.75	2×10^5 to 10^6	±1
Rough cast	4–32 (100–800)	0.3–0.75	2×10^5 to 10^6	±0.7
Rough-welded sheet-iron inlet	8–48 (100–1500)	0.4–0.7	2×10^5 to 10^6	±1.5
Universal venturi tube‡	$\geq 3 \ (\geq 75)$	0.2–0.75	$> 7.5 \times 10^4$	±0.5
Lo-Loss‡	3–120 (75—3000)	0.35–0.85	1.25×10^5 to 3.5×10^6	±1
Nozzle				
ASME	2–16 (50–400)	0.25–0.75	10^4 to 10^7	±2.0
ISA	2–20 (50–500)	0.3–0.6	10^5 to 10^6	±0.8
				$2\beta - 0.4$
Venturi nozzle	3–20 (75–500)	0.3–0.75	2×10^5 to 10^7	±1.2 ± 1.54 β^4
			2×10^5 to 2×10^6	

	Size, in (mm)	β range	Reynolds number	Tolerance, %
Orifice				
Corner, flange, D and $D/2$	2–36 (50–900)	0.2–0.6	10^4 to 10^7	±0.6
		0.6–0.75	10^4 to 10^7	±β
$2\frac{1}{2}D$ and $8D$ (pipe taps)	2–36 (50–900)	0.2–0.75	2×10^3 to 10^4	±0.6 ±β
		0.2–0.5	10^4 to 10^7	±0.8
		0.51–0.7		±1.6
Eccentric				
Flange and vena contracta	4 (100)	0.3–0.75	10^4 to 10^6	±2
	6–14 (150–350)	0.3–0.75	10^4 to 10^6	±1.5
Segmental				
Flange and vena contracta	4–14 (150–350)	0.35–0.75	10^4 to 10^6	±2
Quadrant-edged				
Flange and corner	1–4 (25–100)	0.25–0.6	250 to 6×10^4	±2 – ±2.5
Conical entrance				
Corner		0.1–0.3	25 to 2×10^4	±2 – ±2.5

†ISO 5167 (1980) and ASME *Fluid Meters* (1971) show slightly different values for some devices.
‡The manufacturer should be consulted for recommendations.
Source: Miller (1983).

into the venturi meter, the pressure recovery is relatively large, so that the net friction loss across the entire meter is a relatively small fraction of the measured pressure drop, as indicated in Fig. 10-5. However, since the flow area changes abruptly downstream of the orifice and nozzle, the expansion is uncontrolled, and considerable eddying occurs downstream. This dissipates more energy, resulting in a significantly higher net pressure loss and lower pressure recovery.

The above equations assume that the device is horizontal, i.e. the pressure taps on the pipe are located in the same horizontal plane. If such is not the case, the equations can be easily modified to account for changes in elevation by replacing the pressure P at each point by the total potential $\phi = P + \rho gz$.

The flow nozzle, illustrated in Fig. 10-3, is similar to the venturi meter except that it does not include the diffuser (gradually expanding) section. In fact, one standard design for the venturi meter is basically a flow nozzle with an attached diffuser (see Fig. 10-6). The equations that relate the flow rate and measured pressure drop in the nozzle are the same as for the venturi (e.g. Eqn 10-9), and the nozzle (discharge) coefficient is also shown in Fig. 10-4. It should be noted that the Reynolds number that is used for the venturi coefficient in Fig. 10-4 is based upon the pipe diameter (D), whereas the Reynolds number used for the nozzle coefficient is based upon the nozzle diameter (d) (note that $N_{ReD} = \beta N_{Red}$). There are various "standard" designs for the nozzle, and the reader should consult the literature for details (e.g. Miller, 1983). The discharge coefficient for these nozzles can also be described by Eqn (10-10), with the appropriate parameters given in Table 10-1.

IV. THE ORIFICE METER

The simplest, and most common, device for measuring flow rate in a pipe is the orifice meter, illustrated in Fig. 10-7. This is an "obstruction" meter, which consists of a plate with a hole in it that is inserted into the pipe and the pressure drop across the plate is measured. The major difference between this device and the venturi and nozzle meters is the fact that the fluid stream just downstream of the orifice contracts to an area which is considerably smaller than that of the orifice itself. This is called the *vena contracta*, and it occurs because the fluid has considerable inward radial momentum as it converges into the orifice hole, which causes the fluid to continue to flow "inward" for a distance downstream of the orifice before it starts to expand to fill the pipe. If the pipe diameter is D, the orifice diameter is d, and the diameter of the vena contracta is d_2, the

---→

FIGURE 10-5 Unrecovered (friction) loss in various meters as a percentage of measured pressure drop (Cheremisinoff and Cheremisinoff, 1987).

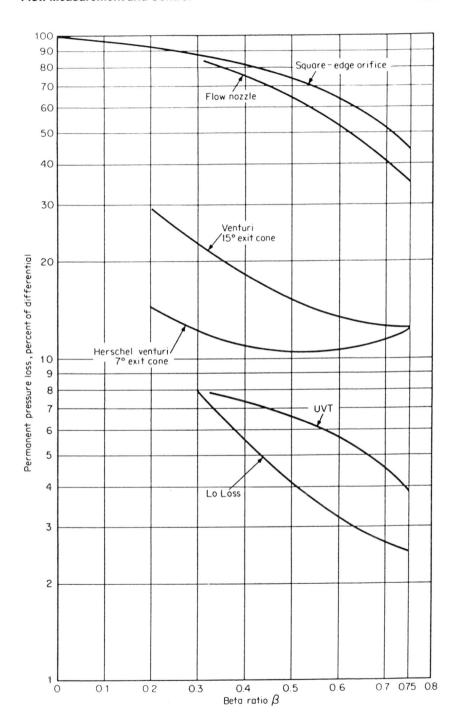

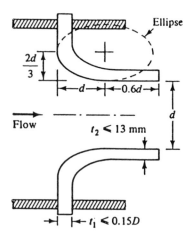

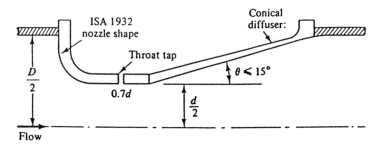

FIGURE 10-6 International standard shapes for nozzle and venturi meter (White, 1994).

contraction ratio for the vena contracta is defined as $C_c = A_2/A_o = (d_2/d)^2$. For highly turbulent flow, $C_c \approx 0.6$.

The complete Bernoulli equation, as applied between point 1 upstream of the orifice where the diameter is D and point 2 in the vena contracta where the diameter is d_2, is

$$\int_{P_2}^{P_1} \frac{dP}{\rho} = \frac{1}{2}(\alpha_2 V_2^2 - \alpha_1 V_1^2) + \frac{K_f}{2} V_1^2 \qquad (10\text{-}11)$$

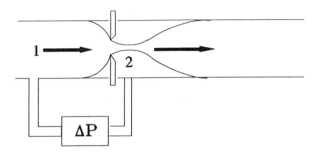

FIGURE 10-7 Orifice meter.

If the continuity equation, along with the contraction ratio, is used to eliminate the upstream velocity from Eqn (10-11), the resulting expression for the mass flow rate through the orifice is

$$\dot{m} = \frac{C_o A_o \rho_2}{\sqrt{1 - \beta^4}} \left[2 \int_{P_2}^{P_1} \frac{dP}{\rho} \right]^{\frac{1}{2}} \tag{10-12}$$

where $\beta = d/D$ and C_o is the orifice coefficient,

$$C_o = \frac{C_c}{\sqrt{\alpha_2}} \left[\frac{1 - \beta^4}{1 - \beta^4 \left(C_c \frac{\rho_2}{\rho_1} \right)^2 \left(\frac{\alpha_1 - K_f}{\alpha_2} \right)} \right]^{\frac{1}{2}} \tag{10-13}$$

C_o is obviously a function of β and K_f (e.g. N_{Re}).

A. INCOMPRESSIBLE FLOW

For incompressible flow, Eqn (10-12) becomes

$$\dot{m} = C_o A_o \sqrt{\frac{2\rho\Delta P}{1 - \beta^4}} \tag{10-14}$$

It is evident that the orifice coefficient incorporates the effects of both friction loss and velocity changes and must therefore depend upon the Reynolds number and beta ratio. This is reflected in Fig. 10-8, in which the orifice (discharge) coefficient is shown as a function of the orifice Reynolds number (N_{Re_d}) and β.

Actually, there are a variety of ''standard'' orifice plate and pressure tap designs (Miller, 1983). Figure 10-9 shows the ASME specifications for the most

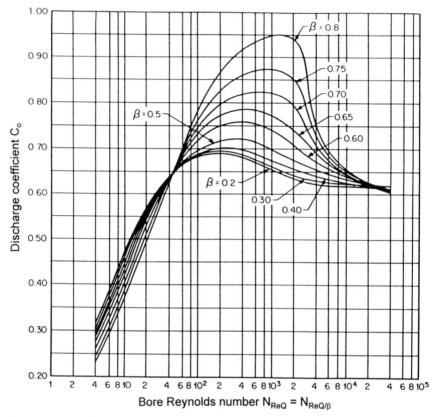

Bore Reynolds number $N_{ReQ} = N_{ReQ/\beta}$

FIGURE 10-8   Orifice discharge coefficient for square-edged orifice and flange, corner, or radius taps (Miller, 1983).

common concentric square-edged orifice. The various pressure tap locations, illustrated in Fig. 10-10 are radius taps (1 D upstream and D/2 downstream); flange taps (1 in. upstream and downstream); pipe taps ($2\frac{1}{2}$ D upstream and 8 D downstream); and corner taps. Radius taps, for which the location is scaled to the pipe diameter, are the most common. Corner taps and flange taps are the most convenient, as they can be installed in the flange which holds the orifice plate and so do not require additional taps through the pipe wall. Pipe taps are less commonly used and essentially measure the total unrecovered pressure drop, or friction loss, in the entire system (which is usually quite a bit lower than the maximum pressure drop across the orifice plate). Vena contracta taps are some-times specified, with the upstream tap 1 D from the plate and the downstream

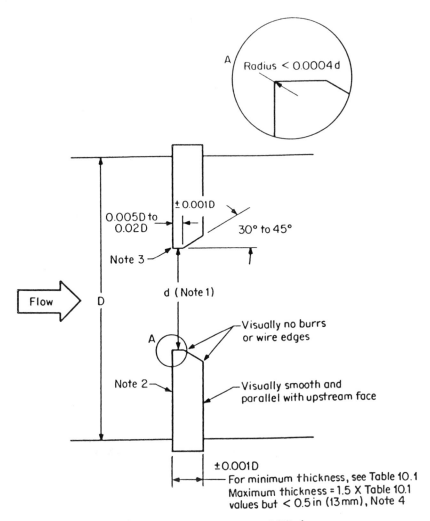

Notes: (1) Mean of four diameters, no diameter > 0.05% of mean diameter. (2) Maximum slope less than 1% from perpendicular; relative roughness < $10^{-4}d$ over a circle not less than 1.5d. (3) Visually does not reflect a beam of light, finish with a fine radial cut from center outward. (4) ANSI/ASME MFC Draft 2 (July 1982), ASME, N.Y. 1982

FIGURE 10-9   Concentric square-edged orifice specifications (Miller, 1983).

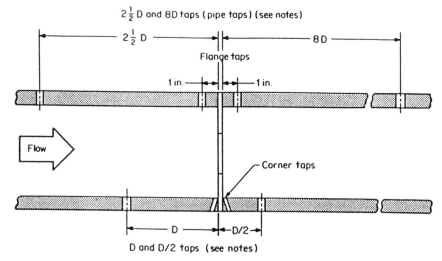

FIGURE 10-10   Orifice pressure tap locations (Miller, 1983).

tap at the vena contracta location, although the latter varies with the Reynolds number and beta ratio and thus is not a fixed position.

The orifice coefficient shown in Fig. 10-8 is valid to about 2–5% (depending upon the Reynolds number) for all pressure tap locations except pipe and vena contracta taps. More accurate values can be calculated from Eqn (10-10), with the parameter expressions given in Table 10-1 for the specific orifice and pressure tap arrangement.

B.   COMPRESSIBLE FLOW

The above relations apply to incompressible fluids, e.g. liquids. For an ideal gas under adiabatic conditions, Eqn (10-12) gives

$$\dot{m} = \frac{C_o A_o P_1}{\sqrt{1 - \beta^4}} \left\{ \frac{2kM}{(k-1)RT_1} \left(\frac{P_2}{P_1}\right)^{\frac{2}{k}} \left[\left(\frac{P_1}{P_2}\right)^{\frac{k-1}{k}} - 1 \right] \right\}^{\frac{1}{2}} \tag{10-15}$$

It is more convenient to express this result in terms of the ratio of Eqn (10-15) to the corresponding incompressible equation, Eqn (10-14), which is called the *expansion factor* Y:

$$\dot{m} = C_o A_o Y \sqrt{\frac{2\rho_1 \Delta P}{1 - \beta^4}} \tag{10-16}$$

where the density ρ_1 is evaluated at the upstream pressure (P_1). For convenience, the values of Y are shown as a function of $\Delta P/P_1$ and β for the square-edged orifice, nozzles, and venturi meters for values of k = c_p/c_v of 1.3 and 1.4 in Fig. 10-11. The lines on Fig. 10-11 for the orifice can be represented by the following equation for radius taps (Miller, 1983):

$$Y = 1 - \frac{\Delta P}{kP_1}(0.41 + 0.35\beta^4) \tag{10-17}$$

and for pipe taps by

$$Y = 1 - \frac{\Delta P}{kP_1}[0.333 + 1.145(\beta^2 + 0.7\beta^5 + 12\beta^{13})] \tag{10-18}$$

V. LOSS COEFFICIENT

The total friction loss in an orifice meter, after all pressure recovery has occurred, can be characterized by the loss coefficient, K_f, which can be derived as follows. With reference to Fig. 10-12, the total friction loss is $P_1 - P_3$. By taking the system to be the fluid in the region between the orifice (P_1) and the downstream position where the stream has filled the pipe (P_3), the momentum balance becomes

$$\sum_{on\ system} F = \dot{m}(V_3 - V_1) = 0 = P_1 A_o + P_2(A_1 - A_o) - P_3 A_1 \tag{10-19}$$

The orifice equation (Eqn 10-14) can be solved for the pressure drop ($P_1 - P_2$) to give

$$P_1 - P_2 = \frac{\rho V_o^2}{2}\left(\frac{1 - \beta^4}{C_o^2}\right) \tag{10-20}$$

Eliminating P_2 from Eqns (10-19) and (10-20) and solving for $P_1 - P_3$ provides a definition for K_f based on the pipe velocity (V_1):

$$P_1 - P_3 = \frac{\rho V_o^2(1 - \beta^4)(1 - \beta^2)}{2C_o^2} = \rho e_f = \frac{\rho V_1^2 K_f}{2} \tag{10-21}$$

Thus the loss coefficient is

$$K_f \approx \frac{(1 - \beta^4)(1 - \beta^2)}{C_o^2 \beta^4} \tag{10-22}$$

(a)

$k = 1.3$ approximately

(for CO_2, SO_2, H_2O, H_2S, NH_3, N_2O, Cl_2, CH_4, C_2H_2, and C_2H_4)

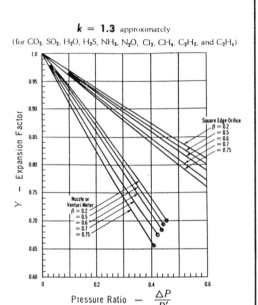

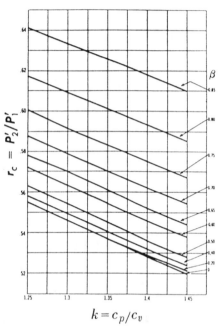

$$k = c_p/c_v$$

(b)

$k = 1.4$ approximately

(for Air, H_2, O_2, N_2, CO, NO, and HCl)

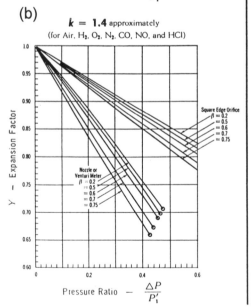

If the loss coefficient is based upon the velocity through the orifice (V_o) instead of the pipe velocity, the β^4 term in the denominator of Eqn (10-22) doesn't appear:

$$K_f = \frac{(1 - \beta^4)(1 - \beta^2)}{C_o^2} \tag{10-23}$$

Equation (10-21) represents the net total (unrecovered) pressure drop due to friction in the orifice. The percentage of the maximum pressure drop that is not recovered is shown in Fig. 10-5.

VI. ORIFICE PROBLEMS

There are three classes of problems involving orifices (or other obstruction meters) that the engineer might encounter that are similar to the types of problems encountered in pipe flows. These are the "unknown pressure drop", "unknown flow rate", and "unknown orifice diameter" problems. Each of these problems involves relationships between the same five basic dimensionless variables: C_d, N_{ReD}, β, $\Delta P/P_1$, and Y. For liquids, this list reduces to four variables, since $Y = 1$ by definition. The basic orifice equation relates these variables:

$$\dot{m} = \frac{\pi D^2 \beta^2 Y C_d}{4} \sqrt{\frac{2\rho_1 P_1}{1-\beta^4}\left(\frac{\Delta P}{P_1}\right)} \tag{10-24}$$

where

$$N_{ReD} = \frac{4\dot{m}}{\pi D \mu} \quad \text{and} \quad \beta = \frac{d}{D} \tag{10-25}$$

and $Y = fn(\beta, P/P_1)$ [as given by Eqn (10-17) or (10-18) or Fig. 10-11], and $C_d = fn(\beta, N_{ReD})$ [as given by Eqn (10-10) or Fig. 10-8]. The procedure for each of these problems is as follows.

A. UNKNOWN PRESSURE DROP

In this case, we want to determine the pressure drop to be expected when a given fluid flows at a given rate through a given orifice:

FIGURE 10-11 Expansion factor for square-edged orifice and nozzle or venturi meter (Crane, 1978). (a) k = 1.3 (for CO_2, SO_2, H_2O, H_2S, NH_3, N_2O, Cl_2, CH_2, C_2H_2, and C_2H_4); (b) k = 1.4 (for Air, H_2, O_2, N_2, CO, NO, and HCl).

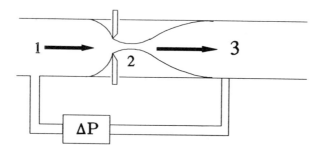

FIGURE 10-12   Pressure loss in orifice

Given: ṁ, μ, ρ_1, D, d (β= d/D), P_1 *Find*: ΔP

The procedure is as follows:

(1) Calculate N_{ReD} and β = d/D from Eqn (10-25).
(2) Get $C_d = C_o$ from Fig. 10-8 or Eqn (10-10).
(3) Assume Y = 1, and solve Eqn (10-24) for $(\Delta P)_1$:

$$(\Delta P)_1 = \left(\frac{4\dot{m}}{\pi D^2 \beta^2 C_o}\right)^2 \left(\frac{1 - \beta^4}{2\rho_1}\right) \tag{10-26}$$

(4) Using $(\Delta P)_1/P_1$ and β, get Y from Eqn (10-17) or (10-18) or Fig. 10-11.
(5) Calculate $\Delta P = (\Delta P)_1/Y^2$.
(6) Use the value of ΔP from step 5 in step 4, and repeat steps 4–6 until there is no change.

B. UNKNOWN FLOW RATE

In this case, the pressure drop across a given orifice is measured for a fluid with known properties, and the flow rate is to be determined:

Given: ΔP, P_1, D, d (β = d/D), μ, ρ_1 *Find*: ṁ

(1) Using $\Delta P/P_1$ and β, get Y from Eqn (10-17) or (10-18) or Fig. 10-11.
(2) Assume C_o = 0.61.
(3) Calculate ṁ from Eqn (10-24).
(4) Calculate N_{ReD} from Eqn (10-25).
(5) Using N_{ReD} and β, get C_o from Fig. 10-8 or Eqn (10-10).
(6) If $C_o \neq 0.61$, use the value from step 5 in step 3, and repeat steps 3–6 until there is no change.

C. UNKNOWN DIAMETER

This is the case in which the proper size orifice (d or β) is to be determined for a specified (maximum) flow rate of a given fluid in a given pipe with a ΔP device having a given (maximum) range:

Given: ΔP, P_1, μ, ρ, D, $\dot{m}$ *Find*: d (i.e. β)

(1) Assume Y = 1 and C_o = 0.61.
(2) Solve Eqn (10-24) for β, i.e.

$$\beta = \left(\frac{X}{1 + X}\right)^{1/4} \text{ where } X = \frac{8}{\rho_1 \Delta P}\left(\frac{\dot{m}}{\pi D^2 Y C_o}\right)^2 \qquad (10\text{-}27)$$

(3) Calculate $N_{Red} = N_{ReD}/\beta$, and get C_o from Fig. 10-8 or Eqn (10-10) and Y from Fig. 10-11 or Eqn (10-17) or (10-18).
(4) Use the results of step 3 in step 2, and repeat steps 2–4 until there is no change.

The required orifice diameter is d = βD.

VII. CONTROL VALVES

Flow control is achieved by control valves, which can be automatically opened or closed and continuously adjusted to achieve a desired flow rate. In this respect, they act much like a "variable resistance" in the flow line, since closing down on the valve is equivalent to increasing the flow resistance (i.e. increasing the K_f) in the line. They are typically computer-controlled and are actuated by pneumatic or electric signals in response to deviations from a set point as measured automatically by a flow meter. The nature of the relation between the valve stem or plug position (which is the manipulated variable) and the flow rate through the valve (which is the desired variable) depends upon the valve characteristics, which include the shape and design of the valve plug and the flow channel through the valve.

A. VALVE CHARACTERISTICS

Different valve plugs are usually available for a given valve, each of which provides a different response characteristic when the valve setting (i.e. the stem position) is changed. The specific valve characteristic is chosen to match the dynamic response of the flow system and control loop.

Figure 10-13 illustrates typical flow characteristic curves for various control valves (Fisher, 1987). The "quick opening" characteristic provides the maximum change in flow rate at low opening or stem travel, with a fairly linear relationship. As the valve approaches wide open, the change in flow with travel

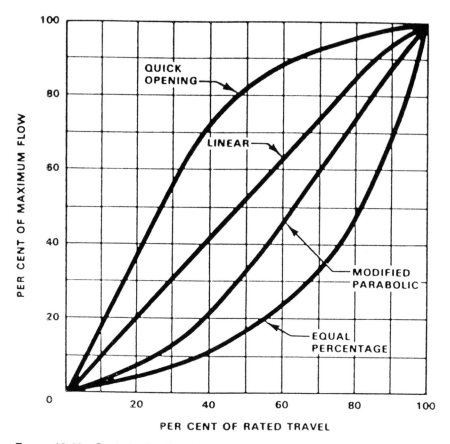

FIGURE 10-13 Control valve flow characterisxtics (Fisher Controls, 1987).

approaches zero. This is best suited for on-off control but is also appropriate where a linear valve is desired. The "linear" flow characteristic has a constant "valve gain", that is, the incremental change in flow rate with change in valve plug position is the same at all flow rates. This is a commonly desired property, particularly for liquid level control. The "equal percentage" flow characteristic provides the same percentage change in flow for equal increments of valve plug position. Thus, with this characteristic the change in flow is always proportional to the value of the flow rate just before the change is made. This characteristic is used in pressure control applications and where a relatively small pressure drop across the valve is required relative to that in the rest of the system. The modified parabolic characteristic is intermediate to the linear and equal percentage characteristics and can be substituted for equal percentage valve plugs in many applications, with some loss in performance.

Some general guidelines for the application of the proper valve charac-teristic are shown in Fig. 10-14. These are "rules of thumb", and the proper valve can only be determined by a complete dynamic analysis of the system in which the valve is to be used (which is beyond the scope of this book—see e.g. Baumann, 1991, for simplified guidelines).

B. VALVE SIZING RELATIONS

1. Incompressible Fluids

Bernoulli's equation relates the pressure drop across a valve and the flow rate through the vaive. If the friction loss is represented by a loss coefficient, K_f, as introduced in Chapter 7, and assuming no change in kinetic or potential energy across the valve, Bernoulli's equation can be solved for the flow rate to give

$$Q = AV = A \sqrt{\frac{2\Delta P}{\rho K_f}} \qquad (10\text{-}28)$$

where A is an appropriate flow area, V is the velocity through that area, $\Delta P = P_1 - P_2$ is the pressure drop across the valve, and K_f is the loss coefficient referred to the velocity V. However, in a control valve the area (and hence V) is a variable, and the internal flow geometry can be quite complex. Also, the pressure drop is not the maximum value in the valve (which would occur if P_2 is at the vena contracta, as for the orifice meter) but is the *net unrecovered* pressure loss corresponding to P_2, which is far enough downstream that any possible pressure recovery has occurred. Combining the flow area and geomet-rical factors and the density of the reference fluid with the friction loss coeffi-cient leads to the following equation for incompressible fluids:

$$Q = C_v \sqrt{\frac{\Delta P}{SG}} \qquad (10\text{-}29)$$

which defines the *flow coefficient*, C_v. The value of C_v is thus different for each valve and also varies with the valve opening (or stem travel) for a given valve. Although Eqn (10-29) is similar to the flow equation for flow meters, the flow coefficient C_v is *not dimensionless*, as is the flow meter discharge coefficient, but has dimensions of $[L^3][L/M]^{1/2}$. More specifically, the "normal engineering" (inconsistent) units of C_v are $[\text{gpm}/(\text{psi})^{1/2}]$. [If the fluid density were included in Eqn (10-29) instead of SG, the dimensions of C_v would be L^2, which follows from the inclusion of the effective valve flow area in the definition of C_v.] The reference fluid for the density is water for liquids and air for gases.

Liquid Level Systems

Control Valve Pressure Drop	Best Inherent Characteristic
Constant ΔP	Linear
Decreasing ΔP with Increasing Load, ΔP at Maximum Load > 20% of Minimum Load ΔP	Linear
Decreasing ΔP with Increasing Load, ΔP at Maximum Load < 20% of Minimum Load ΔP	Equal-Percentage
Increasing ΔP with Increasing Load, ΔP at Maximum Load < 200% of Minimum Load ΔP	Linear
Increasing ΔP with Increasing Load, ΔP at Maximum Load > 200% of Minimum Load ΔP	Quick Opening

Pressure Control Systems

Application	Best Inherent Characteristic
Liquid Process	Equal-Percentage
Gas Process, Small Volume, Less Than 10 ft. of Pipe Between Control Valve and Load Valve	Equal-Percentage
Gas Process, Large Volume (Process has a Receiver, Distribution System or Transmission Line Exceeding 100 ft. of Nominal Pipe Volume) Decreasing ΔP with Increasing Load, ΔP at Maximum Load > 20% of Minimum Load ΔP	Linear
Gas Process, Large Volume, Decreasing ΔP with Increasing Load, ΔP at Maximum Load < 20% of Minimum Load ΔP	Equal-Percentage

Flow Control Processes

FLOW MEASURE-MENT SIGNAL TO CONTROLLER	LOCATION OF CONTROL VALVE IN RELATION TO MEASURING ELEMENT	BEST INHERENT CHARACTERISTIC	
		Wide Range of Flow Set Point	Small Range of Flow but Large ΔP Change at Valve with Increasing Load
Proportional To Flow	In Series	Linear	Equal-Percentage
	In Bypass*	Linear	Equal-Percentage
Proportional To Flow Squared	In Series	Linear	Equal-Percentage
	In Bypass*	Equal-Percentage	Equal-Percentage

*When control valve closes, flow rate increases in measuring element.

FIGURE 10-14   Guidelines for control valve applications (Fisher Controls, 1987).

Valve C_v's are determined by the manufacturer from measurements on each valve type. Because they are not dimensionless, the values will depend upon the specific units used for the variables and parameters in Eqn (10-29). Thus the values of C_v are normally assumed to be defined in terms of typical "engineering units", as follows:

Q = volumetric flow rate (gpm for liquids or scfh for gas or steam)
SG = specific gravity [relative to water for liquids (62.3 lb_m/ft^3) or air at 60°F and 1 atm for gases (0.0764 lb_m/ft^3)]
ρ_1 = density at upstream conditions (lb_m/ft^3)
P_1 = upstream pressure (psia)
ΔP = total (net) pressure drop across valve (psi)

For each valve design, tables for the values of the flow coefficients as a function of valve size and percent of valve opening are provided by the manufacturer. Typical flow coefficient values are shown in Table 10-3. In this table, K_m applies to cavitating and flashing liquids and C_1 applies to critical (choked) compressible flow, as discussed below.

2. Cavitating and Flashing Liquids

The minimum pressure in the valve (P_{vc}) generally occurs at the vena contracta, just downstream of the flow orifice. The pressure then rises downstream to P_2, with the amount of pressure recovery depending upon the valve design. If P_{vc} is less than the fluid vapor pressure (P_v), the liquid will partially vaporize, forming bubbles. If the pressure recovers to a value greater than P_v, these bubbles may collapse suddenly, setting up local shock waves, which can result in considerable damage. This situation is referred to as *cavitation*, as opposed to *flashing* which occurs if the recovered pressure remains below P_v so that the bubbles will not collapse and vaporization continues. After the first vapor cavities form, the flow rate will no longer be proportional to the square root of the pressure difference across the valve, due to the decreasing density of the mixture. If sufficient vapor forms, the flow can become choked, at which point the flow rate will be independent of the downstream pressure as long as P_1 remains constant. The critical pressure ratio ($r_c = P_{2c}/P_v$) at which choking will occur is shown in Fig. 10-15 for water and Fig. 10-16 for other liquids, as a function of the liquid vapor pressure (P_v) relative to the fluid critical pressure (P_c). Table 10-4 lists the critical pressure values for some commonly employed fluids. An equation that represents the critical pressure ratio, r_c, with acceptable accuracy is (Fisher Controls, 1977)

$$r_c = 0.96 - 0.28 \sqrt{\frac{P_v}{P_c}} \tag{10-30}$$

TABLE 10-3   Example Flow Coefficient Values for a Linear Control Valve

Linear — Linear Characteristic

Coefficients	Body Size, Inch	Port Diameter, Inch	Total Travel, Inch	10	20	30	40	50	60	70	80	90	100	$K_m^{(1)}$ and C_1
C_v (Liquid)	2 & 3 x 2	1-7/8	1-1/2	1.69	9.45	21.9	33.4	42.7	50.0	55.6	59.6	61.9	63.6	.72
	3 & 4 x 3	2-7/8	2	3.41	25.4	52.6	76.0	96.4	114	127	133	135	136	.91
	4 & 6 x 4	3-5/8	2	6.89	25.1	50.1	77.9	106	134	157	175	185	188	.86
	6 & 8 x 6	5-3/8	3	9.40	63.8	138	212	282	339	373	389	398	405	.81
C_g (Gas)	2 & 3 x 2	1-7/8	1-1/2	60.8	326	729	1110	1400	1600	1710	1780	1810	1840	28.9
	3 & 4 x 3	2-7/8	2	142	839	1760	2540	3240	3880	4320	4490	4540	4570	33.6
	4 & 6 x 4	3-5/8	2	229	791	1530	2350	3250	4190	5090	5850	6360	6580	35.0
	6 & 8 x 6	5-3/8	3	287	1910	4060	6160	8400	10,600	12,300	13,300	13,800	14,100	34.8
C_s (Steam)	2 & 3 x 2	1-7/8	1-1/2	3.04	16.3	36.5	55.5	70.0	80.0	85.5	89.0	90.5	92.0	28.9
	3 & 4 x 3	2-7/8	2	7.10	42.0	88.0	127	162	194	216	225	227	229	33.6
	4 & 6 x 4	3-5/8	2	11.5	39.6	76.5	118	163	210	255	293	318	329	35.0
	6 & 8 x 6	5-3/8	3	14.4	95.5	203	308	420	530	615	665	690	705	34.8

Equal Percentage — Equal Percentage Characteristic

Coefficients	Body Size, Inch	Port Diameter, Inch	Total Travel, Inch	10	20	30	40	50	60	70	80	90	100	$K_m^{(1)}$ and C_1
C_v (Liquid)	2 & 3 x 2	1-7/8	1-1/8	1.04	1.59	3.52	6.99	12.1	19.7	30.5	40.9	44.6	50.7	.79
	3 & 4 x 3	2-7/8	1-1/2	2.56	51.7	10.80	18.2	28.9	44.9	62.6	82.9	104	117	.91
	4 & 6 x 4	3-5/8	1-1/2	3.44	7.12	13.1	21.8	34.8	54.0	80.4	109	132	154	.71
	6 & 8 x 6	5-3/8	2-1/2	5.27	13.0	22.1	35.3	57	93	141	194	246	308	.64
C_g (Gas)	2 & 3 x 2	1-7/8	1-1/8	41.5	61.2	123	233	401	653	996	1320	1460	1590	31.4
	3 & 4 x 3	2-7/8	1-1/2	88.9	175	381	638	985	1530	2190	2890	3610	4000	34.2
	4 & 6 x 4	3-5/8	1-1/2	134	240	430	700	1080	1650	2480	3440	4210	5140	33.3
	6 & 8 x 6	5-3/8	2-1/2	152	422	673	1020	1710	2730	3990	5490	7350	9220	29.9
C_s (Steam)	2 & 3 x 2	1-7/8	1-1/8	2.08	3.06	6.15	11.7	20.1	32.7	49.8	66.0	73.0	79.5	31.4
	3 & 4 x 3	2-7/8	1-1/2	4.45	8.75	19.1	31.9	49.3	76.5	110	145	181	200	34.2
	4 & 6 x 4	3-5/8	1-1/2	6.70	12.0	21.5	35.0	54.0	82.5	124	172	211	257	33.4
	6 & 8 x 6	5-3/8	2-1/2	7.60	21.1	33.7	51.0	85.5	137	200	275	368	461	29.9

Modified Equal Percentage — Equal Percentage Characteristic

Coefficients	Body Size, Inch	Port Diameter, Inch	Total Travel, Inch	10	20	30	40	50	60	70	80	90	100	$K_m^{(1)}$ and C_1
C_v (Liquid)	2 & 3 x 2	1-7/8	1-1/2	1.07	2.65	6.87	15.1	26.6	38.3	47.6	53.7	57.3	60.4	.73
	3 & 4 x 3	2-7/8	2	3.08	8.63	18.5	34.3	57.8	84.5	108	123	131	135	.88
	4 & 6 x 4	3-5/8	2	4.49	10.7	21.8	41.2	71.0	107	141	166	183	193	.85
	6 & 8 x 6	5-3/8	3	6.67	16.4	29.3	52.0	92.5	151	217	280	346	380	.75
C_g (Gas)	2 & 3 x 2	1-7/8	1-1/2	43.0	95.9	230	493	874	1260	1530	1660	1720	1800	29.8
	3 & 4 x 3	2-7/8	2	105	295	635	1140	1930	2890	3720	4250	4470	4540	33.6
	4 & 6 x 4	3-5/8	2	172	337	663	1280	2240	3380	4470	5480	6460	6670	34.6
	6 & 8 x 6	5-3/8	3	200	298	894	1520	2620	4330	6270	8210	10,700	12,500	32.9
C_s (Steam)	2 & 3 x 2	1-7/8	1-1/2	2.15	4.80	11.5	24.7	43.7	63.0	76.5	83.0	86.0	90.0	29.8
	3 & 4 x 3	2-7/8	2	5.25	14.8	31.8	57.0	96.5	145	186	213	224	227	33.6
	4 & 6 x 4	3-5/8	2	8.60	16.9	33.2	64.0	112	169	224	274	323	334	34.6
	6 & 8 x 6	5-3/8	3	10.0	24.9	44.7	76.0	131	217	314	411	535	625	32.9

1. This column lists the K_m values for the C_v coefficients and the C_1 values for the C_g and C_s coefficients at 100% travel.

Source: Fisher Controls (1987).

With r_c known, the allowable pressure drop across the valve at which cavitation occurs is given by

$$\Delta P_c = K_m(P_1 - r_c P_v) \qquad (10\text{-}31)$$

where K_m is the valve recovery coefficient (which is a function of the valve design). The recovery coefficient is defined as the ratio of the overall net pressure drop $(P_1 - P_2)$ to the maximum pressure drop from upstream to the vena contracta $(P_1 - P_{vc})$:

$$K_m = \frac{P_1 - P_2}{P_1 - P_{vc}} \qquad (10\text{-}32)$$

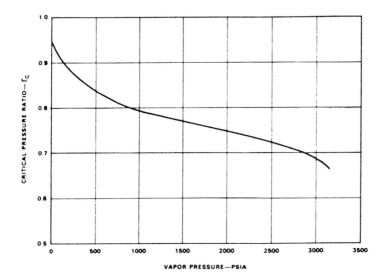

FIGURE 10-15 Critical pressure ratios for water (Fisher Controls, 1987). Enter on the abscissa at the water vapor pressure at the valve inlet. Proceed vertically to intersect the curve. Move horizontally to the left to read the critical pressure ratio, r_c, on the ordinate.

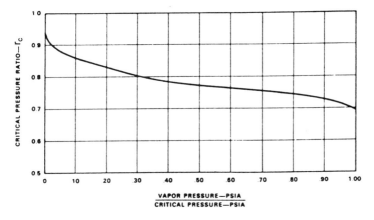

FIGURE 10-16 Critical pressure ratio for cavitating and flashing liquids other than water (Fisher Controls, 1987). Determine the vapor pressure/critical pressure ratio by dividing the liquid vapor pressure at the valve inlet by the critical pressure of the liquid. Enter on the abscissa at the ratio just calculated and proceed vertically to intersect the curve. Move horizontally to the left and read the critical pressure ratio, r_c, on the ordinate.

TABLE 10-4   Critical Pressure of Various Fluids, psia

Ammonia	1636
Argon	705.6
Butane	550.4
Carbon dioxide	1071.6
Carbon monoxide	507.5
Chlorine	1118.7
Dowtherm A	465
Ethane	708
Ethylene	735
Fluorine	808.5
Helium	33.2
Hydrogen	188.2
Hydrogen chloride	1198
Isobutane	529.2
Isobutylene	580
Methane	673.3
Nitrogen	492.4
Nitrous oxide	1047.6
Oxygen	736.5
Phosgene	823.2
Propane	617.4
Propylene	670.3
Refrigerant 11	635
Refrigerant 12	596.9
Refrigerant 22	716
Water	3206.2

Values of K_m for the Fisher Controls Spec D valve are given in the last column of Table 10-3, and representative values for other valves at the full-open condition are given in Table 10-5. If $P > P_c$, the value of P_c is used as the pressure drop in the standard liquid sizing equation to determine Q; otherwise, the value of $P_1 - P_2$ is used:

$$Q = C_v \sqrt{\Delta P_c / SG} \qquad (10\text{-}33)$$

The notation used here is that from the Fisher Controls literature (e.g. Fisher Controls, 1990). The ANSI/ISAS75.01 standard for control valves (e.g.

TABLE 10-5 Representative Full-Open K_m
Values for Various Valves

Body type	K_m
Globe,	
single port, flow opens	0.70–0.80
double port	0.70–0.80
Angle, flow closes	
Venturi outlet liner	0.20–0.25
Standard seat ring	0.50–0.60
Angle, flow opens	
Maximum orifice	0.70
Minimum orifice	0.90
Ball valve:	
V-Notch	0.40
Conventional	0.30
Butterfly valve	
60° open	0.55
90° open	0.30

Source: Hutchison (1971).

Baumann, 1991; Hutchison, 1971) uses the same equations, except that it uses the notation $F_L = \sqrt{K_m}$ and $F_F = r_c$ in place of the factors K_m and r_c.

C. COMPRESSIBLE FLUIDS

1. Subsonic Flow

For relatively low pressure drops, the effect of compressibility is negligible, and the general flow equation (Eqn 10-29) applies. Introducing the conversion factors to give the flow rate in scfh, and the density of air at standard conditions (1 atm, 520°R), this equation becomes

$$Q_{scfh} = 1362\ C_v P_1 \sqrt{\frac{\Delta P}{P_1 (SG) T_1}} \qquad (10\text{-}34)$$

The effect of variable density can be accounted for by an expansion factor Y, as has been done for flow in pipes and meters, in which case Eqn (10-34) can be written

$$Q_{scfh} = 1362\ C_v P_1 Y \sqrt{\frac{X}{(SG) T_1}} \qquad (10\text{-}35)$$

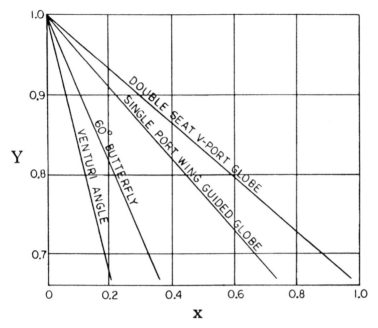

FIGURE 10-17   Expansion factor (Y) as a function of pressure drop ratio (X) for four different types of control valves (Hutchison, 1971).

where

$$X = \frac{\Delta P}{P_1} = \frac{P_1 - P_2}{P_1} \qquad\qquad (10\text{-}36)$$

Deviations from the ideal gas law may be incorporated by multiplying T_1 in Eqn (10-34) or (10-35) by the compressibility factor, Z, for the gas.

The expansion factor Y depends upon the pressure drop X, the dimensions (clearance) in the valve, the gas specific heat ratio k, and the Reynolds number (the effect of which is often negligible). It has been found from measurements (Hutchison, 1971) that the expansion factor for a given valve can be represented, to about ±2%, by the expression

$$Y = 1 - \frac{X}{3X_T} \qquad\qquad (10\text{-}37)$$

where X_T depends upon the specific valve as illustrated in Fig. 10-17.

2. Choked Flow

When the gas velocity reaches the speed of sound, choked flow occurs and the mass flow rate reaches a maximum. From Eqn (10-35), this is equivalent to a maximum in $Y\sqrt{X}$, which occurs at $Y = 0.667$. This corresponds to the terminus of the lines in Fig 10-17. That is, X_T is the pressure ratio across the valve at which choking occurs, and any further increase in X (e.g. P) due to a lower value of P_2 can have no effect on the flow rate.

The flow coefficient C_v is determined by calibration with water, and it is not entirely satisfactory for predicting the flow rate of compressible fluids under choked flow conditions. This has to do with the fact that different valves exhibit different pressure recovery characteristics with gases and hence will choke at different pressure ratios, which is not significant for liquid flows. For this reason, another flow coefficient, C_g, is often determined by calibration with air under critical flow conditions (Fisher Controls, 1977). The corresponding flow equation for gas flow is, therefore,

$$Q_{critical} = C_g P_1 \sqrt{\frac{520}{(SG)T}} \qquad (10\text{-}38)$$

3. Universal Gas Sizing Equation

Equation (10-34), which applies at low pressure drops, and Eqn (10-38), which applies to critical (choked) flow, have been combined into one general "universal" empirical equation by Fisher (1977), by using a sine function to represent the transition between the limits of both of these states:

$$Q_{scfh} = C_g \sqrt{\frac{520}{(SG)T_1}} \; P_1 \, \sin \left[\frac{3417}{C_1} \, \sqrt{\frac{\Delta P}{P_1}} \right]_{deg} \qquad (10\text{-}39)$$

Here, $C_1 = C_g/C_v$ and is determined by measurements on air. For the valve in Table 10-3, values of C_1 are listed in the last column. C_1 is also approximately equal to $40\sqrt{X_T}$ (Hutchison, 1971). For steam or vapor at any pressure, a corresponding equation is

$$Q_{lb/hr} = 1.06 \; C_g \, \sqrt{\rho_1 P_1} \, \sin \left[\frac{3417}{C_1} \, \sqrt{\frac{\Delta P}{P_1}} \right]_{deg} \qquad (10\text{-}40)$$

where ρ_1 is the density of the gas at P_1, in lb_m/ft^3. When the argument of the sine term (in brackets) in Eqn (10-39) or (10-40) is equal to 90° or more, the flow has reached critical flow conditions (choked) and cannot increase above this value without increasing P_1. Under these conditions, the sine term is equal to unity for this and all larger values of ΔP.

The above equations are based upon flow coefficients determined by calibration with air. For application with other gases, the difference between the properties of air and those of the other gas must be considered. The gas density is incorporated into the equations, but a correction must be made for the specific heat ratio (c_p/c_v = k) as well. This can be done by considering the expression for the ideal (isentropic) flow of a gas through a nozzle, which can be written (in "engineering units") as follows:

$$Q_{scfh} = \frac{3.78 \times 10^5 \, A_2 P_1}{SG \, \sqrt{RT}} \sqrt{\frac{k}{k-1} \left[\left(\frac{P_2}{P_1}\right)^{\frac{2}{k}} - \left(\frac{P_2}{P_1}\right)^{\frac{(k+1)}{k}} \right]} \qquad (10\text{-}41)$$

Critical (choked) flow will occur in the nozzle throat when the pressure ratio is

$$r = \frac{P_2}{P_1} = \left(\frac{2}{k+1}\right)^{\frac{k}{k-1}} \qquad (10\text{-}42)$$

Thus, for choked flow, Eqn (10-41) becomes

$$Q_{scfh} = \frac{3.78 \times 10^5 \, A_2 P_1}{SG \, \sqrt{RT}} \sqrt{\left(\frac{k}{k+1}\right)\left(\frac{2}{k+1}\right)^{\frac{2}{k-1}}} \qquad (10\text{-}43)$$

The quantity in the radical, which is a function only of k [fn (k)], represents the dependence of the flow rate on the gas property. Hence this may be used to define a correction factor, C_2, that can be used as a multiplier to correct the flow rate for air to that for any other gas:

$$C_2 = \frac{fn \, (k)_{gas}}{fn \, (k)_{air}} = \frac{\sqrt{\left[\frac{k}{k+1}\right]\left[\frac{2}{k+1}\right]^{\frac{2}{k-1}}}}{0.4839} \qquad (10\text{-}44)$$

A plot of C_2 versus k as given by Eqn (10-44) is shown in Fig. 10-18.

D. VISCOSITY CORRECTION

A correction for fluid viscosity must be applied to the flow coefficient (C_v) for liquids other than water. This viscosity correction factor (F_v) is obtained from Fig. 10-19 by the following procedure, depending upon whether the objective is to find the valve size for a given Q and ΔP, to find Q for a given valve and ΔP, or to find ΔP for a given valve and Q.

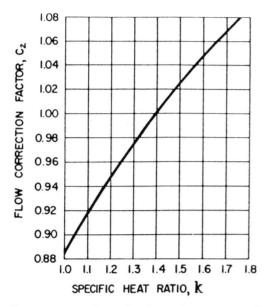

FIGURE 10-18   Correction factor for gas properties.

1.   To Find Valve Size

For the given Q and ΔP, calculate the required C_v as follows:

$$C_v = \frac{Q}{\sqrt{\Delta P/SG}} \qquad (10\text{-}45)$$

Then determine the Reynolds number for the valve from the equation

$$N_{Re} = 17250 \frac{Q}{\sqrt{C_v \nu_{cs}}} \qquad (10\text{-}46)$$

where Q is in gpm, ΔP is in psi, and ν_{cs} is the fluid kinematic viscosity (μ/ρ) in centistokes. The viscosity correction factor, F_v, is then read from the middle line on Fig. 10-19 and used to calculate a corrected value of C_v as follows:

$$C_{v_c} = C_v F_v \qquad (10\text{-}47)$$

The proper valve size and percent opening are then found from the table for the valve flow coefficient (e.g. Table 10-3) at the point where the coefficient is equal to or higher than this corrected value.

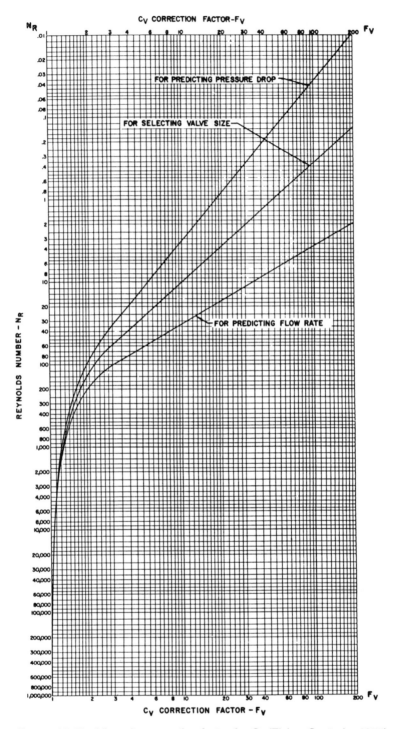

FIGURE 10-19 Viscosity correction factor for C_v (Fisher Controls, 1977).

2. To Predict Flow Rate

For a given valve (i.e. a given C_v) and given ΔP, the maximum flow rate (Q_{max}) is determined as

$$Q_{max} = C_v \sqrt{\Delta P/SG} \qquad (10\text{-}48)$$

The Reynolds number is then calculated from Eqn (10-46), and the viscosity correction factor, F_v, is read from the bottom curve in Fig. 10-19. The corrected flow rate is then

$$Q_c = \frac{Q_{max}}{F_v} \qquad (10\text{-}49)$$

3. To Predict Pressure Drop

For a given valve (C_v) and given flow rate (Q), calculate the Reynolds number as above and read the viscosity correction factor F_v from the top line of Fig. 10-19. The predicted pressure drop across the valve is then

$$\Delta P = SG \left(\frac{QF_v}{C_v} \right)^2 \qquad (10\text{-}50)$$

REFERENCES

Baumann, H. D., *Control Valve Primer*, Instrument Society of America (1991)

Cheremisinoff, N. P., and P. N. Cheremisinoff, *Instrumentation for Process Flow Engineering*, Technomic Publishing Company, Lancaster, Basel (1987)

Crane Company, "Flow of Fluids Through Valves, Fittings, and Pipe," Technical Manual 410 (1978)

Fisher Controls, *Control Valve Handbook*, 2nd Ed., Marshalltown, Iowa (1977)

Fisher Controls, Catalog 10, Ch. 2, Marshalltown, Iowa (1987)

Fisher Controls, *Control Valve Source Book*, Fisher Controls, International, Marshalltown, Iowa (1990)

Hutchison, J. W., *ISA Handbook of Control Valves*, Instrument Society of America (1971)

Miller, R. W., *Flow Measurement Engineering Handbook*, McGraw-Hill (1983)

Murdock, J. W., *Fluid Mechanics and Its Applications*, Houghton Mifflin Company, Boston (1976)

Olson, R. M., *Essentials of Engineering Fluid Mechanics*, 4th Ed., Harper & Row (1980)

White, F. M., *Fluid Mechanics*, 3rd Ed. McGraw-Hill (1994)

PROBLEMS

FLOW MEASUREMENT

1. An orifice meter with a hole of 1 in. diameter is inserted in a $1\frac{1}{2}$ in. sch 40 line carrying SAE 10 lube oil at 70°F (SG = 0.93). A manom-

eter using water as the manometer fluid is used to measure the orifice pressure drop and reads 8 in. What is the flow rate of oil in gpm?

2. An orifice with a 3 in. diameter hole is mounted in a 4 in. diameter pipeline carrying water. A manometer containing a fluid with specific gravity of 1.2 connected across the orifice reads 0.25 in. What is the flow rate in the pipe in gpm?

3. An orifice with a 1 in. diameter hole is installed in a 2 in. sch 40 pipeline carrying SAE 10 lube oil at 100°F. The pipe section where the orifice is installed is vertical, with the flow being upward. Pipe taps are used, which are connected to a manometer containing mercury to measure the pressure drop. If the manometer reading is 3 in., what is the flow rate of the oil, in gpm?

4. The flow rate in a 1.5 in. line can vary from 100 to 1000 bbl/day, so you must install on orifice meter to measure it. If you use a DP cell with a range of 10 in. H_2O to measure the pressure drop across the orifice, what size orifice should you use? After this orifice is installed, you find that the DP cell reads 0.5 in. H_2O. What is the flow rate in bbl/day?

5. A 4 in. sch 80 pipe carries water from a storage tank on top of a hill to a plant at the bottom of the hill. The pipe is inclined at an angle of 20° to the horizontal. An orifice meter with a diameter of 1 in. is inserted in the line, and a mercury manometer across the meter reads 2 in. What is the flow rate in gpm?

6. You must size an orifice meter to measure the flow rate of gasoline (SG = 0.72) in a 10 in. I.D. pipeline at 60°F. The maximum flow rate expected is 1000 gpm, and the maximum pressure differential across the orifice is to be 10 in. of water. What size orifice should you use?

7. A 2 in. sch 40 pipe carries SAE 10 lube oil at 100°F (SG = 0.928). The flow rate may be as high as 55 gpm, and you must select an orifice meter to measure the flow.
   (a) What size orifice should be used if the pressure difference is measured using a DP cell having a full scale range of 100 in. H_2O?
   (b) Using this size orifice, what is the flow rate of oil in gpm when the DP cell reads 50 in. H_2O?

8. A 2 in. sch 40 pipe carries a 35° API distillate at 50°F (SG = 0.85). The flow rate is measured by an orifice meter in the line, which has a diameter of 1.5 in. The pressure drop across the orifice plate is measured by a water manometer connected to flange taps.
   (a) If the manometer reading is 1 in., what is the flow rate of the oil in gpm?
   (b) What would the diameter of the throat of a venturi meter be that would give the same manometer reading at this flow rate?
   (c) Calculate the unrecovered pressure loss for both the orifice and the venturi in psi.

9. An orifice having a diameter of 1 in. is used to measure the flow rate of SAE 10 lube oil (SG = 0.928) in a 2 in. sch 40 pipe at 70°F. The pressure drop across the orifice is measured by a mercury (SG = 13.6) manometer, which reads 2 cm.
   (a) Calculate the volumetric flow rate of the oil in liters/s.
   (b) What is the temperature rise of the oil as it flows through the orifice in °F? [c_v = 0.5 Btu/(lb$_m$ °F)]
   (c) How much power (in horsepower) is required to pump the oil through the orifice? (Note: This is the same as the rate of energy dissipated in the flow.)

10. An orifice meter is used to measure the flow rate of CCl_4 in a 2 in. sch 40 pipe. The orifice diameter is 1.25 in., and a mercury manometer attached to the pipe taps across the orifice reads 1/2 in. Calculate the volumetric flow rate of CCl_4 in ft^3/s (SG of CCl_4 = 1.6). What is the permanent energy loss in the flow above due to the presence of the orifice in ft lb$_f$/lb$_m$? Express this also as a total overall ''unrecovered'' pressure loss in psi.

11. An orifice meter is installed in a 6 in. I.D. pipeline which is inclined upward at an angle of 10° from the horizontal. Benzene is flowing in the pipeline at the flow rate of 10 gpm. The orifice diameter is 3.5 in., and the orifice pressure taps are 9 in. apart.
    (a) What is the pressure drop between the pressure taps in psi?
    (b) What would be the reading of a water manometer connected to the pressure taps?

12. You are to specify an orifice meter for measuring the flow rate of a 35° API distillate (SG = 0.85) flowing in a 2 in. sch 160 pipe at 70°F. The maximum flow rate expected is 2000 gal/hr, and the avail-

able instrumentation for a differential pressure measurement has a limit of 2 psi. What size hole should the orifice have?

13. You must select an orifice meter for measuring the flow rate of an organic liquid (SG = 0.8, = 15 cP) in a 4 in. sch 40 pipe. The maximum flow rate anticipated is 200 gpm, and the orifice pressure difference is to be measured with a mercury manometer having a maximum reading range of 10 in. What size should the orifice be?

14. An oil with SG = 0.9 and viscosity of 30 cP is transported in a 12 in. sch 20 pipeline at a maximum flow rate of 1000 gpm. What size orifice should be used to measure the oil flow rate if a DP cell with a full-scale range of 10 in. of water is used to measure the pressure drop across the orifice? What size venturi would you use in place of the orifice in the above pipeline, everything else being the same?

15. You want to use a venturi meter to measure the flow rate of water, up to 1000 gpm, through an 8 in. sch 40 pipeline. To measure the pressure drop in the venturi, you have a DP cell with a maximum range of 15 in. H_2O pressure difference. What size venturi (i.e. throat diameter) should you specify?

16. Gasoline is pumped through a 2 in. sch 40 pipeline upward into an elevated storage tank at 60°F. An orifice meter is mounted in a vertical section of the line, which uses a DP cell with a maximum range of 10 in. of water to measure the pressure drop across the orifice at radius taps. If the maximum flow rate expected in the line is 10 gpm, what size orifice should you use? If a water manometer with a maximum reading of 10 in. is used instead of the DP cell, what would the required orifice diameter be?

17. You have been asked by your boss to select a flow meter to measure the flow rate of gasoline (SG = 0.85) at 70°F in a 3 in. sch 40 pipeline. The maximum expected flow rate is 200 gpm, and you have a DP cell (which measures differential pressure) with a range of 0 to 10 in. H_2O available.
    (a) If you use a venturi meter, what should the diameter of the throat be?
    (b) If you use an orifice meter, what diameter orifice should you use?
    (c) For a venturi meter with a throat diameter of 2.5 in., what would the DP cell read (in inches of water) for a flow rate of 150 gpm?

(d) For an orifice meter with a diameter of 2.5 in., what would the DP cell read (in inches of water) for a flow rate of 150 gpm?

(e) If the discharge coefficients of the meters reflect the energy dissipated in the meters and nothing else, how much power (in horsepower) is consumed by friction in each of the meters under the conditions of (c) and (d) above?

18. A 2 in. sch 40 pipe is carrying water at a flow rate of 8 gpm. The flow rate is measured by means of an orifice with a 1.6 in. diameter hole. The pressure drop across the orifice is measured using a manometer containing an oil of SG 1.3.

    (a) What is the manometer reading in inches?

    (b) What is the power (in hp) consumed as a consequence of the friction loss due to the orifice plate in the fluid?

19. The flow rate of CO_2 in a 6 in. I.D. pipeline is measured by an orifice meter with a diameter of 5 in. The pressure upstream of the orifice is 10 psig, and the pressure drop across the orifice is 30 in. H_2O. If the temperature is 80°F, what is the mass flow rate of CO_2?

20. An orifice meter is installed in a vertical section of a piping system, in which SAE 10 lube oil is flowing upward (at 100°F). The pipe is 2 in. sch 40, and the orifice diameter is 1 in. The pressure drop across the orifice is measured by a manometer containing mercury as the manometer fluid. The pressure taps are pipe taps ($2^{1}/_{2} \times$ I.D. upstream and $8 \times$ I.D. downstream), and the manometer reading is 3 in. What is the flow rate of the oil in the pipe, in gpm?

21. You must install an orifice meter in a pipeline to measure the flow rate of 35.6° API crude oil, at 80°F. The pipeline diameter is 18 in. sch 40, and the maximum expected flow rate is 300 gpm. If the pressure drop across the orifice is limited to 30 in. of water or less, what size orifice should be installed? What is the maximum permanent pressure loss that would be expected through this orifice, in psi?

22. A 6 inch sch 40 pipeline is designed to carry SAE 30 lube oil at 80°F (SG = 0.87) at a maximum velocity of 10 ft/s. You must install an orifice meter in the line to measure the oil flow rate. If the maximum pressure drop to be permitted across the orifice is 40 inches of water, what size orifice should be used? If a venturi meter is used instead of an orifice, everything else being the same, how large should the throat be?

23. An orifice meter with a diameter of 3 in. is mounted in a 4 in. sch 40 pipeline carrying an oil with a viscosity of 30 cP and a SG of 0.85. A mercury manometer attached to the orifice meter reads 1 in. If the pumping stations along the pipeline operate with a suction (inlet) pressure of 10 psig and a discharge (outlet) pressure of 160 psig, how far apart should the pump stations be, if the pipeline is horizontal?

24. A 35° API oil at 50°F is transported in a 2 in. sch 40 pipeline. The oil flow rate is measured by an orifice meter which is 1.5 in. in diameter, using a water manometer.
    (a) If the manometer reading is 1 in., what is the oil flow rate (in gpm)?
    (b) If a venturi meter is used instead of the orifice meter, what should the diameter of the venturi throat be to give the same reading as the orifice meter at the same flow rate?
    (c) Determine the unrecovered pressure loss for both the orifice and venturi meters.

25. A 6 in. sch 40 pipeline carries a petroleum fraction (viscosity 15 cP, SG 0.85) at a velocity of 7.5 ft/s, from a storage tank at 1 atm pressure to a plant site. The line contains 1500 ft of straight pipe, 25 90° flanged elbows, and 4 open globe valves. The oil level in the storage tank is 15 ft above ground, and the pipeline discharge at a point 10 ft above ground, at a pressure of 10 psig.
    (a) What is the required flow capacity (in gpm) and the head (pressure) to be specified for the pump needed to move the oil?
    (b) If the pump is 85% efficient, what horsepower motor is required to drive it?
    (c) If a 4 in. diameter orifice is inserted in the line to measure the flow rate, what would the pressure drop reading across it be at the specified flow rate?

26. Water drains by gravity out of the bottom of a large tank, through a horizontal 1 cm ID tube, 5 m long, which has a venturi meter mounted in it. The level in the tank is 4 ft above the tube, and a single open vertical tube is attached to the throat of the venturi. What is the smallest diameter of the venturi throat for which no air will be sucked through the tube attached to the throat? What is the flow rate of the water under this condition?

27. Natural gas (CH_4) is flowing in a 6 in. sch 40 pipeline, at 50 psig and 80°F. A 3 in. diameter orifice is installed in the line, which indicates a pressure drop of 20 in. of water. What is the gas flow rate, in lb_m/hr and SCFM?

11

External Flows

I. DRAG COEFFICIENT

When a fluid flows past a solid body, or the body moves through the fluid, the force (F_D) exerted on the body by the fluid can be expressed in terms of a *drag coefficient* (C_D) which is defined by the equation

$$F_D = \frac{C_D}{2} \rho V^2 A \qquad (11\text{-}1)$$

Here, ρ is the density of the fluid, V is the relative velocity between the fluid and the solid body, and A is an appropriate area of the body. For flow past a sphere, as illustrated in Fig. 11-1, this area is taken to be the cross-sectional area normal to the velocity vector V, e.g. $\pi d^2/4$. Note that the definition of the drag coefficient from Eqn (11-1) is analogous to that of the friction factor for flow in a conduit, i.e.

$$\tau_w = \frac{f}{2} \rho V^2 \qquad (11\text{-}2)$$

where τ_w is the force exerted by the moving fluid on the wall of the pipe per unit area. In this case, however, the area is the total contact area between the fluid and the wall instead of the cross sectional area normal to the flow direction.

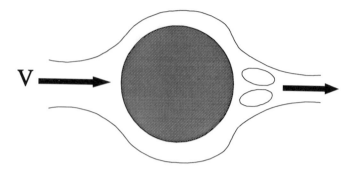

FIGURE 11-1   Drag on sphere.

A. STOKES FLOW

If the relative velocity is sufficiently low, the fluid streamlines can follow the surface of the body almost completely (this is called *creeping flow*). In this case, the microscopic momentum balance equations in spherical coordinates for the two-dimensional flow $[v_r(r,\theta),\ v_\theta(r,\theta)]$ of a Newtonian fluid can be solved for the local stress components. These equations can then be integrated over the surface of the sphere to determine the total drag force. This condition is referred to as *Stokes flow*, and the result is a theoretical expression for the drag coefficient:

$$C_D = \frac{24}{N_{Re}} \tag{11-3}$$

where

$$N_{Re} = \frac{dV\rho}{\mu} \tag{11-4}$$

Equation (11-3) has been found to hold exactly for flow over a sphere for $N_{Re} <$ 0.1 (or to within about 5% for $N_{Re} < 1$). Note the similarity between Eqn (11-3) and the dimensionless Hagen-Poiseuille equation for laminar tube flow, i.e. $f = 16/N_{Re}$.

B. FORM DRAG

As the velocity (hence N_{Re}) increases, it becomes more difficult for the streamlines to follow the back side of the sphere, and at some point they break away from the contour of the surface. This is called *separation*, although it is the smooth streamline that is separating from the surface, not the fluid itself. When separation occurs, eddies form behind the body, as illustrated in Fig. 11-1, which comprise the "wake". Separation is the result of flow into an adverse pressure gradient, which is inherently unstable. As the fluid passes around the front of the sphere it speeds up, since there is less flow area available for the fluid due to the space occupied by the sphere. On the upstream side of the body, the pressure decreases as the velocity increases as a result of the conservation of energy. Conversely, as the fluid passes along the surface on the downstream side of the sphere it must slow down, since the flow area increases, the velocity decreases, and consequently the pressure must increase. This is not unlike the flow in a diffuser or a converging-diverging duct. However, flow from a low-pressure region to a higher pressure region is unstable, and separation results.

As the velocity of the fluid (and N_{Re}) increases, the point of streamline separation behind the sphere moves further upstream, and the wake gets larger. Since the wake region contains circulating eddies of a three-dimensional turbulent nature, it is a region of relatively high velocity and hence low pressure. Since the pressure in the wake is lower than that on the front of the sphere, the product of this pressure difference and the projected area of the wake is a force acting on the sphere in the direction of the flow, i.e. in the same direction as the drag force. This additional force resulting from the low-pressure wake is called *form drag* and is in addition to the viscous forces represented by Stokes drag. The total drag is thus a combination of viscous (Stokes) drag and form (wake) drag, so the drag coefficient is greater than that given by Eqn (11-3) for $N_{Re} > 0.1$, by the additional form drag. This is illustrated in Fig. 11-2, which shows C_D vs N_{Re} for spheres as well as for cylinders and disks (oriented normal to the flow direction). For $1000 < N_{Re} < 1 \times 10^5$, $C_D = 0.45$ (approximately) for spheres. In this region the wake is maximum, and the streamline separation that forms the wake actually occurs slightly ahead of the equator of the sphere. The drag at this point is completely dominated by the wake, which is actually larger in diameter than the sphere (see Fig. 11-4).

C. ALL REYNOLDS NUMBERS

For $N_{Re} > 0.1$ (or > 1, within ~5%), a variety of expressions (largely empirical) have been proposed in the literature. However, a simple and very useful equation which represents the entire range of C_D vs N_{Re} (within experimental error) reasonably well up to about $N_{Re} = 2 \times 10^5$ has been given by Dallavalle (1948):

$$C_D = \left(0.632 + \frac{4.8}{\sqrt{N_{Re}}} \right)^2 \tag{11-5}$$

[Actually, according to Coulson et al. (1991), this equation was first presented by Wadell in 1934.] A comparison of Eqn (11-5) with measured values is shown in Fig. 11-2. A more accurate equation, although a bit more complex, has been developed by Khan and Richardson (1987):

$$C_D = \left[\frac{2.25}{N_{Re}^{0.31}} + 0.358 N_{Re}^{0.06} \right]^{3.45} \tag{11-6}$$

Although Eqn (11-6) is more accurate than Eqn (11-5) at intermediate values of N_{Re}, Eqn (11-5) provides a sufficiently accurate prediction for most applications. And, since it is simpler to manipulate, we will prefer it as an analytical expression for the sphere drag coefficient.

D. CYLINDER DRAG

For flow past a circular cylinder (normal to the cylinder axis) with $L/d \gg 1$, the drag situation is similar to that for the sphere. An equation which adequately represents the drag coefficient over the entire range of N_{Re} (up to about 2×10^5), which is analogous to the Dallavalle equation, is

$$C_D = \left(1.05 + \frac{1.9}{\sqrt{N_{Re}}} \right)^2 \tag{11-7}$$

A comparison of this equation with measured values is also shown in Fig. 11-2.

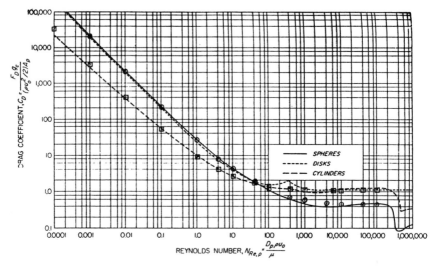

FIGURE 11-2 Drag coefficient for spheres, cylinders, and disks (Perry, 1984). ○ Eqn (11-5) for Spheres. □ Eqn (11-7) for Cylinders.

E. BOUNDARY LAYER EFFECTS

In Fig. 11-2, it is noted that the drag coefficient for the sphere exhibits a sudden drop from 0.45 to about 0.15 (almost 70%) at a Reynolds number of 2×10^5 to 5×10^5. For the cylinder, the drop is from about 1.1 to about 0.35. This drop is a consequence of boundary layer turbulence and can be explained as follows.

As the fluid encounters the solid boundary and proceeds along the surface, a boundary layer forms as illustrated in Fig. 11-3. The boundary layer is the region of the fluid near a boundary in which viscous forces dominate, and the velocity varies with the distance from the wall. Beyond the boundary layer, the fluid velocity is that of the free stream. Near the wall in the boundary layer, where viscous forces dominate, the flow is stable, the velocity is low, and the flow is laminar. However, the boundary layer thickness (δ) grows along the plate (in the x direction), in proportion to $N_{Rex}^{1/2}$, where $N_{Rex} = xV\rho/\mu$. As the boundary layer grows it loses stability until it reaches a point (at $N_{Rex} \approx 2 \times 10^5$) where it becomes unstable and hence turbulent. Within the turbulent boundary layer, the flow streamlines are no longer parallel to the boundary but break up into a three-dimensional eddy structure.

With regard to the flow over an immersed body (e.g. a sphere), the boundary layer grows from the impact (stagnation) point, along the front of the body, and remains laminar until $N_{Rex} \approx 2 \times 10^5$, at which point it becomes turbulent. If the boundary layer is laminar at the point where streamline separation occurs, the earliest separation point can be ahead of the equator of the sphere, resulting in a maximum wake diameter larger than that of the sphere. However, if the boundary layer becomes turbulent before separation occurs, the three-dimensional eddy structure of the boundary layer carries momentum components inward toward the surface, which delays the separation of the streamline. This

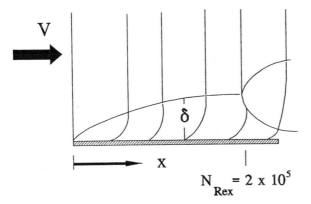

FIGURE 11-3 Boundary layer over a flat plate.

delayed separation results in a smaller wake and a corresponding reduction in form drag, which is the cause of the sudden drop in C_D at $N_{Re} \approx 2 \times 10^5$. This shift in the size of the wake can be rather dramatic, as illustrated in Fig. 11-4, which shows two pictures of a bowling ball falling in water with the wake clearly visible. The bowling ball on the left shows a large wake due to separation from the laminar boundary layer ahead of the equator. The ball on the right has a rougher surface, which promotes turbulence, so that the boundary layer has become turbulent before separation occurs, with a wake that is much smaller due to the delayed separation. Thus, the primary effect of surface roughness on the flow around immersed objects is to delay separation of the stream-lines and thus slightly alter the value of N_{Re} at which the sudden drop (or "kink") in the C_D-N_{Re} curve occurs. This apparent paradox—wherein the promotion of turbulence actually results in lower drag—has been exploited in various ways, such as the dimples on golf balls and the "boundary layer spoilers" on airplane wings and automobiles.

II. FALLING PARTICLES

Many engineering operations involve the separation of solid particles from fluids, in which the motion of the particles is a result of a gravitational (or other potential force) field. To illustrate this, we consider a spherical solid particle

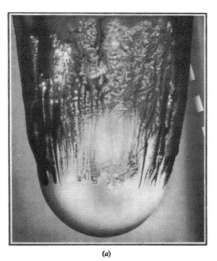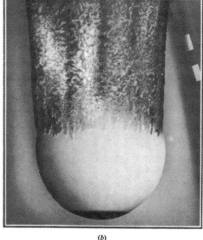

(a)   (b)

FIGURE 11-4   Two bowling balls falling in still water at 25 ft/s. The ball on the left is smooth, and the one on the right has a patch of sand on the nose (DeNevers, 1993).

with diameter d and density ρ_s, surrounded by a fluid of density ρ and viscosity μ, which is released and begins to fall (in the $x = -z$ direction) under the influence of gravity. A momentum balance on the particle is simply $\Sigma F_x = ma_x$, where the forces include gravity acting on the solid (F_g), the buoyant force due to the fluid (F_b), and the drag exerted by the fluid (F_D). The inertial term involves the product of the acceleration ($a_x = dv_x/dt$) and the mass (m). The mass includes that of the solid (m_s) as well as the "virtual mass" (m_f) of the fluid that is displaced by the body and is also accelerated as the body accelerates. It can be shown that the latter is equal to one-half of the mass of fluid displaced by the body, i.e. $m_f = \frac{1}{2}m_s\,(\rho/\rho_s)$. Thus the momentum balance becomes

$$\frac{g(\rho_s - \rho)\pi d^3}{6} - \frac{C_D\rho\pi d^2 V^2}{8} = \frac{\pi d^3(\rho_s + \rho/2)}{6}\frac{dV}{dt} \tag{11-8}$$

At $t = 0$, $V = 0$ and the drag force is zero. As the particle accelerates, the drag force increases, which decreases the acceleration. This process continues until the acceleration drops to zero, at which time the particle falls at a constant velocity resulting from the balance of forces due to drag and gravity. This steady-state velocity is called the *terminal velocity* of the body and is given by the solution of Eqn (11-8) with the acceleration equal to zero:

$$V_t = \left(\frac{4g\Delta\rho d}{3\rho C_D}\right)^{1/2} \tag{11-9}$$

where $\Delta\rho = \rho_s - \rho$. It is evident that the velocity cannot be determined until the drag coefficient is known. If Stokes flow prevails, then $C_D = 24/N_{Re}$, and this can be substituted into Eqn (11-9) to give

$$V_t = \frac{g\Delta\rho d^2}{18\mu} \tag{11-10}$$

However, the criterion for Stokes flow (i.e. $N_{Re} < 1$) cannot be tested until V_t is known, and if it is not valid then Eqn (11-10) will be incorrect. We will address this shortly.

There are several types of problems that we may encounter with falling particles, depending upon what is known and what is to be found. All of these problems involve the two primary dimensionless variables C_D and N_{Re}. The former is determined for gravitation-driven problems by Eqn (11-9), i.e.

$$C_D = \frac{4g\Delta\rho d}{3\rho V_t^2} \tag{11-11}$$

Now C_D and N_{Re} are related by the Dallavalle equation (Eqn 11-5) over the entire practical range of N_{Re}. The following procedures for the various types of problems apply to Newtonian fluids under all flow conditions.

A. UNKNOWN VELOCITY

In this case, the unknown velocity (V_t) appears in both the equation for C_D (Eqn 11-11) and N_{Re}. Hence, a suitable dimensionless group which does not contain the unknown V can be formulated as follows:

$$C_D N_{Re}^2 = \frac{4d^3 \rho g \Delta \rho}{3\mu^2} = \frac{4}{3} N_{Ar} \tag{11-12}$$

where N_{Ar} is the Archimedes number (also sometimes called the Galileo number). The most appropriate set of dimensionless variables to use for this type of problem is thus N_{Ar} and N_{Re}. An equation for N_{Re} can be obtained by multiplying Eqn (11-5) by N_{Re}^2 and solving the result for N_{Re} to give

$$N_{Re} = [(14.42 + 1.827 \sqrt{N_{Ar}})^{1/2} - 3.798]^2 \tag{11-13}$$

The procedure for determining the unknown velocity is therefore as follows:

  Given: d, ρ, ρ_s, μ *Find*: V_t

  (1) Calculate the value of the Archimedes number:

$$N_{Ar} = \frac{d^3 \rho g \Delta \rho}{\mu^2} \tag{11-14}$$

  (2) Insert this value into Eqn (11-13) and calculate N_{Re}.
  (3) Determine V_t from N_{Re}: $V_t = N_{Re} \, \mu/d \, \rho$.

Note that if the value of $N_{Ar} < 15$, then the system is within the Stokes' law range, and the terminal velocity is given by Eqn (11-10).

B. UNKNOWN DIAMETER

It often happens that we know, or can measure, the particle velocity and wish to know the size of the falling particle. In this case, we may formulate a dimensionless group that does not contain d as follows:

$$\frac{C_D}{N_{Re}} = \frac{4\mu \Delta \rho g}{3\rho^2 V_t^3} \tag{11-15}$$

This group can be related to the Reynolds number by dividing Eqn (11-5) by N_{Re}, and then solving the resulting equation for $1/(N_{Re})^{1/2}$ to give

$$\frac{1}{\sqrt{N_{Re}}} = \left(0.00433 + 0.208 \sqrt{\frac{C_D}{N_{Re}}} \right)^{1/2} - 0.0658 \tag{11-16}$$

The two appropriate dimensionless variables are thus C_D/N_{Re} and N_{Re}. The procedure is as follows:

Given: V_t, ρ_s, ρ, μ *Find*: d

(1) Calculate C_D/N_{Re} from Eqn (11-15).
(2) Insert the result into Eqn (11-16) and calculate $1/(N_{Re})^{1/2}$ and hence N_{Re}.
(3) Calculate $d = \mu N_{Re}/V_t\rho$.

If the value of $C_D/N_{Re} > 30$, then the system is within the Stokes' law range, and the diameter can be calculated directly from Eqn (11-10):

$$d = \left(\frac{18\mu V_t}{g\Delta\rho}\right)^{1/2} \tag{11-17}$$

C. UNKNOWN VISCOSITY

The viscosity of a Newtonian fluid can be determined by measuring the terminal velocity of a sphere of known diameter and density if the fluid density is known. If the Reynolds number is low enough ($N_{Re} < 0.1$) for Stokes flow to apply, then the viscosity can be determined directly by rearrangement of Eqn (11-10):

$$\mu = \frac{d^2 g\Delta\rho}{18V_t} \tag{11-18}$$

However, the Stokes flow criterion is rather stringent (e.g. a 1 mm diameter sphere would have to fall at a rate of 1 mm/s or slower in a fluid with a viscosity of 10 cP and SG = 1 to be in the Stokes range, which means that the density of the solid would have to be within 2% of the density of the fluid!).

However, with only a slight loss in accuracy the Dallavalle equation can be used to extend the useful range of this measurement to a much higher Reynolds number, as follows. From known quantities, C_D can be calculated from Eqn (11-11). This can then be equated to the Dallavalle equation (Eqn 11-5), which is then solved for N_{Re}:

$$N_{Re} = \left(\frac{4.8}{C_D^{1/2} - 0.632}\right)^2 \tag{11-19}$$

The viscosity is then determined from the known N_{Re}:

$$\mu = \frac{dV_t\rho}{N_{Re}} \tag{11-20}$$

Note that when $N_{Re} > 1000$, $C_D = 0.45$ (constant). From Eqn (11-19), this gives $\mu = 0$! Although this may seem strange, it is consistent, because in this range the drag is dominated by form (wake) drag, and viscous forces are negligible. However, it should be evident that one cannot extract the viscosity from measurements made under conditions that are insensitive to viscosity, which means that the utility of Eqn (11-19) is limited in practice to approximately $N_{Re} < 300$.

III. CORRECTION FACTORS

A. WALL EFFECTS

All expressions so far have assumed that the particles are surrounded by an infinite sea of fluid, i.e. that the boundaries of the fluid container are far enough from the particle that their influence is negligible. For a falling particle, this might seem to be a reasonable assumption if $d/D < 0.01$ (say), where D is the container diameter. However, the presence of the wall is felt by the particle over a much greater distance than one might expect. This is because as the particle falls, it must displace an equal volume of fluid. This displaced fluid must flow back around the particle to fill the space just vacated by the particle, so the effective "free-stream" (relative) velocity is no longer zero, as it would be for an infinite stagnant fluid. A variety of analyses of this problem have been performed, as reviewed by Chhabra (1992), which result in a wall correction factor (K_w), used to multiply the "infinite fluid" terminal velocity to correct for the wall effect. (This is also equivalent to correcting the Stokes' law drag force by a factor of K_w.) The following equation due to Francis (e.g. Chhabra, 1993) is claimed to be valid for $d/D < 0.97$ and $N_{Re} < 1$:

$$K_{w_0} = \left(\frac{1 - (d/D)}{1 - 0.475 \ d/D} \right)^4 \tag{11-21}$$

For larger Reynolds numbers, the following expression is claimed to be valid for $d/D < 0.8$ and $N_{Re} > 1000$:

$$K_{w_\infty} = 1 - (d/D)^{1.5} \tag{11-22}$$

Although these wall correction factors appear to be independent of Reynolds number for small (Stokes) and large (> 1000) values of N_{Re}, the value of K_w is a function of both N_{Re} and d/D for intermediate Reynolds numbers (Chhabra, 1992).

B. DROPS AND BUBBLES

Very small drops and bubbles are nearly rigid, due to surface tension forces, and behave in a manner similar to rigid particles. However, larger fluid drops or bubbles may experience a considerably different settling behavior. This is because the shear stress on the drop surface can be transmitted to the fluid inside the drop, which in turn results in circulation of the internal fluid. This internal circulation dissipates energy, which is extracted from the energy of the bubble motion, and is equivalent to an additional drag force. For Stokes flow around spherical drops or bubbles (e.g. $N_{Re} < 1$), it has been shown by Hadamard and Rybcznski (see e.g. Grace, 1983) that the drag coefficient can be corrected for this effect as follows:

$$C_d = \frac{24}{N_{Re}} \left(\frac{\kappa + 2/3}{\kappa + 1} \right) \tag{11-23}$$

where $\kappa = \mu_i/\mu_o$, μ_i being the viscosity of dispersed ("inside") fluid and μ_o the viscosity of the continuous ("outside") fluid.

For larger Reynolds numbers ($1 < N_{Re} < 500$), Rivkind and Ryskind (see Grace, 1983) have proposed the following equation for the drag coefficient for spherical drops and bubbles:

$$C_D = \frac{1}{\kappa + 1} \left[\kappa \left(\frac{24}{N_{Re}} + \frac{4}{N_{Re}^{1/3}} \right) + \frac{14.9}{N_{Re}^{0.78}} \right] \tag{11-24}$$

As the drop or bubble gets larger, however, it will become distorted due to the unbalanced forces around it. The viscous shear stresses tend to elongate the shape, whereas the pressure distribution tends to flatten it out normal to the flow direction. Thus the shape tends to progress from spherical to ellipsoidal to a "spherical cap" form as the size increases. Above a certain size, the deformation is so great that the drag force is approximately proportional to the volume, and the terminal velocity becomes nearly independent of size.

IV. NON-NEWTONIAN FLUIDS

The motion of solid particles, drops, or bubbles through non-Newtonian fluid media is encountered frequently and has been the subject of considerable research. A recent monograph by Chhabra (1992) summarizes much of this work, and the interested reader is referred to that work for details of the current state of knowledge. We will present some relations here that are applicable to purely viscous non-Newtonian fluids, although there is also much interest and activity in viscoelastic fluids. Despite the relative large amount of work that has been done in this area, there is still no general agreement as to the "right", or even the "best", description of the drag on a sphere in non-Newtonian fluids. This is due not only to the complexity of the equations which much be solved for the various models, but also to the difficulty in obtaining good, reliable, representative data for fluids with well-characterized, unambiguous, rheological properties.

A. POWER LAW FLUIDS

The usual approach for non-Newtonian fluids is to start with known results for Newtonian fluids and modify them to account for the non-Newtonian properties.

For example, the definition of the Reynolds number for a power law fluid can be obtained by replacing the viscosity in the Newtonian definition by an appropriate shear rate dependent viscosity function. If the characteristic shear rate for flow over a sphere is taken to be V/d, for example, the power law viscosity function becomes

$$\mu \rightarrow \eta(\dot\gamma) \cong m\left(\frac{V}{d}\right)^{n-1} \tag{11-25}$$

and the corresponding expression for the Reynolds number is

$$N_{RePL} = \frac{\rho V^{2-n}d^n}{m} \tag{11-26}$$

The creeping flow drag coefficient is characterized by a correction factor (Y) to the Stokes' law drag coefficient:

$$C_D = Y\,\frac{24}{N_{RePL}} \tag{11-27}$$

A variety of theoretical expressions, as well as experimental values, for the correction factor Y as a function of the power law flow index (n) are summarized by Chhabra (1992). Unfortunately, there is little agreement among either the theoretical or experimental values. One expression, due to Kawase and Moo-Young (Chhabra, 1992), that lies approximately in the mid-range of the various studies is

$$Y = 3^{(3n-3)/2}\left[\frac{-7n^2 - 4n + 26}{5n(n + 2)}\right] \tag{11-28}$$

For high Reynolds number flow, the evidence indicates that the C_D versus N_{RePL} relationship for power law fluids is the same as for Newtonian fluids (within $\pm 30\%$). This means that the Dallavalle equation (Eqn 11-5) can be used as a reasonable estimate for power law fluids by simply using Eqn (11-26) for the Reynolds number.

B. WALL EFFECTS

The wall effect for settling particles in non-Newtonian fluids appears to be significantly smaller than for Newtonian fluids. For power law fluids, the wall correction factor in creeping flow, as well as for very high Reynolds numbers, appears to be independent of Reynolds number. For creeping flow, the wall correction factor given by Chhabra (1992) is

$$K_{w_0} = 1 - 1.6(d/D) \tag{11-29}$$

whereas for high Reynolds numbers he gives

$$K_{w_\infty} = 1 - 3(d/D)^{3.5} \tag{11-30}$$

For intermediate Reynolds numbers, the wall factor depends upon Reynolds number as well as d/D. Over a range of $10^{-2} < N_{RePL} < 10^3$, $0 < d/D < 0.5$, $0.53 < n < 0.95$, the following equation describes the Reynolds number dependence of the wall factor quite well:

$$\frac{(1/K_w) - (1/K_{w_\infty})}{(1/K_{w_0}) - (1/K_{w_\infty})} = [1 + 1.3N_{RePL}^2]^{-0.33} \tag{11-31}$$

C. CARREAU FLUIDS

As discussed in Chapter 3, the Carreau viscosity model is one of the most general and reduces to many of the common two-parameter models (power law, Ellis, Sisko, Bingham, etc.) as special cases. The viscosity model can be written as

$$\eta = \eta_\infty + (\eta_0 - \eta_\infty)[1 + (\lambda\dot\gamma)^2]^{\frac{n-1}{2}} \tag{11-32}$$

where $n = 1 - 2p$, and p is the shear thinning parameter in the original form (Eqn 3-26) of this model. Since the shear conditions surrounding particles virtually never reach the levels corresponding to the high shear viscosity (η_∞), this parameter can be neglected and the parameters reduced to three: η_0, λ, and n. Chhabra and Uhlherr (Chhabra, 1992) have determined the Stokes flow correction factor for this model, which is a function of the dimensionless parameters n and $N_\lambda = \lambda v/d$. The following equation represents their results for the C_D correction factor over a wide range of data, to $\pm 10\%$, for $0.4 < n < 1$ and $0 < N_\lambda < 400$:

$$Y = \frac{1}{[1 + (0.275N_\lambda)^2]^{(1-n)/2}} \tag{11-33}$$

The value of η_0 is used for the viscosity in the Reynolds number.

D. BINGHAM PLASTICS

A particle will not fall through a fluid with a yield stress unless the weight of the particle is sufficient to exceed the yield stress. Because the stress is not uniform around the particle, and the distribution is very difficult to determine, it is not possible to determine the critical "yield" criterion exactly. However, it should be possible to characterize this state by a dimensionless "gravity-yield'' parameter, such as

$$Y_G = \frac{\tau_o}{gd\Delta\rho} \tag{11-34}$$

By equating the vertical component of the yield stress over the surface of the sphere to the weight of the particle, a critical value of $Y_G = 0.212$ can be determined (Chhabra, 1993). Experimentally, however, the results appear to fall into two groups: one for which $Y_G \approx 0.2$ and the other for which $Y_G \approx 0.04$–0.08. There seems to be no consensus as to the correct value, nor is there an adequate explanation of the differences.

With regard to the drag on a sphere which is moving in a Bingham plastic medium, the drag coefficient (C_D) must be a function of the Reynolds number and the Hedstrom number (or the Bingham number, $N_{Bi} = N_{He}/N_{Re} = d\tau_o/\mu V$). If the Stokes drag is considered to consist of a viscous component and a component due to the yield stress, one correlation (Chhabra, 1992) which agrees well with some data is

$$C_D - C_{D0} = 36 \frac{N_{He}}{N_{Re}^2} = 36 \frac{N_{Bi}}{N_{Re}} \tag{11-35}$$

where C_{D0} is the drag coefficient in a Newtonian fluid and the viscosity used in the Bingham plastic Reynolds number is μ_∞.

An alternative approach is to reconsider the Reynolds number from a different perspective, i.e.

$$\text{Newtonian: } N_{Re} = \frac{dV\rho}{\mu} = \frac{\rho V^2}{\mu \frac{V}{D}} \approx \frac{\rho V^2}{\tau} \tag{11-36}$$

$$\text{Bingham plastic: } \frac{\rho V^2}{\tau} \approx \frac{\rho V^2}{\mu_\infty \frac{V}{d} + \tau_o} = \frac{N_{Re}}{1 + N_{Bi}} = N_{ReBP} \tag{11-37}$$

That is, the Reynolds number can be considered a ratio of the momentum flux transported along the flow direction (inertial force) to that transmitted transverse to the flow direction (shear force). Ainsley and Smith (Chhabra, 1992) arrived at a similar group, based upon the slip line theory of soil mechanics:

$$N_{pl} = \frac{N_{Re}}{1 + 7\pi N_{Bi}/24} \tag{11-38}$$

where N_{pl} is called the *plasticity number*. This could be used in place of the Reynolds number in the C_D correlation for a Newtonian fluid to account for the properties of the Bingham plastic. However, instead of the equivalent Stokes relation $(24/N_{Re})$, a better correlation has been found by Chhabra (1992):

$$C_D = \frac{34}{N_{pl}} \quad \text{for} \quad N_{pl} < 20 \quad (11\text{-}39)$$

$$C_D = 0.4 \quad \text{for} \quad N_{pl} > 200 \quad (11\text{-}40)$$

These expressions can be used to represent the entire drag coefficient curve for a sphere in a Bingham plastic by an equivalent Dallavalle-type equation,

$$C_D = \left(0.2 + \frac{5.83}{\sqrt{N_{pl}}}\right)^2 \quad (11\text{-}41)$$

for a sphere in a Bingham plastic.

REFERENCES

Chhabra, R. P., *Bubbles, Drops, and Particles in Non-Newtonian Fluids*, CRC Press (1992)

Chhabra, R. P., and D. De Kee, *Transport Processes in Bubbles, Drops, and Particles*, Hemisphere Publishing Co. (1992)

Coulson, J. M., J. F. Richardson, J. R. Blackhurst, and J. H. Harker, *Chemical Engineering*, Vol. 2, 4th Ed., Pergamon Press (1991)

Dallavalle, J. M., *Micromeritics*, 2nd Ed., Pitman (1948)

DeNevers, N., *Fluid Mechanics*, 2nd Ed., McGraw-Hill (1993)

Grace, J. R., "Hydrodynamics of Liquid Drops in Immiscible Liquids", Ch. 38 in *Handbook of Fluids in Motion*, N. P. Cheremisinoff and R. Gupta, Eds., Ann Arbor Science (1983)

Khan, A. R., and J. F. Richardson, *Chem. Eng. Commun.*, 62, 135 (1987)

Perry, J. H. (Ed.), *Chemical Engineers' Handbook*, 6th Ed., McGraw-Hill (1984)

Wadell, H., *J. Franklin Inst.*, 217, 459 (1934)

PROBLEMS

1. The supports for a tall chimney must be designed to withstand a 120 mph wind. If the chimney is 10 ft in diameter and 40 ft high, what is the wind force on the chimney at this speed? T = 50°F.

2. If your pickup truck has a drag coefficient equivalent to a 5 ft diameter disk and the same projected frontal area, how much horsepower is required to overcome wind drag at 40 mph? What horsepower is required at 70 mph?

3. You take a tumble while water skiing. The handle attached to the tow rope falls beneath the water and remains perpendicular to the direction of the boat's heading. If the handle is 1 in. in diameter and 1 ft long, and the boat is moving at 20 mph, how much horsepower is required to pull the handle through the water?

4. Your new car is reported to have a drag coefficient of about 0.3. If the cross-sectional area of the car is 20 ft^2, how much horsepower is used to overcome wind resistance at 40 mph? 55 mph? 70 mph? 100 mph? (T = 70°F)

5. By careful streamlining, it is possible to reduce the drag coefficient of an automobile from 0.4 to 0.25. How much power would this save at 40 mph and at 60 mph, assuming the effective projected area of the car is 25 ft^2?

6. After blowing up a balloon, you release it without tying off the opening, and it flies out of your hand. If the diameter of the balloon is 6 in., the pressure inside it is 1 psig, and the opening is 1/2 in. in diameter, what is the balloon velocity? You may neglect friction in the escaping air and the weight of the balloon, and assume that an instantaneous steady state (i.e. a pseudo steady state) applies.

7. A speedboat is propelled by a water jet motor that takes water in at the bow through a 10 cm diameter duct and discharges it through an 80 mm diameter nozzle at a rate of 80 kg/s. Neglecting friction in the motor and internal ducts, and assuming that the drag coefficient for the boat hull is the same as for a 1 m diameter sphere, determine:
   (a) The static thrust developed by the motor when it is stationary.
   (b) The maximum velocity attainable by the boat.
   (c) The power (kW) required to drive the motor.
   (Assume seawater density 1030 kg/m^3, viscosity 1.2 cP.)

8. A mixture of titanium (SG = 4.5) and silica (SG = 2.65) particles, with diameters ranging from 50 to 300 μm, is dropped into a tank in which water is flowing upward. What is velocity of the water if all of the silica particles are carried out with the water?

9. A small sample of ground coal is introduced into the top of a column of water 30 cm high, and the time required for the particles to settle out is measured. If it takes 26 s for the first particle to reach the

bottom and 18 hr for all particles to settle, what is the range of particle sizes in the sample? (T = 60°F, SG_{coal} = 1.4)

10. You want to determine the viscosity of an oil which has a specific gravity of 0.9. To do this, you drop a spherical glass bead (SG = 2.7) with a diameter of 0.5 mm into a large vertical column of the oil and measure its settling velocity. If the measured velocity is 3.5 cm/s, what is the viscosity of the oil?

11. A solid particle with a diameter of 5 mm and SG = 1.5 is dropped into a liquid with a viscosity of 10 P and SG = 1. How long will it take for the particle to reach 99% of its terminal velocity?

12. A hot air popcorn popper operates by blowing air through the popping chamber, which carries the popped corn up through a duct and out of the popper, leaving the unpopped grains behind. The unpopped grains weigh 0.15 g, half of which is water, and have an equivalent spherical diameter of 4 mm. The popped corn loses half of the water to steam and has an equivalent diameter of 12 mm. What are the upper and lower limits of the air volumetric flow rate at 200°F over which the popper will operated properly, for a duct diameter of 8 cm?

13. You have a granular solid, with SG = 4, which has particle sizes of 300 μm and smaller. You want to separate out all of the particles with a diameter of 20 μm and smaller by pumping water upward through a slurry of the particles in a column with a diameter of 10 cm. What flow rate is required to ensure that all particles less than 20 μm are swept out of the top of the column? If the slurry is fed to the column through a vertical tube, what should its diameter be to ensure that none of the particles settle out in it?

14. You want to reproduce the experiment shown in the text, which illustrates the wake behind a sphere falling in water, at the point where the boundary layer undergoes transition to a turbulent boundary layer. If the sphere is made of steel with a density of 500 lb_m/ft^3, what should the diameter be?

15. You have a sample of crushed coal containing a range of particle sizes from 1 to 1000 μm in diameter. You wish to separate the particles according to size by entrainment, in which they are dropped into a vertical column of water that is flowing upward. If the water

velocity in the column is 3 cm/s, which particles will be swept out of the top of the column and which will settle to the bottom? (SG of the solid is 2.5)

16. A gravity settling chamber consists of a horizontal rectangular duct 6 m long, 3.6 m wide, and 3 m high. The chamber is used to trap sulfuric acid mist droplets entrained in an air stream. The droplets settle out as the air passes through the duct and may be assumed to behave as rigid spheres. If the air stream has a flow rate of 6.5 m^3/s, what is the diameter of the largest particle that will not be trapped in the duct? (ρ_{acid} = 1.74 g/cm^3; ρ_{air} = 0.01 g/cm^3; μ_{air} = 0.02 cP; μ_{acid} = 2 cP)

17. A small sample of a coal slurry containing particles with equivalent spherical diameters from 1 to 500 μm is introduced into the top of a water column 30 cm high. The particles that fall to the bottom are continuously collected and weighed to determine the particle size distribution in the slurry. If the solid SG is 1.4 and the water viscosity is 1 cP, over what time range must the data be obtained in order to collect and weigh all of the particles in the sample?

12

Fluid–Solid Separations by Free Settling

I. FLUID–SOLID SEPARATIONS

The separation of suspended solids from a carrier fluid is a requirement in many engineering operations. The most appropriate method for achieving this depends upon the specific properties of the system, the most important being the size and density of the solid particles and the solids concentration (e.g. the "solids loading") of the feed stream. For example, for relatively dilute systems (e.g. ~10% or less) of relatively large particles (e.g. ~100 m or more) of fairly dense solids, a gravity settling tank may be appropriate, whereas for more dilute systems of smaller and/or lighter particles, a centrifuge may be more appropriate. For very fine particles, or where a very high separation efficiency is required, a "barrier" system may be needed, such as a filter or membrane. For highly concentrated systems, a gravity thickener may be adequate or, for more stringent requirements, a filter may be needed.

In this chapter, we will consider relatively dilute systems, for which the effects of particle-particle interaction are relatively unimportant (e.g. gravity and centrifugal separation). Situations in which particle-particle interactions are negligible are referred to as *free settling*, as opposed to *hindered settling* in which such interactions are important. Figure 12-1 shows the *approximate* regions of solids concentration and density corresponding to free or hindered settling. In Chapter 14 we will consider systems that are controlled by hindered settling or interparticle interaction (e.g. filtration and sedimentation processes).

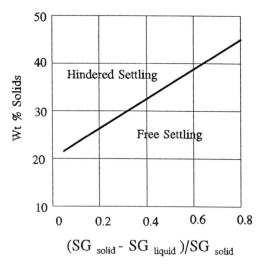

$$(SG_{solid} - SG_{liquid})/SG_{solid}$$

FIGURE 12-1   Regions of hindered and free settling.

II. GRAVITY SETTLING

Solid particles can be removed from a dilute suspension by passing the suspension through a vessel which is large enough that the vertical component of the fluid velocity is lower than the terminal velocity of the particles and the residence time is sufficiently long to allow the particles to settle out. A typical gravity settler is illustrated in Fig. 12-2. If the upward velocity of the liquid (Q/A) is less than the terminal velocity of the particles (V_t) they will settle to the bottom; otherwise, they will be carried out with the overflow. If Stokes flow is applicable (i.e. N_{Re} < 1), the diameter of the smallest particle that will settle out is

$$d = \left(\frac{18\mu Q}{g\Delta\rho A}\right)^{1/2} \tag{12-1}$$

If Stokes flow is not applicable (or even if it is), the Dallavalle equation in the form of Eqn (11-16) can be used to determine the Reynolds number, and hence the diameter, of the smallest setting particle:

$$\frac{1}{\sqrt{N_{Re}}} = \left(0.00433 + 0.208 \sqrt{\frac{C_D}{N_{Re}}}\right)^{1/2} - 0.0658 \tag{12-2}$$

where

$$\frac{C_D}{N_{Re}} = \frac{4\mu\Delta\rho g}{3\rho^2 V_t^3}, \qquad d = \frac{N_{Re}\mu A}{Q\rho} \tag{12-3}$$

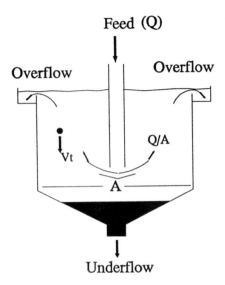

Feed (Q)

Overflow

Overflow

V_t

Q/A

A

Underflow

FIGURE 12-2 Gravity settling tank.

Alternatively, it may be desired to determine the maximum capacity (e.g. flow rate, Q) at which particles of a given size, d, will (or won't) settle out. This is obtained directly from the Dallavalle equation in the form of Eqn (11-13), by solving for the unknown flow rate:

$$Q = \left(\frac{\mu A}{d\rho}\right)[(14.42 + 1.827 \sqrt{N_{Ar}})^{1/2} - 3.798]^2 \tag{12-4}$$

where

$$N_{Ar} = \frac{d^3 \rho g \Delta \rho}{\mu^2} \tag{12-5}$$

III. CENTRIFUGAL SEPARATION

A. FLUID–SOLID SEPARATION

For very small particles or low-density solids, the terminal velocity may be too low to enable separation by gravity settling in a reasonably sized tank. However, the separation may be carried out in a centrifuge, which operates on the same principle as the gravity settler but employs the (radial) acceleration in a rotating system (e.g. $\omega^2 r$) in place of gravitational acceleration as the driving force. Centrifuges can be designed to operate at very high rotating speeds, which may be equivalent to many ''g's'' of acceleration.

A simplified diagram of a particle in a centrifuge is illustrated in Fig. 12-3. A radial force (momentum) balance on the particle can be written

$$F_{cf} - F_b - F_D = m_e \frac{dV_r}{dt} \tag{12-6}$$

where F_{cf} if the centrifugal force on the particle, F_b is the buoyant force (equal to the centrifugal force acting on the displaced fluid), F_D is the drag force, and m_e is the "effective" mass of the particle, which includes the solid particle and the "virtual mass" of the displaced fluid, which is half the actual mass of displaced fluid. Equation (12-6) thus becomes

$$(\rho_s - \rho)\left(\frac{\pi d^3}{6}\right)\omega^2 r - \rho V_r^2 C_D\left(\frac{\pi d^2}{8}\right) = \left(\rho_s + \frac{\rho}{2}\right)\left(\frac{\pi d^3}{6}\right)\frac{dV_r}{dt} \tag{12-7}$$

When the particle reaches its (radial) terminal velocity, $dV_r/dt = 0$ and Eqn (12-7) can be solved for V_{rt}, the radial terminal velocity:

$$V_{rt} = \left(\frac{4\Delta\rho d\omega^2 r}{3\rho C_d}\right)^{1/2} \tag{12-8}$$

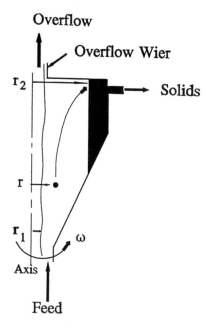

FIGURE 12-3   Particle in a centrifuge.

If $N_{Re} < 1$, Stokes' law holds and $C_D = 24/N_{Re}$, in which case Eqn (12-8) yields

$$V_{rt} = \frac{dr}{dt} = \frac{\Delta\rho d^2\omega^2 r}{18\mu} \qquad (12\text{-}9)$$

This shows that the terminal velocity is not a constant but increases with r, since the (centrifugal) driving force increases with r. Assuming that all of the fluid is rotating at the same speed as the centrifuge, integration of Eqn (12-9) gives

$$\ln(r_2/r_1) = \frac{\Delta\rho d^2\omega^2}{18\mu} t \qquad (12\text{-}10)$$

where t is the time required for the particle to travel a radial distance from r_1 to r_2. The time available for this to occur is the residence time of the particle in the centrifuge, i.e. $t = V/Q$, where V is the volume of fluid in the centrifuge. If the region occupied by the fluid is cylindrical, then $V = \pi(r_2^2 - r_1^2)L$. Conversely, if d is the smallest particle that will travel from the surface of the fluid (r_1) to the wall (r_2) in time t:

$$d = \left(\frac{18\mu Q \ln(r_2/r_1)}{\Delta\rho\omega^2 V}\right)^{1/2} \qquad (12\text{-}11)$$

Rearranging Eqn (12-11) to solve for Q gives

$$Q = \frac{\Delta\rho d^2\omega^2 V}{18\mu \ln(r_2/r_1)} = \left(\frac{\Delta\rho g d^2}{18\mu}\right)\left(\frac{V\omega^2}{g \ln(r_2/r_1)}\right), \qquad (12\text{-}12)$$

which can also be written

$$Q = V_t \sum, \qquad \sum = \left(\frac{V\omega^2}{g \ln(r_2/r_1)}\right) \qquad (12\text{-}13)$$

Here, V_t is the terminal velocity of the particle in a gravitational field and $\sum$ is the cross-sectional area of the gravity settling tank that would be required to remove the same size particles.

This analysis is based on the assumption that Stokes' law applies, i.e. $N_{Re} < 1$. This is frequently a bad assumption, since many industrial centrifuges operate under conditions where $N_{Re} > 1$. If such is the case, an analytical solution to the problem is still possible using the Dallavalle equation for C_D, rearranged to solve for N_{Re} as follows:

$$N_{Re} = \left(\frac{d\rho}{\mu}\right)\frac{dr}{dt} = \left[\left(14.42 + 1.827 \sqrt{N_{Ar}}\right)^{1/2} - 3.797\right]^2 \qquad (12\text{-}14)$$

where

$$N_{Ar} = \frac{d^2\rho\omega^2 r\Delta\rho}{\mu^2} \qquad (12\text{-}15)$$

Equation (12-14) can be integrated from r_1 to r_2 to give

$$t = N_{1-2} \frac{\mu}{d^2 \omega^2 \Delta \rho} \qquad (12\text{-}16)$$

where

$$N_{1-2} = 0.599(N_{Re2} - N_{Re1}) + 13.65(\sqrt{N_{Re2}} - \sqrt{N_{Re1}}) \qquad (12\text{-}17)$$
$$+ 17.29 \ln\left(\frac{N_{Re2}}{N_{Re1}}\right) + 48.34\left(\frac{1}{\sqrt{N_{Re2}}} - \frac{1}{\sqrt{N_{Re1}}}\right)$$

and the values of N_{Re2} and N_{Re1} are computed from Eqn (12-14) and the values of N_{Ar2} and N_{Ar1} at r_1 and r_2, respectively. Since $t = V/Q$, Eqn (12-16) can be rearranged to solve for Q:

$$Q = \frac{\Delta \rho d^2 \omega^2 V}{\mu N_{1-2}} = \frac{\Delta \rho d^2 \omega^2 V}{18 \mu \ln(r_2/r_1)} \left(18 \frac{\ln(r_2/r_1)}{N_{1-2}}\right) \qquad (12\text{-}18)$$

where the term in the large parentheses is a ''correction factor'' that can be applied to the Stokes flow solution to account for non-Stokes conditions.

For separating very fine solids, emulsions, and immiscible liquids, a disk-bowl centrifuge is frequently used, in which the settling occurs in the spaces between a stack of conical disks as illustrated in Fig. 12-4. The advantage of this arrangement is that the particles have a much smaller distance to travel before striking a wall and being trapped. The disadvantage is that the carrier fluid circulating between the disks has a higher velocity in the restricted spaces, which can retard the settling motion of the particles. Separation will occur only when $V_{rt} > V_{rf}$, where V_{rt} is the radial terminal velocity of the particle and V_{rf} is the radial velocity component of the carrier fluid in the region where the fluid flow is in the inward radial direction.

B. SEPARATION OF IMMISCIBLE LIQUIDS

The problem of separating immiscible liquids in a centrifuge can best be understood by first considering the static gravity separation of immiscible liquids, as illustrated in Fig. 12-5, where the subscript 1 represents the lighter liquid and 2 represents the heavier liquid. In a continuous system, the static head of the heavier liquid in the overflow pipe must be balanced by the combined head of the lighter and heavier liquids in the separator:

$$\rho_2 z g = \rho_2 z_2 g + \rho_1 z_1 g \qquad (12\text{-}19)$$

or

$$z = z_2 + z_1 \frac{\rho_1}{\rho_2} \qquad (12\text{-}20)$$

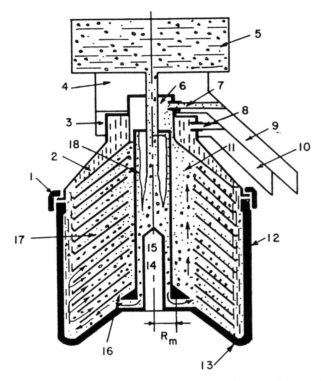

FIGURE 12-4 Schematic of disk-bowl centrifuge: (1) Ring; (2) bowl; (3,4) collectors for products of separation; (5) feed tank; (6) tube; (7,8) discharge nozzles; (9,10) funnels for collectors; (11) through channels; (12) bowl; (13) bottom; (14) thick-walled tube; (15) hole for guide; (16) disk fixator; (17) disks; (18) central tube (Azbel and Cheremisinoff, 1983).

In a centrifuge, the position of the overflow weir is similarly determined by the relative amounts of the heavier and lighter liquids and their densities, along with the size and speed of the centrifuge. The feed stream may consist of either the lighter liquid, 1, dispersed in the heavier liquid, 2, or vice versa. An illustration of the overflow weir positions is shown in Fig. 12-6. Since there is no slip at the interface between the liquids, the axial velocity must be the same at that point for both fluids, i.e.

$$V_1 = \frac{Q_1}{A_1} = \frac{Q_1}{\pi(r_i^2 - r_1^2)}$$

$$V_2 = \frac{Q_2}{A_2} = \frac{Q_2}{\pi(R^2 - r_i^2)}$$

(12-21)

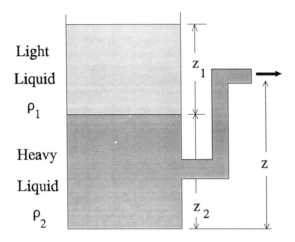

FIGURE 12-5 Gravity separation of immiscible liquids.

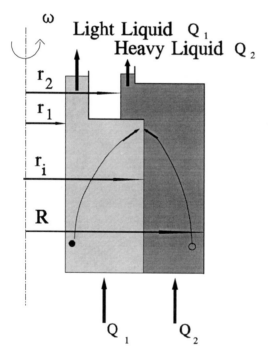

FIGURE 12-6 Centrifugal separation of immiscible liquids.

or

$$\frac{Q_1}{Q_2} = \frac{r_i^2 - r_1^2}{R^2 - r_i^2} \tag{12-22}$$

This provides a relation between the locations of the interface (r_i) and the inner weir (r_1) and the relative feed rates of the two liquids. Likewise, the residence time for each of the two liquids in the centrifuge is the same, i.e.

$$t = \frac{V_T}{Q_T} = \frac{\pi L(R^2 - r_i^2)}{Q_2 + Q_1} \tag{12-23}$$

For drops of the lighter liquid (1) dispersed in the heavier liquid (2), assuming that Stokes flow applies, the time required for the drops to travel from the maximum radius (R) to the interface (r_i) is

$$t = \frac{18\mu_2}{\Delta\rho d^2\omega^2} \ln\left(\frac{R}{r_i}\right) = \frac{\pi L(R^2 - r_i^2)}{Q_2 + Q_1} \tag{12-24}$$

For the case of drops of the heavier liquid dispersed in the lighter liquid, the corresponding time required for the maximum radial travel from the surface (r_1) to the interface (r_i) is

$$t = \frac{18\mu_1}{\Delta\rho d^2\omega^2} \ln\left(\frac{r_i}{r_1}\right) = \frac{\pi L(R^2 - r_i^2)}{Q_2 + Q_1} \tag{12-25}$$

Equations (12-22) and (12-24) or (12-25) determine the locations of the light liquid weir (r_1) and the interface (r_i) for given feed rates, centrifuge size, and operating conditions.

The proper location of the heavy liquid weir (r_2) can be determined by a balance of the radial pressure difference through the liquid layers, which is analogous to the gravity head balance in the gravity separator in Fig. 12-5. The radial pressure gradient due to centrifugal force is

$$\frac{dP}{dr} = \rho\omega^2 r \qquad \text{or} \qquad \Delta P = \frac{1}{2}\rho\omega^2\Delta r^2 \tag{12-26}$$

Since both the heavy liquid surface at r_2 and the light liquid surface at r_1 are at atmospheric pressure, the sum of the pressure differences from r_1 to R to r_2 must be zero:

$$\frac{1}{2}\rho_2\omega^2(r_2^2 - R^2) + \frac{1}{2}\rho_2\omega^2(R^2 - r_i^2) + \frac{1}{2}\rho_1\omega^2(r_i^2 - r_1^2) = 0 \tag{12-27}$$

which can be rearranged to give

$$\frac{\rho_2}{\rho_1} = \frac{r_i^2 - r_1^2}{r_i^2 - r_2^2} \tag{12-28}$$

Solving for r_2:

$$r_2^2 = \frac{\rho_1}{\rho_2}\left[r_i^2\left(\frac{\rho_2}{\rho_1} - 1\right) + r_1^2\right] \tag{12-29}$$

Equations (12-22), (12-24) or (12-25), and (12-29) thus determine the three design parameters r_i, r_1, and r_2. These equations can be arranged in the following dimensionless form. From Eqn (12-22):

$$\left(\frac{r_1}{R}\right)^2 = \left(\frac{r_i}{R}\right)^2\left(1 + \frac{Q_1}{Q_2}\right) - \frac{Q_1}{Q_2} \tag{12-30}$$

For drops of the light liquid in the heavy liquid, Eqn (12-24) becomes

$$\left(\frac{r_1}{R}\right)^2 = 1 - \frac{18\mu_2(Q_1 + Q_2)}{\pi LR^2 \Delta\rho d^2 \omega^2}\ln\left(\frac{R}{r_i}\right) \tag{12-31}$$

For drops of the heavy liquid in the light liquid, Eqn (12-25) becomes

$$\left(\frac{r_1}{R}\right)^2 = 1 - \frac{18\mu_1(Q_1 + Q_2)}{\pi LR^2 \Delta\rho d^2 \omega^2}\ln\left[\frac{r_i}{R}\left(\frac{R}{r_1}\right)\right] \tag{12-32}$$

Also, Eqn (12-29) is equivalent to

$$\left(\frac{r_2}{R}\right)^2 = \frac{\rho_1}{\rho_2}\left[\frac{r_i^2}{R^2}\left(\frac{\rho_2}{\rho_1} - 1\right) + \frac{r_1^2}{R^2}\right] \tag{12-33}$$

These three equations can be solved simultaneously (by iteration) for r_1/R, r_2/R, and r_i/R.

IV. CYCLONE SEPARATIONS

A. GENERAL CHARACTERISTICS

Centrifugal force can also be used to separate solid particles from fluids by inducing the fluid to undergo a rotating flow pattern in a stationary vessel (e.g. a cyclone), instead of rotating the equipment as is done in the centrifuge. Cyclones are widely used to remove small particles from gas streams (e.g. "aerocyclones") but can also be used to remove suspended solids from liquid streams (e.g. "hydrocyclones").

A typical cyclone is illustrated in Fig. 12-7. The suspension enters through a rectangular duct tangential to the cylindrical separator, which has a conical bottom. The flow generates a rotating vortex motion which imparts centrifugal force to the particles, which are thrown outward to the walls of the vessel where they fall by gravity to the conical bottom and are removed. The carrier fluid spirals inward to the cylindrical exit duct (also referred to as the "vortex

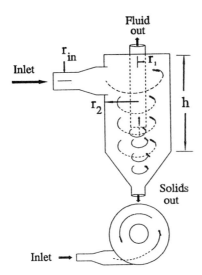

FIGURE 12-7 Cyclone separator.

finder''), where it leaves the vessel. The separation is not perfect, and some solid particles leave in the overflow as well as the underflow. The particle size for which 50% leaves in the overflow and 50% leaves in the underflow is called the *cut size*.

For hydrocyclones, the cyclone diameter can range from 10 mm to 2.5 m, cut sizes from 2 to 250 μm, and flow rate (capacities) from 0.1 to 7200 m³/hr. Pressure drop can range from 0.3 to 6 atm (Svarovsky, 1984). For aerocyclones, very little fluid leaves with the solids underflow, although for hydrocyclones the underflow solids content is typically 45–50% by volume, or less.

Advantages of the hydrocyclone include (Svarosky, 1984):

(1) Versatility—virtually any slurry or suspension can be concentrated, liquids degassed, or the solids classified by size, density, or shape.
(2) Simplicity and economy—they have no moving parts and little maintenance.
(3) Small—low residence times and relatively fast response.
(4) High shear forces—can break up agglomerates, etc.

The primary disadvantages are:

(1) Inflexibility—a given design is not easily adapted to a range of conditions. Performance is strongly dependent upon flow rate and feed composition, and the turndown ratio (i.e. range of operation) is small.

(2) Limited separation performance—in terms of the sharpness of the cut, range of cut size, etc.

(3) Susceptibility to erosion.

(4) High shear prevents utilization of flocculents to aid the separation, as can be done in gravity settlers.

An increase in any one operating parameter generally increases all others as well. For example, increasing the flow rate will increase both separation efficiency and pressure drop, and vice versa.

B. ANALYSIS OF OPERATION

Although the dominant velocity component in the cyclone is in the angular (tangential) direction, the swirling flow field includes significant velocity components in the radial and axial directions as well, which complicate the situation and make a rigorous analysis impossible. This complex flow field also results in significant particle-particle collisions, which cause some particles of a given size to be carried out in both the overhead and underflow discharge, thus affecting the separation efficiency.

There are a variety of approaches to the analysis of cyclone performance, as discussed by Svarovsky (1984). The one we present here is that of Bhonet (1983). In the centrifuge, it is assumed that the particles rapidly reach their terminal velocity in the radial direction (V_{tr}). If Stokes flow prevails, this is

$$V_{rt} = \frac{\omega^2 r d^2 \Delta\rho}{18\mu} \tag{12-34}$$

Since $\omega = V_\theta/r$, this is equivalent to

$$V_{rt} = \frac{d^2 g \Delta\rho}{18\mu}\left(\frac{V_\theta^2}{gr}\right) = V_t \frac{V_\theta^2}{gr} \tag{12-35}$$

where the terminal velocity in a gravitational field, V_t, is taken to be the reference velocity. The critical particle diameter corresponding to the "cut diameter" is designated d*, and the corresponding gravity settling velocity is V_t^*, as given by Stokes' law, i.e.

$$d_p^* = \left[\frac{18\mu V_t^*}{\Delta\rho g}\right]^{1/2} \tag{12-36}$$

It is assumed that for this size particle the radial (outward) centrifugal force just balances the radial drag resistance, so that all particles larger than this will be deposited on the cyclone wall and removed, and all smaller particles will be carried out of the cyclone. The problem is that it is not possible to rigorously determine the corresponding value of V_θ in Eqn (12-35) because of the flow complexity.

As has been noted, a radial pressure gradient is generated in a rotating flow field. If the contribution to the radial pressure gradient due to radial flow is neglected, this is

$$\frac{\partial P}{\partial r} = \rho \omega^2 r = \rho \frac{V_\theta^2}{r} \tag{12-37}$$

Measurements of the radial pressure distribution can be used with Eqn (12-37) to determine the variation of V_θ with r. The results show (Svarosky, 1984) that

$$V_\theta r^n = \text{constant} \tag{12-38}$$

where the empirical exponent n has been found to be between 0.5 and 0.9. (Note that if the fluid were undergoing a uniform rotation, i.e. $\overline{\omega}$ = const., then n = -1. For a free vortex in inviscid flow, n = 1, i.e. $\omega \sim 1/r^2$.) The maximum value of V_θ occurs in the vicinity of the outlet or exit duct (vortex finder) at $r = r_1$, so this is taken to be the reference point for the critical value of the tangential velocity, V_θ. It is this quantity which must be estimated in order to determine the cut diameter, as follows.

From continuity, assuming that the volumetric flow rate (Q) is the same through any cross section in the flow field, we have

$$Q = A_{in}V_{in} = A_{out}V_{out} = V_{rt}A_r \tag{12-39}$$

where

$$A_{in} = \pi r_{in}^2 \text{ (round)} \qquad \text{or} \qquad A_{in} = ba \text{ (rectangular)}$$

$$A_{out} = \pi r_1^2, \qquad A_r = 2\pi r_1 h$$

so

$$Q = V_{in}A_{in} = V_{rt}2\pi r_1 h \tag{12-40}$$

Using this to eliminate V_{rt} from Eqn (12-35) gives

$$V_t = V_t^* = V_{in}\frac{A_{in}}{2\pi r_1 h}\left(\frac{gr_1}{V_\theta^2}\right) \tag{12-41}$$

which can be written in the following form, since $A_{out} = \pi r_1^2$:

$$V_t^* = \left(\frac{V_\theta}{V_{in}}\right)^{-2}\left(\frac{h}{r_1}\right)^{-1}\left(\frac{A_{in}}{A_{out}}\right)\left(\frac{r_1 g}{2V_{in}}\right) \tag{12-42}$$

in which everything on the right is known except V_θ, the tangential velocity. According to Bohnet (1983), V_θ can be approximated by assuming that all friction losses occur at a specific point in the cyclone. A momentum balance incorprating this assumption leads to

$$\frac{V_\theta}{V_{in}} = \left[\alpha\, \frac{r_1}{r_{in}} + \frac{A_{out}}{A_{in}} \left(\frac{h}{r_1}\right) k_f \right]^{-1} \qquad (12\text{-}43)$$

Here, r_{in} is the radius of the inlet pipe, α is a correction factor for the contraction of the fluid stream at the inlet that depends on the geometry of the inlet duct, and k_f is a friction loss coefficient for the cyclone, for which Bohnet (1983) recommends a value of 0.02. The correction factor α is shown in Fig. 12-8 for two different fluid inlet geometries. For a given cyclone (i.e. known dimensions), the "cut diameter" can be determined from the above relations. The procedure is as follows:

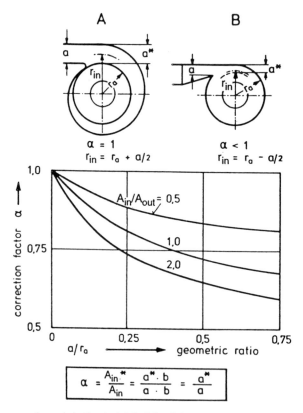

where b is the height of the inlet slit

Figure 12-8   Inlet correction factor α, for two different inlet geometries (Bhonet, 1983).

(1) Determine V_θ/V_{in} from Eqn (12-43).
(2) Insert this into Eqn (12-42) to give V_t^*.
(3) Calculate the cut diameter, d_p^*, from Eqn (12-36) knowing the properties of the fluid and solid.

C. SEPARATION EFFICIENCY

The efficiency of a cyclone is defined by the "fractional separation efficiency", η_F, which is the fraction of the particles in the feed that are separated into the underflow. A value of $\eta_F = 0.5$ for a certain particle diameter means that 50% of the particles of that size are removed. Ideally, the $\eta_F = 0.5$ cutoff point should occur at the critical particle diameter, given by Eqn (12-36), although this ideal situation does not always occur in practice. Actual fractional separation efficiency curves, as a function of d_p/d_p^*, for various aero- and hydrocyclones are shown in Fig. 12-9.

The fractional efficiency may be determined from the particle size distribution in the feed, the overflow, and the underflow. If F is the percent of solids in the feed above a specified size, F_o the corresponding value for the overflow, and F_u the corresponding value for the underflow, the separation efficiency for that particle size is given by

$$\eta_F = \frac{F - F_o}{F_u - F_o} \tag{12-44}$$

The total efficiency, η_T, for aerocyclones is given by

$$\eta_T = \frac{M_U}{M_F} \tag{12-45}$$

where M_F is the mass flow rate of the feed and M_U is the mass flow rate of the underflow. This assumes that the underflow is essentially all solids. For hydrocyclones, the underflow typically includes 50% or more of the fluid along with the solids, in which case the total efficiency must be modified as follows:

$$\eta_T \Rightarrow \frac{\eta_T - R}{1 - R} \tag{12-46}$$

where $R = M_U/M_F$ is the underflow-to-feed ratio.

Also, for hydrocyclones the interaction between particles can reduce the separation efficiency if the entering volumetric concentration of solids (c_v) exceeds about 10^{-4}. In such cases, the efficiency determined from the fractional separation curve must be diminished by multiplying by a factor C, as shown in Fig. 12-10 (Bohnet, 1983), where c_{vmax} is the highest solids volume concentration

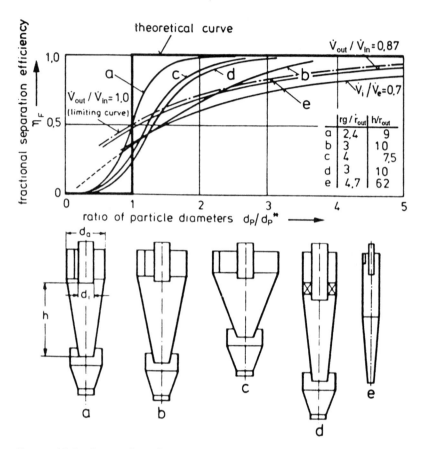

FIGURE 12-9   Separation efficiency for several aero- and hydrocyclones (Bhonet, 1983).

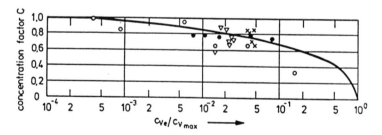

FIGURE 12-10   Concentration correction factor for separation efficiency (Bhonet, 1983).

at which the suspension still flows. Svarovsky (1984) gives a correction factor equal to $(1 - c)^{4.4}$, which is to be multiplied by V_t^* to account for hindered settling.

D. PRESSURE LOSS

The pressure loss in the cyclone occurs primarily in two regions: the inlet loss and friction within the cyclone and loss in the outlet pipe. Equations for these losses are given by Bhonet (1983):

$$\Delta P = (K_{fi} + K_{fo}) \frac{\rho V_\theta^2}{2} \tag{12-47}$$

where

$$K_{fi} = \frac{r_1}{r_a} \left\{ \left[1 + \frac{V_\theta}{V_{in}} \left(\frac{A_{out}}{A_{in}} \right) \left(\frac{h}{r_1} k_f \right) \right]^{-2} - 1 \right\} \tag{12-48}$$

and

$$K_{fo} = \left\{ K \left[\frac{V_\theta}{V_{in}} \left(\frac{A_{out}}{A_{in}} \right) \right]^{-2/3} + 1 \right\} \tag{12-49}$$

Here, k_f can be approximated as 0.02, and $K = 4.4$ for sharp-edged outlet pipes or $K = 3.4$ for rounded outlet pipes. The value of r_a is the (minimum) radius of the inside wall of the cyclone (see Fig. 12-8). Equation (12-49) is said to be valid only if $(V_\theta/V_{in})(A_{out}/A_{in}) \geq 1$. Otherwise, the following equation should be used for K_{fo}:

$$K_{fo} = \left\{ K \left[\frac{V_\theta}{V_{in}} \left(\frac{A_{out}}{A_{in}} \right) \right]^{-2/3} + 1 \right.$$
$$\left. + K_o \left[\frac{V_\theta}{V_{in}} \left(\frac{A_{out}}{A_{in}} \right) \right]^{-1} \left\{ \left[\frac{V_\theta}{V_{in}} \left(\frac{A_{out}}{A_{in}} \right) \right]^{-1} - 1 \right\} \right\} \tag{12-50}$$

where $K_o = 2.0$ for sharp-edged outlet pipes and $K_o = 1.1$ for rounded outlet pipes.

E. OPTIMAL DESIGN

The optimum cyclone not only separates particles above the critical particle size but also considers the pressure loss. The optimization of cyclone design can be done on the basis of either area or space (Bhonet, 1983). Bhonet (1983) gives a procedure for optimizing on the needed space that is based on a characteristic dimension defined as follows:

$$r_a^* = (r_a^2 h)^{1/3} \qquad (12\text{-}51)$$

which is used in the definition of a characteristic velocity

$$V^* = \frac{Q}{\pi r_a^{*2}} \qquad (12\text{-}52)$$

These parameters are used to define a dimensionless pressure loss coefficient:

$$\xi^* = \frac{\Delta P}{\rho \dfrac{V^{*2}}{2}} = (K_{fi} + K_{fo})\left[\frac{V_\theta}{V_{in}}\left(\frac{A_{out}}{A_{in}}\right)\right]^2 \left[\left(\frac{r_a}{r_1}\right)^2 \left(\frac{h}{r_1}\right)\right]^{4/3} \qquad (12\text{-}53)$$

Also, a dimensionless separation number, B^*, is defined as

$$B^* = \frac{V_t V^*}{2 r_a^* g} = \left[2\left(\frac{r_a}{r_1}\right)\left(\frac{h}{r_1}\right)\left(\frac{V_\theta}{V_{in}}\right)\left(\frac{A_{out}}{A_{in}}\right)\right]^{-2} \qquad (12\text{-}54)$$

A design diagram showing ξ^* versus B^* is presented in Fig. 12-11, where the numbers on the plot refer to the various cyclone designs shown in Fig. 12-12. The optimum design corresponds to curve B on Fig. 12-11. For hydrodynamic cyclones, Fig. 12-11 applies for operation with a closed apex, i.e. zero under-flow, and $Q_{in} = Q_{out}$. According to Bohnet (1983), the length of the cyclones determined from Fig. 12-11 is relatively large for aerocyclones, most of which are shorter. He therefore recommends that the design be such as to not exceed curve C of Fig. 12-11.

F. ALTERNATIVE DESIGN METHODS

A different approach to the analysis of hydrocyclones has been presented by Svarovsky (1984). He deduced that the system can be described in terms of three dimensionless groups, in addition to dimensionless geometric parameters. These groups are the Stokes number,

$$N_{Sk50} = \frac{V_{tr}}{V} = \frac{\Delta \rho d_{50}^2 Q}{36 \pi \mu r_2^3} \qquad (12\text{-}55)$$

the Euler number,

$$N_{Eu} = \frac{\Delta P}{\rho V^2/2} = \frac{2\pi^2 \Delta P r_2^4}{\rho Q^2} \qquad (12\text{-}56)$$

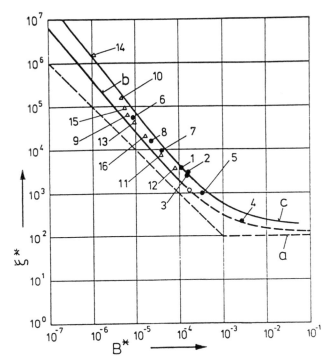

$\textsc{Figure}$ 12-11   Design diagram for aero- and hydrocyclones (Bhonet, 1983). Data points correspond to design numbers in Fig. 12-12.

and the Reynolds number,

$$N_{Re} = \frac{DV\rho}{\mu} = \frac{4Q\rho}{\pi D\mu} \tag{12-57}$$

In each of these groups, the characteristic length is the cyclone diameter, $D = 2r_2$, and the characteristic velocity is $V = 4Q\rho/\pi D\mu$. The relationship between these groups was found to be

$$N_{Sk50}N_{Eu} = C \tag{12-58}$$

and

$$N_{Eu} = K_p N_{Re}^n \tag{12-59}$$

The quantities C, K_p, and n are empirical constants, with the same values for a given family of geometrically similar cyclones. The values of $N_{Sk50}N_{Eu}$ range from 0.06 to 0.33, the exponent n varies from zero to 0.8, and K_p ranges from

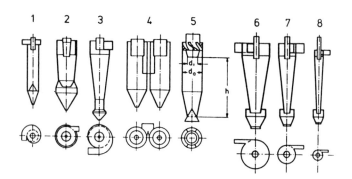

aerocyclone		1	2	3	4	5*	6	7	8
geometric dimensions	r_i m	0,450	0,770	0,260	0,635	0,045	0,050	0,050	0,050
	r_a m	1,12	2,00	0,588	1,00	0,075	0,250	0,150	0,150
	h m	7,35	6,70	2,41	3,40	0,650	0,900	0,900	1,80
	r_a/r_1 –	2,49	2,60	2,26	1,58	1,67	5,0	3,0	3,0
	r_{in}/r_1 –	3,73	1,95	3,08	1,14	1,44	4,48	2,48	2,48
	h/r_i –	16,3	8,70	9,27	5,35	14,4	18,0	18,0	36,0
	A_{in}/A_{out} –	1,97	1,10	1,00	1,25	0,852	1,00	1,00	1,00
	a/r_a –	1,00	0,482	0,723	0,553	-	0,205	0,341	0,341
	α –	1,00	0,720	1,00	0,700	1,00	0,840	0,790	0,790
	k_f –	0,02	0,02	0,02	0,02	0,02	0,02	0,02	0,02
velocity ratio (Eqn 12-43)	V_o/V_{in} –	2,30	1,89	1,96	1,43	0,94	1,83	1,47	0,963
vortex finder coefficients	K –	4,4	4,4	3,4	4,4	4,4	3,4	3,4	3,4
pressure loss coefficient	K_0 –	-	-	-	-	-	-	-	1,1
(Eqn 12-47) $(K_{fi} + K_{fo})$ –		5,61	4,50	3,83	5,22	5,76	4,78	4,81	7,74
volume based pressure loss coefficient (Eqn 12-53)	ξ^*	$3,62 \cdot 10^3$	$3,03 \cdot 10^3$	$2,51 \cdot 10^3$	$2,15 \cdot 10^2$	$1,03 \cdot 10^3$	$5,53 \cdot 10^4$	$9,16 \cdot 10^3$	$1,60 \cdot 10^4$
volume based separation number (Eqn 12-54)	B^*	$1,11 \cdot 10^4$	$1,65 \cdot 10^4$	$1,48 \cdot 10^4$	$2,69 \cdot 10^3$	$3,33 \cdot 10^4$	$9,22 \cdot 10^6$	$3,97 \cdot 10^5$	$2,31 \cdot 10^5$

hydrocyclone		9	10	11	12	13	14	15	16
geometric dimensions	r_i m	0,01	0,01	0,01	0,01	0,01	0,00255	0,0053	0,0075
	r_a m	0,05	0,05	0,025	0,025	0,05	0,025	0,025	0,025
	h m	0,33	1,0	0,33	0,15	0,33	0,332	0,329	0,304
	r_a/r_i –	5,0	5,0	2,5	2,5	5,0	9,81	4,71	3,33
	r_{in}/r_i –	4,49	4,49	1,99	1,99	4,28	9,29	4,25	2,79
	h/r_i –	33	100	33	15	33	130	62	40,5
	A_{in}/A_{out} –	1	1	1	1	2	1,02	0,93	1,05
	a/r_a –	0,2	0,2	0,4	0,4	0,289	0,104	0,200	0,328
	α –	0,855	0,855	0,766	0,766	0,731	0,909	0,864	0,786
	k_f –	0,02	0,02	0,02	0,02	0,02	0,02	0,02	0,02
velocity ratio (Eqn 12-43)	V_o/V_{in} –	1,18	0,457	0,957	1,46	2,00	0,378	0,651	0,950
vortex finder coefficients	K –	3,4	3,4	3,4	3,4	3,4	3,4	3,4	3,4
	K_0 –	-	1,1	1,1	-	-	1,1	1,1	1,1
pressure loss coefficient	$(K_{fi} + K_{fo})$ –	5,42	21,26	5,79	3,81	4,73	36,81	11,14	6,52
volume based pressure loss coefficient	ξ^* –	$5,84 \cdot 10^4$	$1,51 \cdot 10^5$	$6,46 \cdot 10^3$	$3,46 \cdot 10^3$	$3,66 \cdot 10^4$	$1,47 \cdot 10^6$	$8,35 \cdot 10^4$	$1,83 \cdot 10^4$
volume based separation number	B^* –	$6,59 \cdot 10^6$	$4,79 \cdot 10^6$	$4,01 \cdot 10^5$	$8,34 \cdot 10^5$	$9,18 \cdot 10^6$	$1,12 \cdot 10^6$	$5,98 \cdot 10^6$	$1,68 \cdot 10^5$

*blade angle $\beta = 31°$

FIGURE 12-12 Dimensions and flow values for various aero- and hydrocyclones (Bhonet, 1983).

TABLE 12-1 Operational Parameters for Some Known Hydrocyclone Designs

Cyclone type and size of hydrocyclone	Geometrical proportions					Scale-up constants			Running cost criterion
	D_I/D	D_o/D	l/D	L/D	Angle θ, degrees	$N_{Stk50} \cdot N_{Eu}$	K_p	n_p	$N_{Stk50}^{4/3} \cdot N_{Eu}$
Rietema's design (optimum separation), D = 0.075 m	0.28	0.34	0.4	5	20	0.0611	316	0.134	2.12
Bradley's design, D = 0.038 m	0.133 (1/7.5)	0.20 (1/5)	0.33 (1/3)	6.85	9	0.1111	446.5	0.323	2.17
Mozley cyclone, D = 0.022 m	0.154 (1/6.5)	0.214 (3/14)	0.57 (4/7)	7.43	6	0.1203	6381	0	3.20
Mozley cyclone, D = 0.044 m	0.160 (1/6.25)	0.25 (1/4)	0.57 (4/7)	7.71	6	0.1508	4451	0	4.88
Mozley cyclone, D = 0.044 m	0.197 (1/5)	0.32 (1/3)	0.57 (4/7)	7.71	6	0.2182	3441	0	8.70
Warman 3″ Model R, D = 0.076 m	0.29 (1/3.5)	0.20 (1/5)	0.31	4.0	15	0.1079	2.618	0.8	2.07
RW 2515 (AKW), D = 0.125 m	0.20 (1/5)	0.32 (1/3)	0.8	6.24	15	0.1642	2458	0	6.66
Hi-Klone model 2 D = 0.097 m	0.175	0.25	0.92 (0.59)*	5.6	10		873.5	0.2	
Hi-Klone model 3 D = 0.1125 m	0.15	0.20	0.80 (0.51)*	5.4	10		815.5	0.2	
Demco D = 0.051 m	0.217	0.50	1.0	4.7	25				
Demco D = 0.102 m	0.244	0.313	0.833	3.9	20				

*represents a modification; blank spaces in table are where the figures are not known.
Source: Svarovsky (1984).

2.6 to 6300. A summary of these parameters corresponding to some known hydrocyclone designs is given in Table 12-1 (from Svarovsky, 1984). The references in this table are found in Svarovsky's book, and the notation is as follows: $D = 2r_2$, $D_i = 2r_i$, $D_o = 2r_1$, l is the length of the vortex finder, and L is the total length of the hydrocyclone. These equations can be used to predict the performance of a given cyclone, for example, as follows. Equation (12-59) can be solved for the capacity, Q, to give

$$Q^{2+n} = \frac{2\pi^2 \Delta P r_1^4}{\rho K_p} \left(\frac{\pi r_1 \mu}{2\rho} \right)^n \qquad (12\text{-}60)$$

and the cut size obtained from Eqn (12-58):

$$d_{50}^2 = \frac{36\pi N_{Sk50} N_{Eu} r_1^3 \mu}{K_p Q \Delta \rho} \left(\frac{\pi r_1 \mu}{2Q\rho} \right)^n \qquad (12\text{-}61)$$

REFERENCES

Azbel, D. S., and N. P. Cheremisinoff, *Fluid Mechanics and Unit Operations*, Ann Arbor Science (1983)

Bohnet, M., ''Design Methods for Aerocyclones and Hydrocyclones'', Chapter 32 in *Handbook of Fluids in Motion*, N. P Cheremisinoff, and R. Gupta, Eds., Ann Arbor Science (1983)

Coulson, J. M., J. F. Richardson, J. R. Blackhurst, and J. H. Harker, *Chemical Engineering*, Vol. 2, 3rd Ed., Pergamon Press (1980)

McCabe, W. L., J. C. Smith, and P. Hariott, *Unit Operations of Chemical Engineering*, McGraw-Hill (1993)

Svarovsky, L., *Hydrocyclones*, Technomic Publ. Co. (1984)

PROBLEMS

FREE SETTLING FLUID-PARTICLE SEPARATIONS

1. A slurry containing solid particles having a density of 2.4 g/cm^3, ranging in diameter from 0.001 in. to 0.1 in., is fed to a setting tank 10 ft in diameter. Water is pumped into the tank at the bottom and overflows the top, carrying some of the solid with it. If it is desired to separate out all particles of diameter 0.02 in. and smaller, what flow rate of the water in gpm is required?

2. A handful of sand and gravel is dropped into a tank of water 5 ft deep. The time required for the solids to reach the bottom is measured and found to vary from 3 to 20 s. If the solid particles behave as equivalent spheres and have a specific gravity of 2.4, what is the range of equivalent particle diameters?

3. It is desired to determine the size of pulverized coal particles by measuring the time it takes them to fall a given distance in a known fluid. It is found that the coal particles (SG = 1.35) take a time ranging from 5 s to 1000 min to fall 23 cm through a column of methanol (SG = 0.785, μ = 0.88 cP). What is the size range of the particles in terms of their equivalent spherical diameters? Assume the particles are falling at their terminal velocities at all times.

4. A water slurry containing coal particles (SG = 1.35) is pumped into the bottom of a large tank (10 ft diameter, 6 ft high), at a rate of 500 gal/hr, and overflows the top. What is the largest coal particle that will be carried over in the overflow? If the flow rate is increased to 5000 gal/hr, what size particles would you expect in the overflow? The slurry properties can be taken to be the same as for water.

5. In order to determine the settling characteristics of a sediment, you drop a sample of the material into a column of water. You measure the time it takes for the solids to fall a distance of 2 ft and find that it ranges from 1 to 20 s. If the solid has SG = 2.5, what is the range of particle sizes in the sediment, in terms of the diameters of equivalent spheres?

6. You want to separate all the coal particles having a diameter of 100 μm or larger from a slurry. To do this, the slurry is pumped into the bottom of the large tank. It flows upward and overflows over the top of the tank, where it is collected in a trough. If the solid coal has SG = 1.4, and the total flow rate is 250 gpm, how big should the tank be?

7. A gravity settling chamber consists of a horizontal rectangular duct 6 m long, 3.6 m wide, 3 m high. The duct is used to trap sulfuric acid mist droplets entrained in an air stream. The droplets settle out as the air passes through the duct and may be assumed to behave as rigid spheres. If the air stream has a flow rate of 6.5 m^3/s, what is the diameter of the largest particle that will not be trapped by the duct? (Acid: ρ = 1.75 g/cm^3, μ = 3 cP; Air: = 0.01 g/cm^3, μ = 0.02 cP)

8. Solid particles of diameter 0.1 mm and density 20 g/cm^3 are to be separated from air in a settling chamber. If the air flow rate is 100 ft^3/s and the maximum height of the chamber is 4 ft, what should its minimum length and width be? (Air: ρ = 0.75 lb$_m$/ft^3, μ = 0.018 cP)

9. A sedimentation tank contains solid particles which have a wide range of sizes. Water is pumped into the tank from the bottom and overflows the top, at a rate of 10,000 gal/hr. If the tank diameter is 3 ft, what separation of particle size is achieved? (That is, what size particles are carried out the top of the tank, assuming the particles to be spherical?) Solid density = 150 lb_m/ft^3.

10. You want to use a viscous Newtonian fluid to transport small granite particles through a horizontal 1 in. I.D. pipeline, 100 ft long. The granite particles have a diameter of 1.5 mm and SG = 4.0. The specific gravity of the fluid can be assumed to be 0.95. The fluid should be pumped as fast as possible to minimize settling of the particles in the pipe but must be kept in laminar flow, so you design the system to operate at a pipe Reynolds number of 1000. The flow rate must be fast enough that the particles will not settle a distance greater than half the I.D. of the pipe from the entrance to the exit. What should the viscosity of the fluid be, and what should the flow rate be (in gpm) at which it is pumped through the pipe?

11. An aqueous slurry containing particles with the size distribution shown below is fed to a 20 ft diameter settling tank [see McCabe (1993) or Perry's Handbook for definition of mesh size]:

Tyler mesh size	% of Total solids in feed
8/10	5.0
10/14	12.0
14/20	26.0
20/28	32.0
28/35	21.0
35/48	4.0

    The feed enters near the center of the tank, and the liquid flows upward and overflows the top of the tank. The solids loading of the feed is 0.5 lb of solids per gallon of slurry, and the feed rate is 50,000 gpm. What is the total solids concentration and particle size distribution in the overflow? Density of solids is 100 lb_m/ft^3. Assume: (1) The particles are spherical; (2) the particles in the tank are unhindered; (3) the feed and overflow have the same properties as water.

12. A water stream contacts a bed of particles, which have diameters ranging from 1 to 1000 μm and SG = 2.5. The water stream flows

upward at a rate of 3 cm/s. What size particles will be carried out by the stream, and what size will be left behind?

13. An aqueous slurry containing particles with a range of sizes up to 300 μm flows upward through a small tube into a larger vertical chamber with a diameter of 10 cm. You want the liquid to carry all of the solids through the small tube, but you want only those particles with diameters less than 20 μm to be carried out the top of the larger chamber. What should the flow rate of the slurry be (in gpm), and what size should the smaller tube be?

14. A dilute aqueous $CaCO_3$ slurry is pumped into the bottom of a classifier, at a rate of 0.4 m^3/s, and overflows the top. The solid density is 2.71 g/cm^3.
    (a) What should the diameter of the classifier be if the overflow is to contain no particles larger than 0.2 mm in diameter?
    (b) The same slurry as in (a) is sent to a centrifuge, which operates at 5000 rpm. The centrifuge diameter is 20 cm, its length is 30 cm, and the liquid layer thickness is 20% of the centrifuge radius. What is the maximum flow rate that the centrifuge can handle and achieve the same separation as the classifier?

15. A centrifuge that is 40 cm I.D. and 30 cm long has an overflow weir that is 5 cm wide. The centrifuge operates at a speed of 3600 rpm.
    (a) What is the maximum capacity of the centrifuge (in gpm) for which particles with a diameter of 25 μm and SG = 1.4 can be separated from the suspension?
    (b) What would be the diameter of a settling tank that would do the same job?
    (c) If the centrifuge I.D. was 30 cm, how fast would it have to rotate to do the same job, everything else being equal?

16. Solid particles with a diameter of 10 μm and SG = 2.5 are to be removed from an aqueous suspension in a centrifuge. The centrifuge inner radius is 1 ft, the outer radius is 2 ft, and the length is 1 ft. If the required capacity of the centrifuge is 100 gpm, what should the operating speed (in rpm) be?

17. A centrifuge is used to remove solid particles with a diameter of 5 μm and SG = 1.25 from a dilute aqueous stream. The centrifuge rotates at 1200 rpm and is 3 ft high, the radial distance to the liquid surface is 10 in., and the radial distance to the wall is 14 in.

(a) Assuming that the particles must strike the centrifuge wall to be removed, what is the maximum capacity of this centrifuge, in gpm?

(b) What is the diameter, in feet, of the gravity settling tank that would be required to do the same job?

18. A dilute aqueous slurry containing solids with a diameter of 20 μm and SG = 1.5 is fed to a centrifuge rotating at 3000 rpm. The radius of the centrifuge is 18 in., its length is 24 in., and the overflow weir is 12 in. from the centerline.

(a) If all of the solids are to be removed in the centrifuge, what is the maximum capacity that it can handle (in gpm)?

(b) What is the diameter of the gravity settling tank that would be required for this separation, at the same flow rate?

19. A centrifuge with a radius of 2 ft and a length of 1 ft has an overflow weir located 1 ft from the centerline. If particles with diameters of 10 μm and less are to be removed from an aqueous suspension at a flow rate of 100 gpm, what should the operating speed of the centrifuge be (in rpm)?

20. A centrifuge with a diameter of 20 in. operates at a speed of 1800 rpm. If there is a water layer 3 in. thick on the centrifuge wall, what is the pressure exerted on the wall?

21. A vertical centrifuge, operating at 100 rpm, contains an aqueous suspension of solid particles with SG = 1.3 and radius 1 mm. When the particles are 10 cm from the axis of rotation, determine the direction in which they are moving relative to a horizontal plane.

13

Flow in Porous Media

I. DESCRIPTION OF POROUS MEDIA

By "porous medium" is meant a solid, or collection of solid bodies, with sufficient open space in or around the solids to enable a fluid to pass through or around them. There are various conceptual ways of describing a porous medium.

One concept is a continuous solid with holes in it, such as a brick or a region of sandstone. Such a medium is referred to as *consolidated*, and the holes may be unconnected ("closed cell" or impermeable) or connected ("open cell", or permeable). Another concept is a collection (or "pile") of solid particles in a packed bed, where the fluid can pass through the voids between the particles, which is referred to as *unconsolidated*. A schematic representation is shown in Fig. 13-1. Either of these concepts may be valid, depending upon the specific medium under consideration, and both have been used as the basis for developing the equations which describe fluid flow behavior. In practice, porous media may range from a "tight" oil-bearing rock formation to a packed column containing relatively large packing elements.

The "pile of solid particles" concept is useful for either consolidated or unconsolidated media as a basis for analyzing the flow process, since many consolidated media are actually made up of individual particles that are just stuck together (e.g. sandstone). One of the key properties of a porous medium

Consolidated Medium

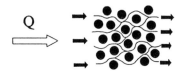

Unconsolidated Medium

FIGURE 13-1 Porous media. (a) Consolidated medium; (b) unconsolidated medium.

is the porosity ε, or void fraction, which is defined by

$$\varepsilon = \frac{\text{Total vol.} - \text{Vol. solids}}{\text{Total volume}} = 1 - \frac{A_{\text{solid}}}{A} = \frac{A_{\text{voids}}}{A} \tag{13-1}$$

where A_{Solid} is the area of the solid phase in a cross section of area A.

We also distinguish between the velocity of approach, or the "superficial" velocity of the fluid,

$$V_s = \frac{Q}{A} \tag{13-2}$$

and the "interstitial" velocity, which is the actual velocity within the pores or voids,

$$V_i = \frac{Q}{\varepsilon A} = \frac{V_s}{\varepsilon} \tag{13-3}$$

A. HYDRAULIC DIAMETER

Since the fluid in a porous medium follows a tortuous path through channels of varying size, one method of describing the flow behavior in the pores is to consider the flow path as a "noncircular conduit". This requires an appropriate definition of the hydraulic diameter, which is

$$D_h = 4\frac{A_i}{W_p} = 4\frac{A_iL}{W_pL} = 4\frac{\text{Flow volume}}{\text{Wetted surface area}}$$

$$= 4\frac{\varepsilon \times (\text{Bed volume})}{(\text{No. particles})\,(\text{Surface area/Particle})} \tag{13-4}$$

The medium, with overall dimensions AL, is assumed to be made up of a collection of individual particles and may be either consolidated or unconsolidated. The number of particles in the medium can be expressed as

$$\text{No. particles} = \frac{(\text{Bed volume})\,(\text{Fraction of solids in bed})}{\text{Volume/Particle}}$$

$$= \frac{(\text{Bed volume})\,(1 - \varepsilon)}{\text{Volume/Particle}} \tag{13-5}$$

Substitution of this into Eqn (13-4) leads to

$$D_h = 4\left(\frac{\varepsilon}{1 - \varepsilon}\right)\left(\frac{1}{a_s}\right) \tag{13-6}$$

where a_s = (particle surface area)/(particle volume). If the particles are spherical, with diameter d, then $a_s = 6/d$. Thus, for a medium composed of uniform spherical particles,

$$D_h = \frac{2d\varepsilon}{3(1 - \varepsilon)} \tag{13-7}$$

If the particles are not spherical, the parameter d may be replaced by

$$d = \psi d_s = \frac{6}{a_s} \tag{13-8}$$

where ψ is the *sphericity factor*, defined by

$$\psi = \frac{\text{Surface area of a sphere with same volume as the particle}}{\text{Surface area of the particle}} \tag{13-9}$$

and, d_s is the diameter of the sphere with the same volume as the particle.

B. POROUS MEDIUM FRICTION FACTOR

The expressions for the hydraulic diameter and the superficial velocity can be incorporated into the definition of the friction factor to give an equivalent expression for the porous medium friction factor, as follows:

$$f \equiv \frac{e_f}{(4L/D_h)(V_i^2/2)} = \frac{e_f d\varepsilon}{3L(1 - \varepsilon)V_i^2} = \frac{e_f d\varepsilon^3}{3L(1 - \varepsilon)V_s^2} \qquad (13\text{-}10)$$

Most references use Eqn (13-10) without the numerical factor of 3 as the definition of the porous medium friction factor, i.e.

$$f_{PM} \equiv \frac{e_f d\varepsilon^3}{L(1 - \varepsilon)V_s^2} \qquad (13\text{-}11)$$

C. POROUS MEDIUM REYNOLDS NUMBER

In a like fashion, the hydraulic diameter and the superficial velocity can be introduced into the definition of the Reynolds number to give

$$N_{Re} = \frac{D_h V_i \rho}{\mu} = \frac{2d\varepsilon V_i \rho}{3(1 - \varepsilon)\mu} = \frac{2dV_s \rho}{3(1 - \varepsilon)\mu} \qquad (13\text{-}12)$$

Here, again, the usual porous medium Reynolds number is defined by Eqn (13-12), without the numerical factor (2/3):

$$N_{RePM} = \frac{dV_s \rho}{(1 - \varepsilon)\mu} \qquad (13\text{-}13)$$

II. FRICTION LOSS IN POROUS MEDIA

A. LAMINAR FLOW

By analogy with laminar flow in a tube, the friction factor in laminar flow is

$$f = \frac{16}{N_{Re}} \quad \text{or} \quad f_{PM} = \frac{72}{N_{RePM}} \qquad (13\text{-}14)$$

However, this expression does not properly account for the deviation of the flow path from a straight line through the medium, nor does it account for the resistance due to extensional deformation as the fluid moves through channels of varying cross-sectional area. The latter results in a stress component normal to the flow direction, which is proportional to the product of the extension (stretch) rate and the extensional viscosity (which, for a Newtonian fluid, is three times the shear viscosity). If, for example, the flow path is assumed to deviate from a straight line by 45°, then the equation should be modified by substituting $\sqrt{2}L$ for L, and $\sqrt{2}V_i$ for V_i. The result is $f = 144/N_{RePM}$ instead of Eqn (12-14). In practice, a value of 150 to 180 instead of 144 is in closer agreement with observations at low Reynolds numbers, i.e.

$$f_{PM} = \frac{180}{N_{RePM}} \quad \text{for} \quad N_{RePM} < 10 \tag{13-15}$$

Equation (13-15) is known as the *Blake-Kozeny equation* and, as noted, applies for $N_{RePM} < 10$.

B. TURBULENT FLOW

At high Reynolds numbers (high turbulence levels), the flow is dominated by inertial forces and "wall roughness". The flow in a porous medium can be considered "extremely rough", with $\varepsilon/d \sim 1$. Thus, the flow at sufficiently high Reynolds numbers should be fully turbulent, and the friction factor should be constant. This has been confirmed by observations, with the value of the constant equal to approximately 1.75:

$$f_{PM} = 1.75 \quad \text{for} \quad N_{RePM} > 1000 \tag{13-16}$$

This is known as the *Burke-Plummer equation* and, as noted, applies for $N_{RePM} > 1000$.

C. ALL REYNOLDS NUMBERS

An expression that adequately represents the porous medium friction factor over all values of Reynolds number is

$$f_{PM} = 1.75 + \frac{180}{N_{RePM}} \tag{13-17}$$

This equation (with a value of 150 instead of 180) is called the *Ergun equation* and is simply the sum of Eqns (13-15) and (13-16). Obviously, for $N_{RePM} < 10$, the first term is very small relative to the second, and the Ergun equation reduces to the Blake-Kozeny equation. Likewise, for $N_{RePM} > 1000$ the first term is much larger than the second, and the equation reduces to the Burke-Plummer equation.

If the definitions of f_{PM} and N_{RePM} are inserted into the Ergun equation, the resulting expression for the frictional energy loss (dissipation) per unit mass of fluid in the medium is

$$e_f = 1.75 \frac{V_s^2}{d} \left(\frac{1 - \varepsilon}{\varepsilon^3}\right) L + 150 \frac{V_s \mu (1 - \varepsilon)^2 L}{d^2 \varepsilon^3 \rho} \tag{13-18}$$

III. PERMEABILITY

The "permeability" of a porous medium (K) is defined as the constant which relates the flow rate through the medium in proportion to the pressure drop and the cross-sectional area and inversely proportional to the fluid viscosity and length of the medium, i.e.

$$Q = K \frac{-\Delta PA}{\mu L} \tag{13-19}$$

Equation (13-19), which constitutes the definition of permeability, is known as *Darcy's Law*, and the most common unit for the permeability is the "darcy", which is defined as the flow rate, in cm^3/s, that results when a pressure drop of 1 atm is applied to a porous sample or region that is 1 cm^2 in cross-sectional area and 1 cm long, for a fluid with viscosity of 1 cP. It should be evident that the dimensions of the darcy are L^2, and the conversion factors are (approximately) 10^{-8} cm^2/darcy $\doteq 10^{-11}$ ft^2/darcy. The flow properties of crude-oil-bearing rock formations are often described in permeability units of millidarcies.

If the Blake-Kozeny equation for laminar flow through the medium is used to determine the friction loss, which is then equated to $\Delta P/\rho$ (from the Bernoulli equation), the resulting expression for the flow rate is

$$Q = \frac{-\Delta PA}{\mu L} \left(\frac{d^2 \varepsilon^3}{180(1 - \varepsilon)^2} \right) \tag{13-20}$$

It is evident that the permeability is identical to the term in brackets and is thus determined by the particle size and porosity of the medium. Since the permeability is defined by a laminar flow relation, it has no meaning under turbulent flow conditions.

IV. MULTIDIMENSIONAL FLOW

Flow in a porous medium in two or three dimensions is important in situations such as the production of crude oil from reservoir formations. Thus, it is of interest to consider this situation briefly and to point out some characteristics of the governing equations.

Consider the flow of an incompressible fluid through a two-dimensional porous medium, as illustrated in Fig. 13-2. Assuming that the kinetic energy is negligible and that the flow is laminar as characterized by Darcy's law, the Bernoulli equation becomes

$$\frac{\Delta P}{\rho} + g\Delta z = -e_f = \frac{\mu V_s L}{K\rho} \tag{13-21}$$

or

$$\Delta \left(\frac{\Phi}{\rho} \right) = -\frac{\mu V_s L}{K\rho} \tag{13-22}$$

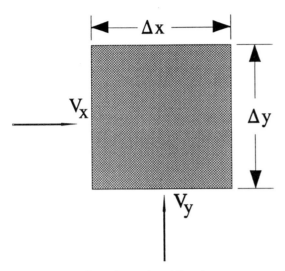

FIGURE 13-2   Two-dimensional flow in a porous medium.

where the density cancels out if the fluid is incompressible. Equation (13-22) can be applied in both the x and y directions, where L = Δx for the x direction, and L = Δy for the y direction, as follows:

$$\frac{\Delta\Phi}{\Delta x} = -\frac{\mu V_x}{K} = \frac{\partial\Phi}{\partial x} \tag{13-23}$$

and

$$\frac{\Delta\Phi}{\Delta y} = -\frac{\mu V_y}{K} = \frac{\partial\Phi}{\partial y} \tag{13-24}$$

If Eqn (13-23) is differentiated with respect to x, and Eqn (13-24) is differentiated with respect to y, and the results are added assuming μ and K to be constant, we get

$$\frac{\partial^2\Phi}{\partial x^2} + \frac{\partial^2\Phi}{\partial y^2} = -\frac{\mu}{K}\left(\frac{\partial V_x}{\partial y} + \frac{\partial V_y}{\partial y}\right) = 0 \tag{13-25}$$

The term in brackets is zero as a result of the conservation of mass (e.g. the microscopic continuity equation) for an incompressible fluid. Equation (13-25) can be generalized as

$$\nabla^2\Phi = 0 \tag{13-26}$$

which is known as the *Laplace equation*. This equation also governs the three-dimensional (potential) flow of an inviscid fluid. Note that the Laplace equation follows from Eqn (13-25) either for an incompressible viscous fluid, by virtue of the continuity equation, or for any flow with negligible viscosity effects (e.g. compressible flow outside the boundary layer, far from a solid boundary). It is interesting that the same equation governs both of these extreme cases.

The Laplace equation also applies to the distribution of electrical potential and current flow in a conducting medium, as well as the temperature distribution and heat flow in a thermally conducting medium. For example, if $\Phi \to E$, $V \to i$, and $\mu/K \to r_e$, Eqn (13-22) becomes *Ohm's law*, where r_e is the electrical resistivity ($r_e = RA/\Delta x$):

$$\frac{\partial E}{\partial x} = -r_e i_x \quad , \quad \nabla^2 E = 0 \quad , \quad \text{and} \quad \left(\frac{\partial i_x}{\partial x} + \frac{\partial i_y}{\partial y} \right) = 0 \qquad (13\text{-}27)$$

Also, with $\Phi \to T$, $V \to q$, and $K/\mu \to k$, where k is the thermal conductivity, the same equations govern the flow of heat in a thermally conducting medium (e.g. *Fourier's law*):

$$\frac{\partial T}{\partial X} = -\frac{1}{k} q_x \quad , \quad \nabla^2 T = 0 \quad , \quad \text{and} \quad \left(\frac{\partial q_x}{\partial x} + \frac{\partial q_y}{\partial y} \right) = 0 \qquad (13\text{-}28)$$

By making use of these analogies, electrical analog models can be constructed that can be used to determine the pressure and flow distribution in a porous medium from measurements of voltage and current distribution in a conducting medium, for example. The process becomes more complex, however, when the local permeability varies with position within the medium, which is often the case.

V. PACKED COLUMNS

At the other end of the spectrum from "porous rock" are unconsolidated media composed of beds of relatively large scale packing elements. These include a variety of shapes, such as rings, saddles, grids, and meshes, which are typically used to provide a large gas-liquid interface for promoting mass transfer in such operations as distillation or absorption or for liquid-liquid extraction. A typical application might be the removal of an impurity from a gas stream by selective absorption by a solvent in an absorption column filled with packing. The gas (or lighter liquid, in the case of liquid-liquid extraction) typically enters the bottom of the column, and the heavier liquid enters the top, the flow being countercurrent, as illustrated in Fig. 13-3.

For single-phase flow through packed beds, the pressure drop can generally be predicted adequately by the Ergun equation. However, because the flow is normally two-phase through packed columns, this situation is more complex. The effect of increasing the liquid mass flow rate (L) on the pressure drop through the column for a given gas mass flow rate (G), starting with dry packing,

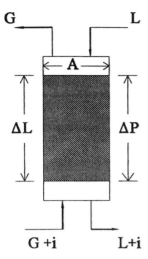

FIGURE 13-3 Schematic of packed column.

is illustrated in Fig. 13-4. The pressure drop for wet, drained, packing is higher than for dry packing, since the liquid occupies some of the void space between packing elements even in the "drained" condition. As the liquid flow rate increases, the liquid occupies an increasing portion of the void space, so that the area available to the gas is reduced and the total pressure drop is increased. As the liquid flow rate increases, the curve of ΔP vs G becomes increasingly non-

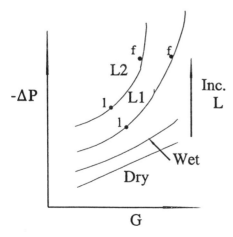

FIGURE 13-4 Effect of liquid rate on ΔP.

linear. The points labeled "l" in Fig. 13-4 are referred to as the "loading" points and indicate the point where the liquid flow begins to have a marked effect on the gas flow. At this point, the interaction between the gas and liquid is optimum, and thus this is the desired operation point for the column. The points labeled "f" on Fig. 13-4 are called the "flooding" points. At these points, the pressure drop through the column (e.g. in. of water per ft of column height) has reached a magnitude comparable to the static head of liquid in the column. At this point, the liquid will no longer drain through the packing by gravity, and the column is said to be "flooded". It is obviously undesirable to operate at or near the flooding point, since a slight increase in gas flow at this point will carry the liquid out of the top of the column.

The pressure drop through packed columns, and the flooding conditions, can be estimated from the generalized correlation of Leva (1992), shown in Fig. 13-5. The pressure drop in (in. of water)/(ft of packed height) is the parameter for the curves, and interpolation is usually necessary to determine the pressure

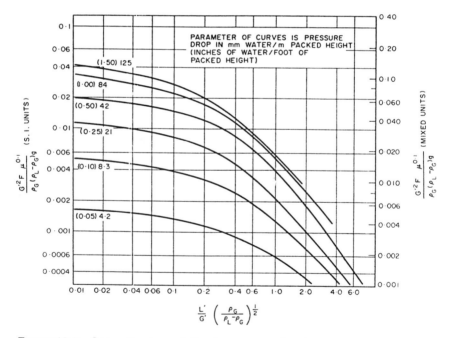

FIGURE 13-5 Generalized correlation for pressure drop in packed columns (Coulson et al., 1991). L' = liquid rate, lb/(s ft^2), kg/(m^{2}s); G' = gas rate, lb/(s ft^2); kg/(m^{2}s); ρ_L = liquid density, lb/ft^3; kg/m^3; ρ_G = gas density, lb/ft^3; kg/m^3; F = packing factor (see Table 13-1); μ = viscosity of liquid, centipoise (mNs/m^2); g = gravitational acceleration = 9 · 81 m/s^2 (32 · 2 ft/s^2).

drop (note that the pressure drop is not linearly proportional to the spacing between the curves). Correction factors for liquid density and viscosity, which are to be applied to the Y axis of this correlation are also shown. The parameter F in this correlation is called the *packing factor*. Values of F are given in Table 13-1, which shows the dimensions and physical properties of a variety of types of packing. Note that in Table 13-1 the term S_B is equal to $a_s(1 - \varepsilon)$, where a_s is the surface area per unit volume of the packing element. The packing factor F is equivalent to the term S_B/ε^3 in the definition of f_{PM} but is an empirical factor which characterizes the packing better than S_B/ε^3.

VI. FILTRATION

A common method of separating suspended solids from a liquid carrier is by filtration. The slurry is passed through a filter medium, usually a cloth or screen, which provides a support for the solid particles removed from the slurry. In actuality, the pores in the filter medium are frequently larger than the particles, which must penetrate some distance into the medium before being trapped. The layer of solids that builds up on the surface of the medium is called the *cake*, and it is the cake that provides the actual filtration. The pressure-flow charac- teristics of the porous cake mainly determine the performance of the filter.

A. GOVERNING EQUATIONS

A schematic of the flow through the cake and filter medium is shown in Fig. 13-6. The slurry flow rate is Q, and the total volume of filtrate that has passed through the filter is V. The flow through the cake and filter medium is inevitably laminar, so the resistance can be described by the permeability, K, and Darcy's law:

$$\frac{-\Delta P}{L} = \frac{Q\mu}{KA} \tag{13-29}$$

Applying this across the cake and filter medium in series:

$$P_1 - P_2 = \left(\frac{L}{K}\right)_{cake} \frac{Q\mu}{A} \tag{13-30}$$

$$P_2 - P_3 = \left(\frac{L}{K}\right)_{FM} \frac{Q\mu}{A} \tag{13-31}$$

The total pressure drop across the filter is the sum of these:

$$P_1 - P_3 = \frac{Q\mu}{A}\left[\left(\frac{L}{K}\right)_{cake} + \left(\frac{L}{K}\right)_{FM}\right] \tag{13-32}$$

TABLE 13-1 Design Data for Various Column Packings

	Size		Wall thickness		Number		Bed density		Contact surface S_B		Free space %	Packing factor	
	in.	mm	in.	mm	/ft³	/m³	lb/ft³	kg/m³	ft²/ft³	m²/m³	(100 ε)	ft²/ft³	m²/m³
Ceramic Raschig Rings	0.25	6.35	0.03	0.76	85,600	3,022,600	60	961	242	794	62	1600	5250
	0.38	9.65	0.05	1.27	24,700	872,175	61	977	157	575	67	1000	3280
	0.50	12.7	0.07	1.78	10,700	377,825	55	881	112	368	64	640	2100
	0.75	19.05	0.09	2.29	3090	109,110	50	801	73	240	72	255	840
	1.0	25.4	0.14	3.56	1350	47,670	42	673	58	190	71	160	525
	1.25	31.75			670	23,660	46	737			71	125	410
	1.5	38.1			387	13,665	43	689			73	95	310
	2.0	50.8	0.25	6.35	164	5790	41	657	29	95	74	65	210
	3.0	76.2			50	1765	35	561			78	36	120
Metal Raschig Rings	0.25	6.35	0.03125	0.794	88,000	3,107,345	133	2131			72	700	2300
	0.38	9.65	0.03125	0.794	27,000	953,390	94	1506			81	390	1280
	0.50	12.7	0.03125	0.794	11,400	402,540	75	1201	127	417	85	300	980
	0.75	19.05	0.03125	0.794	3340	117,940	52	833	84	276	89	185	605
(N.B. Bed densities	0.75	19.05	0.0625	1.59	3140	110,875	94	1506			80	230	750
are for mild steel;	1.0	25.0	0.03125	0.794	1430	50,494	39	625	63	207	92	115	375
multiply by 1.105,	1.0	25.0	0.0625	1.59	1310	46,260	71	1137			86	137	450
1.12, 1.37, 1.115 for	1.25	31.75	0.0625	1.59	725	25,600	62	993			87	110	360
stainless steel,	1.5	38.1	0.0625	1.59	400	14,124	49	785			90	83	270
copper, aluminum,	2.0	50.8	0.0625	1.59	168	5932	37	593	31	102	92	57	190
and monel	3.0	76.2	0.0625	1.59	51	1800	25	400	22	72	95	32	105
respectively)													
Carbon Raschig Rings	0.25	6.35	0.0625	1.59	85,000	3,001,410	46	737	212	696	55	1600	5250
	0.50	12.7	0.0625	1.59	10,600	374,290	27	433	114	374	74	410	1350
	0.75	19.05	0.125	3.175	3140	110,875	34	545	75	246	67	280	920
	1.0	25.0	0.125	3.175	1325	46,787	27	433	57	187	74	160	525
	1.25	31.75			678	23,940	31	496			69	125	410
	1.5	38.1			392	13,842	34	545			67	130	425

	2.0	50.8	0.250	6.35	166	5862	27	433	29	95	74	65	210
	3.0	76.2	0.312	7.92	49	1730	23	368	19	62	78	36	120
Metal Pall Rings	0.625	15.9	0.018	0.46	5950	210,098	37	593	104	341	93	70	230
(N.B. Bed densities	1.0	25.4	0.024	0.61	1400	49,435	30	481	64	210	94	48	160
are for mild steel)	1.25	31.75	0.030	0.76	375	13,240	24	385	39	128	95	28	92
	2.0	50.8	0.036	0.915	170	6003	22	353	31	102	96	20	66
	3.5	76.2	0.048	1.219	33	1165	17	273	20	65.6	97	16	52
Plastic Pall Rings	0.625	15.9	0.03	0.762	6050	213,630	7.0	112	104	341	87	97	320
(N.B. Bed densities	1.0	25.4	0.04	1.016	1440	50,848	5.5	88	63	207	90	52	170
are for	1.5	38.1	0.04	1.016	390	13,770	4.75	76	39	128	91	40	130
polypropylene)	2.0	50.8	0.06	1.524	180	6356	4.25	68	31	102	92	25	82
	3.5	88.9	0.06	1.524	33	1165	4.0	64	26	85	92	16	52
Ceramic Intalox	0.25	6.35			117,500	4,149,010	54	865			65	725	2400
Saddles	0.38	9.65			49,800	1,758,475	50	801			67	330	1080
	0.50	12.7			18,300	646,186	46	737			71	200	660
	0.75	19.05			5640	199,150	44	705			73	145	475
	1.0	25.4			2150	75,918	42	673			73	92	300
	1.5	38.1			675	23,835	39	625	59	194	76	52	170
	2.0	50.8			250	8828	38	609			76	40	130
	3.0	76.2			52	1836	36	577			79	22	72
Plastic Super Intalox	No. 1				1620	57,200	6.0	96	63	207	90	33	108
	No. 2				190	6710	3.75	60	33	108	93	21	210
	No. 3				42	1483	3.25	52	27	88.6	94	16	52
Intalox Metal		25			4770	168,425					96.7	41	135
		40			1420	50,140					97.3	25	82
		50			416	14,685					97.8	16	52
		70			131	4625					98.1	13	43
Hy-Pak	No. 1				850	30,014	19	304			96	43	140
(N.B. Bed densities	No. 2				107	3778	14	224			97	18	59
are for mild steel)	No. 3				31	1095	13	208			97	15	49
Plastic Cascade Mini	No. 1											25	82
Rings	No. 2											15	49
	No. 3											12	39

TABLE 13-1 Continued

	Size		Wall thickness		Number		Bed density		Contact surface S_B		Free space %	Packing factor	
	in.	mm	in.	mm	/ft³	/m³	lb/ft³	kg/m³	ft²/ft³	m²/m³	(100 ε)	ft²/ft³	m²/m³
Metal Cascade Mini Rings	No. 0											55	180
	No. 1											34	110
	No. 2											22	72
	No. 3											14	46
	No. 4											10	33
Ceramic Cascade Mini Rings	No. 2											38	125
	No. 3											24	79
	No. 5											18	59

Note: The packing factor F replaces the term S_B/ε^3. Use of the given value of F in Fig. 13-5 permits more predictable performance of designs incorporating packed beds since the values quoted are derived from operating characteristics of the packings rather than from their physical dimensions.
Source: Coulson et al. (1991).

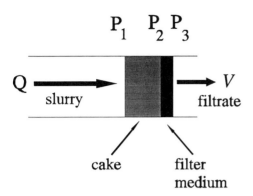

FIGURE 13-6 Schematic of flow through filter cake and medium.

The term $(L/K)_{cake}$ is the resistance of the cake, and $(L/K)_{FM}$ is the resistance of the filter medium. The latter is higher for a ''dirty'' filter medium than for a clean one, but once the initial particles become embedded in the medium and the cake starts to build up, it remains constant. The cake resistance, on the other hand, continues to increase with time as the cake thickness increases. The cake thickness is directly proportional to the volume of solids that has been deposited from the slurry, and inversely proportional to the area:

$$L_{cake} = \frac{V_{cake}}{A} = \frac{V_{solids}}{A(1 - \varepsilon)} = \frac{M_{solids}}{A\rho_s(1 - \varepsilon)} \qquad (13\text{-}33)$$

Now $M_{solids}/V)$ is the mass of solids per unit volume of liquid in the slurry feed (e.g. the ''solids loading''), where V is the volume of liquid (filtrate) that has passed to deposit M_{solids} on the cake. Thus, the cake thickness is given by

$$L_{cake} = \left(\frac{M_{solids}}{V}\right)\left(\frac{V}{A}\right)\left(\frac{1}{\rho_s(1 - \varepsilon)}\right) = W\frac{V}{A} \qquad (13\text{-}34)$$

where $W = (M_{solids}/V)/\rho_s(1 - \varepsilon)$ is a property of the specific slurry/cake. The density of the cake is given by

$$\rho_c = (1 - \varepsilon)\rho_s + \varepsilon\rho_{liq} \qquad (13\text{-}35)$$

Substituting Eqn (13-34) into Eqn (13-32) and rearranging results in the basic equation governing the filter performance:

$$\frac{Q}{A} = \frac{1}{A}\frac{dV}{dt} = \frac{P_1 - P_3}{\mu\left(\dfrac{VW}{AK} + a\right)} \qquad (13\text{-}36)$$

where a is the filter medium resistance, i.e. $a = (L/K)_{FM}$.

It should be evident that the operation of a filter is an unsteady cyclic process. As the cake builds up and its resistance increases with time, either the flow rate (Q) will drop or the pressure drop (ΔP) will increase with time. The specific behavior depends upon how the filter is operated, as follows.

B. CONSTANT-PRESSURE OPERATION

If the slurry is fed to the filter by a centrifugal pump, which delivers approximately a constant head, or if the filter is operated by drawing a vacuum downstream, the pressure drop will remain essentially constant during operation and the flow rate will drop as the cake thickness (resistance) increases. In this case, Eqn (13-36) can be integrated keeping the pressure constant to give

$$C_1 \left(\frac{V}{A}\right)^2 + C_2 \left(\frac{V}{A}\right) = (-\Delta P)t \tag{13-37}$$

where $C_1 = \mu W/2K$ and $C_2 = \mu a$, both being assumed to be independent of pressure (we will consider compressible cakes later). In Eqn (13-37), t is the time required to pass volume V of filtrate through the filter.

Since C_1 and C_2 are unique properties of a specific slurry/cake system, it is usually more appropriate to determine their values from laboratory tests using samples of the specific slurry and filter medium that are to be evaluated in the plant. For this purpose, it is more convenient to rearrange Eqn (13-37) in the following form:

$$\left(\frac{-\Delta Pt}{V/A}\right) = C_2 \left(\frac{V}{A}\right) + C_2 \tag{13-38}$$

If V is measured as a function of t in a lab experiment, for given values of ΔP and A, the data can be arranged in the form of Eqn (13-38). When the left-hand side is plotted versus V/A, the result should be a straight line with slope C_1 and intercept C_2 (which are easily determined by linear regression).

C. CONSTANT-FLOW OPERATION

If the slurry is fed to the filter by a positive displacement pump, the flow rate will be constant regardless of the pressure drop, which will then increase with time. In this case, noting that $V = Qt$, Eqn (13-36) can be rearranged to give

$$-\Delta P = 2C_2 \left(\frac{Q}{A}\right)^2 t + C_2 \left(\frac{Q}{A}\right) \tag{13-39}$$

This shows that, for given Q and A, the plot of ΔP versus t should be straight and the system constants C_1 and C_2 can be determined from the slope $2C_1(Q/A)^2$ and intercept $C_2(Q/A)$.

It is evident that the filter performance is governed by the system constants C_1 and C_2, regardless of whether the operation is at constant pressure or constant flow rate. Thus, these constants can be evaluated from laboratory data taken under either type of operation and used to analyze the performance of the plant filter for either type of operation.

D. CYCLE TIME

As has been mentioned, the operation of a filter is cyclic. The filtration process proceeds (and the pressure increases or the flow rate drops) until either the cake has built up to fill the space available for it or the pressure drop reaches the operational limit. At that point, the filtration must cease, and the cake is removed. There is often a wash cycle prior to removal of the cake, in order to remove the slurry carrier liquid from the pores of the cake using a clean liquid. The pressure-flow behavior during the wash period is a steady-state operation, controlled by the maximum cake and filter medium resistance, since no solids are deposited during this period. The cake can be removed by physically disassembling the filter and taking out the cake and the medium (as for a plate-and-frame filter), then reassembling the filter and starting the cycle over. Or, as in the case of a rotary drum filter, the cake removal is part of the rotating drum cycle, which is continuous although the filtration operation is still cyclic (this will be discussed below).

Thus, the t variable in the above equations is the actual time that the filtrate has been passing through the medium (t_{filter}) and is only part of the total time of the cycle (t_{cycle}). The rest of the cycle, which may include wash time, disassembly and assembly time, cleaning time, etc., we shall call "dead" time (t_{dead}):

$$t_{cycle} = t_{filter} + t_{dead} \qquad (13\text{-}40)$$

The filter capacity is determined by the amount of slurry processed during the total cycle time, not just the "filter" time, and is represented by the average flow rate ($\overline{Q}$):

$$\overline{Q} = \left(\frac{V_{cycle}}{t_{cycle}}\right)_{const\ \Delta P} = \left(\frac{Qt_{filter}}{t_{cycle}}\right)_{const\ Q} \qquad (13\text{-}41)$$

E. PLATE-AND-FRAME FILTERS

A plate-and-frame filter press consists of alternate solid plates and hollow frames in a "sandwich" arrangement. The open frames are covered by the filter medium (e.g. the filter cloth), and the slurry enters through the frames and deposits the cake on the filter medium. The operation is "batch", in that the filter must be disassembled when the cake fills the frame space, cleaned, and reassembled, after which the entire process is repeated. A schematic of a plate-and-frame

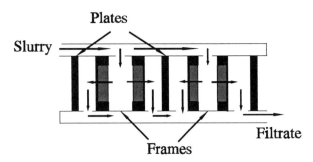

FIGURE 13-7 Plate-and-frame filter.

press is shown in Fig. 13-7. In the arrangement shown, all of the frames are in parallel, and the total filter area (which appears in the equations) is

$$A = 2nA_f \tag{13-42}$$

where n is the number of frames and A_f is the filter area of (one side) of the frame. The "dead time" in the cycle may include washing, disassembly, cleaning, and reassembly.

There are a variety of arrangements that operate in the same manner as the plate-and-frame filter, such as the "leaf filter". The leaf filter may consist of one or more "frames", covered by the filter medium, which are immersed in the slurry. They are often operated by a vacuum, which draws the filtrate through the filter, and the cake collects on the filter medium on the outside of the frame.

F. ROTARY DRUM FILTERS

The rotary drum filter is a "continuous" process, since it does not have to be shut down during part of the cycle. The operation is still cyclic, however. A schematic is shown in Fig. 13-8. The drum rotates at a rate N (rpm), and the total drum surface is the filter area, i.e. $A = \pi DL$. However, if the fraction of the drum that is in contact with the slurry is f, then the length of time in the cycle during which any one point on the surface is actually filtering is f/N. Thus:

$$t_{cycle} = \frac{1}{N} \quad , \quad t_{filt} = \frac{f}{N} \tag{13-43}$$

G. COMPRESSIBLE CAKE

The equations presented so far all assume that the cake is incompressible, i.e. the permeability and density are constant. For many cakes, this is not so, and

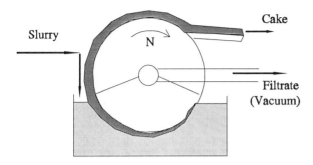

FIGURE 13-8 Rotary drum filter.

the cake properties vary with pressure (e.g. flocs, gels, fibers, pulp, etc.). For such cases, the basic filter equation, Eqn (13-36), can be expressed in the form

$$\frac{Q}{A} = \frac{(-\Delta P)^{1-s}}{\mu \left[\alpha \left(\frac{V}{A}\right)\left(\frac{M_s}{V}\right) + a \right]} = \frac{1}{A}\frac{dV}{dt} \tag{13-44}$$

where the parameters s and α represent the pressure-dependent properties of the cake. There are several modes of performance of the filter, depending upon the value of s, e.g.:

(1) If $s = 0$, then $Q \sim \Delta P$ (e.g. incompressible cake).
(2) If $s = 1$, then Q is independent of ΔP (compressible).
(3) If $s < 1$, then Q increases as ΔP increases (slightly compressible).
(4) If $s > 1$, then Q decreases as P increases (highly compressible).

In case 4, the increasing pressure compresses the cake to such an extent that it actually ''chokes off'' the flow, so that the higher the pressure, the lower the flow. This situation can be compensated for by adding a ''filter aid'' to the slurry, which is a dispersed solid that forms an incompressible cake (diatomaceous earth, sand, etc.). This will provide ''rigidity'' to the cake and enhance its permeability, thus increasing the filter capacity (it may seem like a paradox that adding more solids to the slurry feed actually increases the filter performance, but it works!).

The equations that apply for a compressible cake are as follows:
Constant Pressure Drop:

$$C_1 \left(\frac{V}{A}\right)^2 + C_2 \left(\frac{V}{A}\right) = t(-\Delta P)^{1-s} \tag{13-45}$$

Constant Flow Rate:

$$2C_1\left(\frac{Q}{A}\right)^2 t + C_2\left(\frac{Q}{A}\right) = (-\Delta P)^{1-s} \qquad (13\text{-}46)$$

where $C_1 = (\mu\alpha/2)(M_s/V)$ and $C_2 = \mu a$. There are now three parameters that must be determined empirically from laboratory measurements—C_1, C_2, and s. The easiest way to do this would be to utilize the constant-pressure mode in the laboratory (a Buchner funnel, for example, with a vacuum pressure difference) and obtain several sets of data for V as a function of t, each set at a different value of ΔP. For each data set, the plot of t/V versus V should yield a straight line, with a slope of $C_1/[A^2(-\Delta P)^{1-s}]$ and intercept of $C_2/[A(-\Delta P)^{1-s}]$. Thus, a log-log plot of either the slope or the intercept versus ΔP should have a slope of $s - 1$, which determines s.

REFERENCES

Branan, C. R., *Rules of Thumb for Chemical Engineers*, Gulf Publishing Co. (1994)

Coulson, J. M., J. F. Richardson, J. R. Blackhurst, and J. H. Harker, *Chemical Engineering*, Vol. 2, 3rd Ed., Pergamon Press (1991)

Leva, M., "Reconsider Packed Tower Pressure Drop Correlations", *Chem. Eng. Prog.*, **88**, 65–72 (1992)

PROBLEMS

POROUS MEDIA

1. A packed bed is composed of crushed rock with a density of 175 lb_m/ft^3 having a size and shape such that the average ratio of surface area to volume for the particles is 50 in.^{2}/in.^{3}. The bed is 6 ft deep, has a porosity of 0.3, and is covered by a 2 ft deep layer of water which drains by gravity through the bed. Calculate the flow rate of water through the bed in gpm/ft^2, assuming it exits at 1 atm pressure.

2. An impurity in very small concentration in a water stream is to be removed in a charcoal trickle bed filter. The filter is in a cylindrical column which is 2 ft in diameter, and the bed is 4 ft deep. The water is kept at a level which is 2 ft above the top of the bed, and it trickles through by gravity flow. If the charcoal particles have a geometric surface area to volume ratio of 48 in.$^{-1}$, and they pack with a porosity of 0.45, what is the flow rate of water through the column, in gpm?

3. A trickle bed filter is composed of a packed of broken rock that has a shape such that the average ratio of the surface area to volume for the rock particles is 30 in.$^{-1}$. The bed is 2 ft deep, has a porosity of 0.3, and is covered by a layer of water that is 2 ft deep and drains by gravity through the bed.
   (a) Determine the volume flow rate of the water through the bed per unit bed area (in gpm/ft^2).
   If the water is pumped upward through the bed (e.g. to flush it out):
   (b) Calculate the flow rate (in gpm/ft^2 of bed area) which would be required to fluidize the bed.
   (c) Calculate the corresponding flow rate which would sweep the rock particles away with the water. The rock density is 120 lb$_m$/ft^3.

PACKED COLUMNS

4. A packed column that is 3 ft in diameter with a packing height of 25 ft is used to absorb an impurity from a methane gas stream, using an amine solution absorbent. The gas flow rate is 2000 scfm, and the liquid has a density of 1.2 g/cm^3 and a viscosity of 2 cP. If the column operates at 1 atm and 80°F, determine the liquid flow rate at which flooding would occur in the column and the pressure drop at 50% of the flooding liquid rate for the following packings:
   (a) 2 in. ceramic Raschig rings.
   (b) 2 in. plastic Pall rings.

5. A packed column is used to scrub SO$_2$ from air using water. The gas flow rate is 500 scfm/ft^2, and the column operates at 90°F and 1 atm. If the packing is No. 1 plastic Intalox packing, what is the maximum liquid flow rate (per unit cross section of column) that could be used without flooding?

6. A stripping column packed with 2 in. metal Pall rings uses hot air at 5 psig, 200°C, to strip an impurity from an absorber oil (SG = 0.9, viscosity = 5 cP). If the flow rate of the oil is 500 lb$_m$/min and that of the air is 20 lb$_m$/min:
   (a) What is the minimum column diameter that can be used without flooding?
   (b) If the column diameter is 50% greater than the minimum size, what is the pressure drop per ft of column height?

7. A packed column, which is 0.6 m in diameter and 4 m high containing 25 mm Raschig rings, is used in a gas absorption process to remove an inpurity from the gas stream by absorbing it in a liquid solvent. The liquid, which has a viscosity of 5 cP and SG = 1.1, enters the top of the column at a rate of 2.5 kg/(sm^2), and the gas, which may be assumed to have the same properties as air, enters the bottom of the column at a rate of 0.6 kg/(sm^2). The column operates at atmospheric pressure and 25°C. Determine:
   (a) The pressure drop through the column, in inches of water.
   (b) How high the liquid rate could be increased before the column would flood.

FILTRATION

8. A fine aqueous suspension containing 1 lb$_m$ of solids per ft^3 of suspension is to be filtered in a constant-pressure filter. It is desired to filter at an average rate of 100 gpm, and the filter cake must be removed when it gets 2 in. thick. What filter area is required? Data: $-\Delta P$ = 10 psi, ρ(wet cake) = 85 lb$_m$/ft^3, K (permeability) = 0.118 darcies, a = 2 × 10^9 ft^{-1}.

9. An aqueous slurry containing 1.5 lb$_m$ of solid per gallon of liquid is pumped through a filter cloth by a centrifugal pump. If the pump provides a constant pressure drop of 150 psig, how long will it take for the filter cake to build up to a thickness of 2 in.? The density of the filter cake is 30 lb$_m$/ft^3 and its permeability is 0.01 darcies.

10. A packed bed is to be used to filter solids from an aqueous slurry. To determine the filter properties, you test a small section of the bed, which is 6 in. in diameter and 6 in. deep, in the lab. When the slurry is pumped through this test model at a constant flow rate of 30 gpm, the pressure drop across the bed rises to 2 psia after 10 min. How long will it take to filter 100,000 gal of water from the slurry in a full-sized bed which is 10 ft in diameter and 2 ft deep, if the slurry is maintained at a depth of 2 ft over the bed and drains by gravity through the bed?

11. A slurry containing 1 lb$_m$ of solids per gallon of water is to be filtered in a plate-and-frame filter having a total filtering area of 60 ft^2. The slurry is fed to the filter by a centrifugal pump which develops a head of 20 psig. How long would it take to build up a layer of filter cake 4 in. thick on the filter medium? Laboratory data were taken

on the slurry using a positive displacement pump operating at 5 gpm and 1 ft^2 of filter medium. It was found that the pressure drop increased linearly with time from an initial value of 0.2 psi, reaching a value of 50 psi after 1 min. The density of the filter cake was found to be 0.85 g/cm^3.

12. A rotary drum filter which is 6 ft in diameter and 8 ft long is to be used to filter a slurry. The drum rotates at 0.5 rpm, and one-third of the drum's surface is submerged under the slurry at any given time. A vacuum is drawn in the drum so that the constant pressure drop of 10 psi is maintained across the drum and filter cake. You test the slurry in the lab by pumping it at a constant filtrate rate of 20 gpm through 1 ft^2 of the drum filter screen, and find that after 1 min the pressure drop is 8 psi and after 3 min the pressure drop is 12 psi. How long will it take to filter 100,000 gal of filtrate from the slurry using the rotary drum?

13. A plate-and-frame filter press contains 16 frames and operates at a constant flow rate of 30 gpm. Each filter frame has an active filtering area of 4 ft^2, and it takes 15 min to disassemble, clean, and reassemble the press. The press must be shut down for disassembly when the pressure difference builds up to 10 psi. What is the total net filtration rate in gpm for a slurry having properties determined by the following lab test? A sample of the slurry is pumped at a constant pressure differential of 5 psi through 0.25 ft^2 of the filter medium. After 3 min, 1 gal of filtrate had been collected. The resistance of the filter medium may be neglected.

14. A rotary drum filter is used to filter a slurry. The drum rotates at a rate of 3 min/cycle, and 40% of the drum surface is submerged beneath the slurry. A constant pressure drop at 3 psi is maintained across the filter. If the drum is 5 ft in diameter and 10 ft long, calculate the total net filtration rate in gpm that is possible for a slurry having properties as determined by a lab test as follows. A sample of the slurry was pumped at a constant flow rate of 1 gpm through 0.25 ft^2 of the filter medium. After 10 min, the pressure difference across the filter had risen to 2.5 psi. The filter medium resistance may be neglected.

15. You must filter 1000 lb_m/min of an aqueous slurry containing 40% solids by weight using a rotary drum filter with a diameter of 4 m and a length of 4 m, which operates at a vacuum of 25 in. of mercury with 30% of its surface submerged beneath the slurry. A lab test is

run on a sample of the slurry using 200 cm^2 of the same filter medium, under a vacuum of 25 in. Hg. During the first minute of operation, 300 cm^3 of filtrate is collected, and during the second minute an additional 140 cm^3 is collected.

(a) How fast should the drum be rotated?

(b) If the drum is rotated at 2 rpm, what would the filter capacity be in pounds of slurry filtered per minute?

16. A rotary drum filter is to be used to filter a lime slurry. The drum rotates at a rate of 0.2 rpm, and 30% of the drum surface is submerged beneath the slurry. The filter operates at a constant ΔP of 10 psi. The slurry properties were determined from a lab test at a constant flow rate of 0.5 gpm using 1/2 ft^2 of the filter medium. The test results indicated that the pressure drop rose to 2 psi after 10 s and to 10 psi after 60 s. Calculate the net filtration rate per unit area of the drum under these conditions, in gpm/ft^2.

17. A plate-and-frame filter press operating at a constant ΔP of 150 psi is to be used to filter a sludge containing 2 lb$_m$ of solids per ft^3 of water. The filter must be disassembled and cleaned when the cake thickness builds up to 1 in. The frames have a projected area of 4 ft^2, and the downtime for cleaning is 10 min/frame. The properties of the sludge and cake were determined in a lab test operating at a constant flow rate of 0.2 gpm of filtrate and a filter area of 1/4 ft^2. The test results show that the pressure drop rises to 3 psi after 20 s, and to 8 psi after 60 s. Calculate the overall net filtration rate per frame in the filter in gpm of filtrate, accounting for the downtime. The density of the cake was found to be 150 lb$_m$/ft^3.

18. A packed bed composed of crushed rock having a density of 175 lb$_m$/ft^3 is to be used as a filter. The size and shape of the rock particles is such that the average surface area to volume ratio is 50 in.^{2}/in.^{3}, and the bed porosity is 0.3. A lab test using the slurry to be filtered is run on a small bed of the same particles, which is 6 in. deep and 6 in. in diameter. The slurry is pumped through this bed at a constant filtrate rate of 10 gpm, and it is found that after 5 min the pressure drop is 5 psi, while after 10 min it is 8 psi.

(a) Calculate how long it would take to filter 100,000 gal of filtrate from the slurry in a full-scale bed which is 10 ft in diameter and 2 ft deep, if the slurry is maintained at a depth of 2 ft above the bed and drains through it by gravity. Assume the slurry densities to be the same as water.

(b) The filter is cleaned by back-flushing with water, pumped up-ward through the bed. At what flow rate of water (in gpm) would the bed become fluidized?

19. A rotary drum filter is available, which has a diameter of 6 ft and a length of 8 ft and rotates at a rate of 30 s/cycle. The filter operates at a vacuum of 500 mm Hg, with 30% of its surface submerged. The slurry to be filtered is tested in the lab, using 0.5 ft^2 of the drum filter medium, in a filter funnel operating at 600 mm Hg vacuum. After 5 min of operation, 250 cm^3 of filtrate is collected through the funnel, and after 10 min, a total of 400 cm^3 is collected.
    (a) What would be the net (average) filtration rate of this slurry in the rotary drum filter, in gpm?
    (b) How much could this filtration rate be increased by increasing the speed (i.e. rotation rate) of the drum?

20. A rotary drum filter, 10 ft in diameter and 8 ft long, is to be used to filter a slurry of incompressible solids. The drum rotates at 1.2 rpm, and 40% of its surface is submerged beneath the slurry at all times. A vacuum in the drum maintains a constant pressure drop of 10 psi across the drum and filter cake. The slurry is tested in the lab by pumping it at a constant rate of 5 gpm through 0.5 ft^2 of the drum filter screen. After 1 min, the pressure drop is 9 psi, and after 3 min it has risen to 15 psi. How long will it take to filter 1 million gallons of filtrate from the slurry using the rotary drum? How long would it take if the drum rotated at 3 rpm?

21. A slurry is being filtered at a net rate of 10,000 gal/day by a plate-and-frame filter with 15 frames, having an active filtering area of 1.5 ft^2 each, fed by a positive displacement pump. The pressure drop varies from 2 psi at start-up to 25 psi after 10 min, at which time it is shut down for cleanup. It takes 10 min to disassemble, clean out, and reassemble the filter. Your boss decides that it would be more economical to replace this filter by a rotary drum filter, using the same filter medium. The rotary filter operates at a vacuum of 200 mm Hg with 30% of its surface submerged, and rotates at a rate of 5 min/rev. If the drum length is equal to its diameter, how big should it be?

22. You want to select a rotary drum filter to filter a coal slurry at a rate of 100,000 gal of filtrate per day. The filter operates at a differential pressure of 12 psi, and 30% of the surface is submerged in the slurry

at all times. A sample of the slurry is filtered in the lab through a 6 in. diameter sample of the filter medium, at a constant rate of 1 gpm. After 1 min the pressure drop across this filter is 3 psi, and after 5 min it is 10 psi. If the drum rotates at a rate of 3 rpm, what total filter area is required?

23. A slurry is delivered to a rotary drum filter which is 4 ft in diameter and 6 ft long and operates at a vacuum of 25 in. Hg. A lab test is run with a 50 cm² sample of the filter medium and the slurry, at a constant flow rate of 200 cm³/min. After 1 min, the pressure across the lab filter is 6 psi, and after 3 min it is 16 psi. If 40% of the rotary drum is submerged under the slurry, how fast should it be rotated (rpm) in order to filter the slurry at the same rate that it is delivered?

24. A slurry is to be filtered with a rotary drum filter which is 5 ft in diameter, 8 ft long, rotates once every 10 s, and has 20% of its surface immersed in the slurry. The drum operates with a vacuum of 20 in. Hg. A lab test was run on a sample of the slurry using 1/4 ft² of the filter medium, at a constant flow rate of 40 cm³/s. After 20 s, the pressure drop was 30 psi across the lab filter, and after 40 s it was 35 psi. How many gallons of filtrate can be filtered per day in the rotary drum?

25. A slurry of $CaCO_3$ in water at 25°C, containing 20% solids by weight, is to be filtered in a plate-and-frame filter. The slurry and filter medium are tested in a constant-pressure lab filter, having an area of 0.0439 m², at a pressure drop of 338 kPa. It is found that 10^{-3} m³ of filtrate is collected after 9.5 s, and 5×10^{-3} m³ is collected after 107.3 s. The plate-and-frame filter has 20 frames, with 0.873 m² of filter medium per frame, and operates at a constant flow rate of 0.00462 m³ of slurry per second. The filter is operated until the pressure drop reaches 500 kPa, at which time it is shut down for cleaning. The downtime is 15 min per cycle. Determine how much filtrate passes through the filter in each 24 hr period of operation (SG of $CaCO_3$ is 1.6).

26. An algal sludge is to be clarified by filtering. A lab test is run on the sludge using an area A of the filter medium. At a constant pressure drop of 40 kN/m², a plot of the time required to collect a volume V of the filtrate times $\Delta PA/V$ versus V/A gives a straight line with a slope of 1.2×10^6 kN s/m⁴ and an intercept of 6.0×10^4 kN s/m³.

A repeat of the data at a pressure drop of 200 kN/m^2 also gave a straight line on the same type of plot, with the same intercept but with a slope of 2.1 $\times$ 10^6 kN s/m^4. When a filter aid was added to the sludge in an amount equal to 20% of the algae by weight, the lab test gave a straight line with the same intercept but with a slope of 1.4 $\times$ 10^6 kN s/m^4.

(a) What does this tell you about the sludge?

(b) The sludge is to be filtered using a rotary drum filter, with a diameter of 4 ft and a length of 6 ft, operating at a vacuum of 700 mm Hg with 35% of the drum submerged. If the drum is rotated at a rate of 2 rpm, how many gallons of filtrate will be collected in a day, with and without the filter aid?

(c) What would the answer to (b) be if the drum speed was 4 rpm?

27. A slurry containing 0.2 kg of solids per kg water is filtered through a rotary drum filter, operating at a pressure difference of 65 kN/m^2. The drum is 0.6 m in diameter and 0.6 m long, rotates once every 350 s, and has 20% of its surface submerged below the slurry.

(a) If the overall average filtrate flow rate is 0.125 kg/s, the cake is incompressible with a porosity of 50%, and the solids SG = 3.0, determine the maximum thickness of the cake on the drum (you may neglect the filter medium resistance).

(b) The filter breaks down, and you want to replace it with a plate-and-frame filter having the same overall capacity, which operates at a pressure difference of 275 kN/m^2. The frames are 10 cm thick, so the maximum cake thickness at which the filter will still operate properly is 4 cm. It will take 100 s to disassemble the filter, 100 s to clean it out, and 100 s to reassemble it. If the frames are 0.3 m square, how many frames should the filter contain?

14

Fluidization and Sedimentation

I. FLUIDIZATION

When a fluid is passed upward through a bed of particles, the pressure drop increases as the fluid velocity increases. The pressure drop times the bed cross-sectional area is the net upward force on the bed, and when this force becomes equal to the weight of the bed (solids and fluid), the bed becomes suspended by the fluid. In this state, the particles move freely within the "bed", which thus behaves much like a boiling liquid. Under these conditions the bed is said to be "fluidized", and the situation is illustrated in Fig. 14-1. The fluid will tend to entrain the particles and carry them out of the bed. However, if the fluid velocity above the bed is less than the terminal velocity of the particles, they will fall back and remain in the bed. This "freely flowing" or bubbling behavior results in a high degree of mixing in the bed, which provides a great advantage for heat or mass transfer efficiency as compared with a fixed bed.

A. GOVERNING EQUATIONS

The Bernoulli equation relates the pressure drop across the bed to the fluid flow rate and the bed properties:

$$\frac{-\Delta P}{\rho_f} - gh = e_f = \frac{f_{PM} h V_s^2}{d} \left(\frac{1 - \varepsilon}{\varepsilon^3} \right) \tag{14-1}$$

389

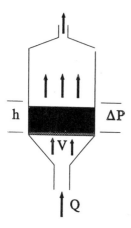

FIGURE 14-1 Fluidized bed.

where the porous medium friction factor is given by the Ergun equation

$$f_{PM} = 1.75 + \frac{180}{N_{RePM}} \tag{14-2}$$

and the porous medium Reynolds number is

$$N_{RePM} = \frac{dV_s\rho}{(1 - \varepsilon)\mu} \tag{14-3}$$

Now the criterion for incipient fluidization is that the force due to the pressure drop must balance the weight of the bed, i.e.

$$-\Delta P = \frac{\text{Bed wt.}}{A} = \rho_s(1 - \varepsilon)gh + \rho_f\varepsilon gh \tag{14-4}$$

where the first term on the right is the weight of the solids and the second the weight of the fluid in the bed. When the pressure drop is eliminated between Eqn (14-4) and Eqn (14-1), an equation for the "minimum fluidization velocity" (V_{mf}) results:

$$(\rho_s - \rho_f)(1 - \varepsilon)g = \frac{\rho e_f}{h} = 1.75 \frac{\rho_f V_{mf}^2(1 - \varepsilon)}{d\varepsilon^3}$$

$$+ 180 \frac{V_{mf}\mu(1 - \varepsilon)^2}{d^2\varepsilon^3} \tag{14-5}$$

This can be written in dimensionless form as follows:

$$N_{Ar} = 1.75 \frac{\hat{N}_{Re}^2}{\varepsilon^3} + 180\left(\frac{1-\varepsilon}{\varepsilon^3}\right)\hat{N}_{Re} \tag{14-6}$$

where

$$N_{Ar} = \frac{\rho_f g \Delta \rho d^3}{\mu^2}, \qquad \hat{N}_{Re} = \frac{dV_s \rho_f}{\mu} \tag{14-7}$$

Equation (14-6) can be solved for the Reynolds number to give

$$\hat{N}_{Re} = \sqrt{C_1^2 + C_2 N_{Ar}} - C_1 \tag{14-8}$$

where

$$C_1 = \frac{180(1-\varepsilon)}{3.5}, \qquad C_2 = \frac{\varepsilon^3}{1.75} \tag{14-9}$$

Actually, the values of C_1 and C_2 that give the best results have been found, from empirical observations, to be

$$C_1 = 27.2, \qquad C_2 = 0.0408 \tag{14-10}$$

B. MINIMUM BED VOIDAGE

Before the bed can become fluidized, the particles must dislodge from their "packed" state, and it must expand. Thus, the porosity (ε) in Eqns (14-5)– (14-9) is not the initial "packed bed" porosity, but the "expanded bed" porosity at the point of minimum fluidization (ε_{mf}). This is the "minimum bed voidage" that the bed will exhibit just prior to fluidization. By comparing the empirical values of C_1 and C_2 (Eqn 14-10) with Eqns (14-9), the C_1 value of 27.2 is equivalent to $\varepsilon_{mf} = 0.471$, while the C_2 value of 0.0408 is equivalent to $\varepsilon_{mf} = 0.415$. In actuality, the value of ε_{mf} may vary considerably with the nature of the solid particles, as shown in Fig. 14-2.

C. NONSPHERICAL PARTICLES

Many particles are not spherical and so will not have the same drag properties as spherical properties. The effective diameter for such particles is often char- acterized by the equivalent *Stokes diameter*, which is the diameter of the sphere that has the same terminal velocity as the particle. This can be determined from a direct measurement of the settling rate of the particles and provides the best value of equivalent diameter for use in the equations derived assuming spherical particles.

An alternative description of nonspherical particles is to define a "sphe- ricity factor" ψ, which is the number that, when multiplied by the diameter of

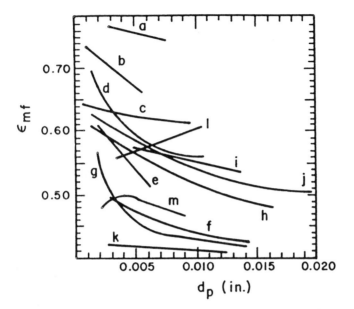

FIGURE 14-2 Values of ε_{mf} for various solids. (a) Soft brick; (b) absorption carbon; (c) broken Raschig rings; (d) coal and glass powder; (e) carborundum; (f) sand; (g) round sand, $\psi = 0.86$; (h) sharp sand, $\psi = 0.67$; (i) Fischer-Tropsch catalyst, $\psi = 0.58$; (j) anthracite coal, $\psi = 0.63$; (k) mixed round sand, $\psi = 0.86$; (l) coke; (m) carborundum. (ψ is the particle shape factor.)

a sphere with the same volume as the particle (d_s), gives the particle effective diameter (d_p):

$$d_p = \psi d_s \tag{14-11}$$

The sphericity factor is defined by

$$\psi = \frac{\text{Surface area of the sphere with same volume as particle}}{\text{Surface area of the particle}} \tag{14-12}$$

Thus:

$$\psi = \frac{A_s}{A_p} = \frac{A_s/V_s}{A_p/V_p} = \frac{6/d_s}{a_s} \tag{14-13}$$

Equations (14-11) and (14-13) show that $d_p = 6/a_s$, where a_s is the surface-to-volume ratio for the particle (A_p/V_p), as previously deduced. Since $V_p = V_s$, by definition, the equivalent definition of ψ is

$$\psi = \frac{6}{d_s} \left(\frac{V_p}{A_p} \right) = (6^2 \pi)^{1/3} \frac{V_p^{2/3}}{A_p} = 4.84 \frac{V_p^{2/3}}{A_p} \tag{14-14}$$

The minimum bed expansion porosity at incipient fluidization for nonspherical particles can be estimated from

$$\varepsilon_{mf} \cong (14\psi)^{-1/3} \tag{14-15}$$

For spherical particles ($\psi = 1$), Eqn (14-15) reduces to $\varepsilon_{mf} = 0.415$.

II. SEDIMENTATION

Sedimentation, or thickening, involves increasing the solids content of a slurry or suspension by gravity settling in order to effect separation (or partial separation) of the solids and the fluid. It differs from the gravity settling process that we previously considered in that the solids fraction is relatively high in such systems, so the particle settling rates are strongly influenced by the presence of the surrounding particles. This situation is called *hindered settling*. Fine particles tend to behave differently than larger or coarse particles, since the fine particles usually exhibit a high degree of flocculation due to the dominance of surface forces and high surface areas.

A. HINDERED SETTLING

A mixture of particles of different sizes can settle in two different ways, according to Coulson et al. (1991). These two modes are illustrated in Fig. 14-3. Case (a) corresponds to a suspension with a range of particle sizes less than about 6:1. In this case, all of the particles settle at about the same velocity, in the "constant-composition zone" (B), leaving a layer of clear liquid above. As the sediment (D) builds up, however, the liquid that is "squeezed out" of this layer serves to further retard the particles just above it, resulting in a zone of variable composition (C). Case (b) in Fig. 14-3 is less common and corresponds to a broad particle size range, in which the larger particles settle at a rate significantly greater than that of the smaller ones, and consequently there is no constant-composition zone.

The settling characteristics of hindered settling systems differ significantly from those of freely settling particles in several ways:

(1) The large particles are hindered by the small particles, which effectively increase the viscosity of the suspending medium for the large particles. At the same time, however, the small particles tend to be "dragged down" by the large particles, so that all particles fall at

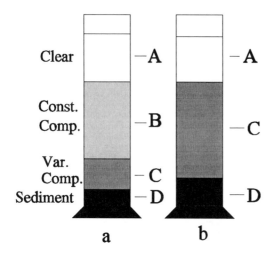

FIGURE 14-3   Two modes of settling. (a) Narrow size range; (b) broad size range.

about the same rate (unless the size range is very large—greater than 6:1 or so).

(2) The upward velocity of the displaced fluid flowing in the interstices between the particles is appreciable, so the apparent settling velocity (relative to a fixed point) is significantly lower than the particle velocity relative to the fluid.

(3) The velocity *gradients* in the suspending fluid flowing upward between the particles are increased considerably, resulting in greater shear forces.

(4) Because of the high surface area to volume ratio of the small particles, surface forces are important, resulting in flocculation and ''clumping'' of the smaller particles into larger effective particle groups. This effect is more pronounced in a highly ionic (conducting) fluid, because the electrostatic surface forces that would cause the particles to be repelled are ''shorted out''.

There are essentially three different approaches to describing hindered settling. One approach is to define a ''correction factor'' to Stokes' law for free settling in an infinite Newtonian fluid (which we will designate V_o), as a function of the solids loading. A second approach is to consider the suspending fluid properties (e.g. viscosity and density) to be modified by the presence of the fine particles. A third approach is to consider the collection or ''swarm'' of particles equivalent to a moving porous bed, the resistance to flow through the bed being determined by an equivalent to the Kozeny equation. There is not sufficient hard evidence to say that any one of these approaches is any better or worse than the others. For many systems, they may all give comparable results, whereas for others one of these methods may be better or worse than the others.

If all of the solids are relatively fine and/or the slurry is sufficiently concentrated that settling is extremely slow, the slurry can usually be approximated as a uniform continuous medium with properties (viscosity and density) that depend upon the solids loading, particle size and density, and interparticle forces (surface charges, conductivity, etc.). Such systems are most generally quite non-Newtonian, with properties described by the Bingham plastic or power law models. If a significant fraction of the particles are fines (e.g. less than about 30 μm or so), the suspending fluid plus fines can be considered to be a continuous medium, with increased viscosity and density, through which the larger particles must move. Such systems may or may not be non-Newtonian, depending on solids loading, etc. If the solids loading is relatively low (e.g. below about 10% solids by volume) and/or the particle size and/or density are relatively large, the system will be heterogeneous and the larger particles will settle readily. Such systems are usually Newtonian. A summary of the flow behavior of these various systems has been presented by Darby (1986).

B. FINE PARTICLES

For fine suspensions, or systems containing a significant amount of fines, the suspending fluid can be considered to be homogeneous, with density and viscosity modified by the presence of the fines. These properties depend primarily upon the solids loading of the suspension, which may be described in terms of either the porosity or void fraction (ε) or, more commonly, the volume fraction of solids, ϕ ($\phi = 1 - \varepsilon$). The buoyant force on the particles is due to the difference in density between the solid (ρ_s) and the surrounding suspension (ρ_ϕ), which is

$$\rho_s - \rho_\phi = \rho_s - [\rho_s(1 - \varepsilon) + \rho\varepsilon]$$
$$= \varepsilon(\rho_s - \rho) = (1 - \phi)(\rho_s - \rho) \quad (14\text{-}16)$$

where ρ is the density of the suspending fluid.

The viscosity of the suspension (μ_ϕ) is modified by the presence of the solids. For uniform spheres at a volumetric fraction of 2% or less, Einstein (1906) showed that

$$\mu_\phi = \mu(1 + 2.5\phi) \quad (14\text{-}17)$$

where μ is the viscosity of the suspending fluid. For more concentrated suspensions, a wide variety of expressions have been proposed in the literature (see e.g. Darby, 1986). Vand (1948) proposed the expression

$$\mu_\phi = \mu \, \exp\left(\frac{2.5\phi}{1 - 0.609\phi}\right) \qquad (14\text{-}18)$$

although Mooney (1951) concluded that the constant 0.609 could actually range from 0.75 to 1.5, depending upon the system. Equations (14-16) and (14-18) (or equivalent) may be used to modify Stokes' law, which is

$$\text{Stokes' law: } V_o = \frac{(\rho_s - \rho)gd^2}{18\mu} \qquad (14\text{-}19)$$

In this equation, V_o is the relative velocity between the particle and the fluid. However, in a suspension this velocity is increased by the velocity of the displaced fluid, which flows back up through the suspension, in a cross-sectional area equal to the void space between the particles. Thus, if V_s is the (superficial) settling velocity of the suspension and V_L is the velocity of the fluid, the total flux of solids and liquid is $[\phi V_s + (1 - \phi)V_L]$. Now the relative velocity between the fluid and solids in the "swarm" is $V_r = V_s - V_L$. If the total net flux is zero (e.g. "batch" settling in a closed-bottom container with no outflow), elimination of V_L gives

$$V_r = \frac{V_s}{1 - \phi} \qquad (14\text{-}20)$$

This also shows that $V_L = -\phi V_s/(1 - \phi)$, i.e. V_L is negative (of the opposite sign) relative to V_s.

From Eqns (14-16), (14-18), and (14-20), it is seen that the ratio of the settling velocity of the suspension (V_s) to the terminal velocity of a single freely settling sphere (V_o) is

$$\frac{V_s}{V_o} = \frac{(1 - \phi)^2}{\exp\left(\dfrac{2.5\phi}{1 - k_2\phi}\right)} \qquad (14\text{-}21)$$

where the constant k_2 may lie between 0.61 and 1.5, depending upon the system. However, Coulson et al. (1991) remark that the use of a modified viscosity for the suspending fluid is more appropriate for the settling of large particles through a suspension of fines than for the uniform settling of a "swarm" of particles with a narrow size distribution. They state that in the latter case the increased resistance is due to the higher velocity gradients in the interstices rather than to an increased viscosity. However, the net effect may be very similar for either mechanism. This approach, as well as the other two mentioned above, all result in expressions of the general form

$$\frac{V_s}{V_o} = \varepsilon^2 \, fn(\varepsilon) \qquad \text{where } \varepsilon = 1 - \phi \tag{14-22}$$

A widely used empirical expression for the function in Eqn (14-22) is that of Richardson and Zaki (1954):

$$fn(\varepsilon) = \varepsilon^{2.65} \tag{14-23}$$

An alternative expression, due to Davies et al. (1977), is

$$\frac{V_s}{V_o} = \exp(-k_1\phi) \tag{14-24}$$

which agrees very well with Eqn (14-23) for $k_1 = 5.5$. Another expression for $fn(\varepsilon)$, deduced by Steinour (1944) from settling data on tapioca in oil, is

$$fn(\varepsilon) = 10^{-1.82(1-\varepsilon)} \tag{14-25}$$

Barnea and Mizrahi (1973) considered the effects of the modified density and viscosity of the suspending fluid, as represented by Eqn (14-21), as well as crowding'' or hindrance effect that decreases the effective space around the particles and increases the drag. This introduces an additional ''crowding factor'' $(1 + k_2\phi^{1/3})$ in Eqn (14-21), so the modified Stokes velocity thus becomes

$$\frac{V_s}{V_o} = \frac{(1 - \phi)^2}{(1 + \phi^{1/3}) \exp\left[\dfrac{5\phi}{3(1 - \phi)}\right]} \tag{14-26}$$

where the constant 2.5 in Eqn (14-21) has been replaced by 5/3, and the constant k_2 set equal to unity, based upon settling observations.

C. COARSE PARTICLES

Coarser particles (e.g. ~100 μm or larger) have a specific surface that is much smaller, so flocculation is not common. Also, the suspending fluid surrounding the particles is the liquid phase rather than the ''pseudo-continuous'' phase of fines in suspension, which modify the fluid viscosity and density properties. Thus, the properties of the continuous phase are taken to be those of the pure fluid, unaltered by the presence of fine particles. In this case, it can be shown by dimensional analysis that the dimensionless settling velocity V_s/V_o must be a function of the particle drag coefficient, which in turn is a unique function of the particle Reynolds number, N_{Re}, the void fraction (porosity), $\varepsilon = 1 - \phi$, and the ratio of the particle diameter to container diameter, d/D. Since we have seen that there is a unique relationship between the drag coefficient, the Reynolds number, and the Archimedes number for settling particles, the result can be expressed by a function of the form

$$\frac{V_s}{V_o} = fn \left\{ N_{Ar}, \frac{d}{D}, \varepsilon \right\} \tag{14-27}$$

Empirically, it has been found that this relationship can be represented by the following expression (Coulson et al., 1991):

$$\frac{V_s}{V_o} = \varepsilon^n \left[1 + 2.4 \frac{d}{D} \right]^{-1} \tag{14-28}$$

where the exponent n is given by

$$\frac{4.8 - n}{n - 2.4} = 0.043 N_{Ar}^{0.57} \left[1 - 2.4 \left(\frac{d}{D} \right)^{0.27} \right] \tag{14-29}$$

or

$$n = \frac{4.8 + 2.4X}{X + 1} \tag{14-30}$$

where X represents the right-hand side of Eqn (14-29).

D. ALL FLOW REGIMES

The above expressions give the suspension velocity (V_s) relative to the single-particle settling velocity, V_o, which has been referred to as the Stokes velocity. However, it is not necessary that the settling particle conditions correspond to the Stokes regime to use these equations. As shown in Chapter 11, the Dallavalle equation can be used to calculate the single-particle terminal velocity V_o under any flow conditions from a known value of the Archimedes number:

$$V_o = \frac{\mu}{\rho d} [(14.42 + 1.827 \sqrt{N_{Ar}})^{1/2} - 3.798]^2 \tag{14-31}$$

where

$$N_{Ar} = \frac{d^3 \rho g \Delta \rho}{\mu^2} \tag{14-32}$$

This result applies directly to coarse-particle systems. For fine-particle systems, the suspending fluid properties are assumed to be modified by the fines in suspension, which necessitates modifying the definitions of the Reynolds and Archimedes numbers accordingly. Furthermore, since the particle drag is a direct function of the local relative velocity between the fluid and the solid (i.e. the interstitial velocity, V_r), it is this velocity that must be used in the drag equations (e.g. the modified Dallavalle equation). Since $V_r = V_s/(1 - \phi) = V_s/\varepsilon$, the appropriate definitions for the Reynolds number and drag coefficient for the suspension (e.g. the particle "swarm") are (after Barnea and Mizrahi, 1973)

$$N_{Re\phi} = \frac{dV_r\rho}{\mu_\phi} = N_{Re}\left(\frac{V_s}{V_o}\right)\frac{1}{(1 - \phi)\exp[5\phi/3(1 - \phi)]} \qquad (14\text{-}33)$$

and

$$C_{D\phi} = C_D\left(\frac{V_o}{V_s}\right)^2\left(\frac{(1 - \phi)^2}{1 + \phi^{1/3}}\right) \qquad (14\text{-}34)$$

where $N_{Re} = dV_o\rho/\mu$ and $C_D = 4gd(\rho_s - \rho)/(3\rho V_o^2)$ are the Reynolds number and drag coefficient for a single particle. Data presented by Barnea and Mizrahi (1973) show that the "swarm" dimensionless groups $N_{Re\phi}$ and $C_{D\phi}$ are related in the same way as the corresponding groups for single particles, e.g. by the Dallavalle equation,

$$C_{D\phi} = \left(0.6324 + \frac{4.8}{N_{Re\phi}^{1/2}}\right)^2 \qquad (14\text{-}35)$$

Thus, the settling velocity, or the terminal velocity of the "swarm", may be determined from

$$N_{Re\phi} = [(14.42 + 1.827N_{Ar\phi}^{1/2})^{1/2} - 3.798]^2 \qquad (14\text{-}36)$$

where

$$N_{Ar\phi} = \frac{3}{4}C_{D\phi}N_{Re\phi}^2 = \left(\frac{d^3\rho g(\rho_s - \rho)}{\mu^2(1 + \phi^{1/3})}\right)\exp\left(\frac{-10\phi}{3(1 - \phi)}\right) \qquad (14\text{-}37)$$

Once $N_{Re\phi}$ is determined from Eqns (14-36) and (14-37), the settling velocity, V_s, is determined from Eqn (14-33).

III. GENERALIZED SEDIMENTATION/FLUIDIZATION

The above relations all apply to hindered settling of a suspension (or "swarm") of particles in a stagnant suspending medium. Barnea and Mizrahi (1973) showed that these generalized relations may be applied to fluidization as well, since a fluidized bed may be considered a particle "swarm" suspended by the continuous fluid flowing upward at the terminal velocity of the swarm. In this case the above equations apply, with V_s replaced by the velocity V_f, the superficial velocity of the fluidizing medium. Barnea and Mizrahi (1973) presented data for both settling and fluidization which cover a very wide range of the dimensionless parameters, as shown in Fig. 14-4.

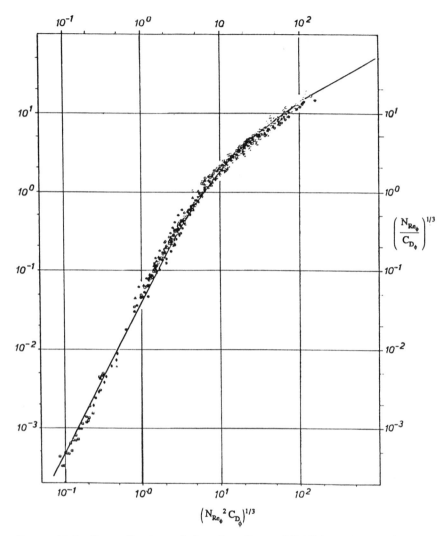

FIGURE 14-4 Generalized correlation of settling and fluidizing velocities (Barnea and Mizrahi, 1973).

IV. THICKENING

The process of thickening involves the concentration of a slurry, suspension, or sludge, usually by gravity settling. Since concentrated suspensions, and/or fine particle dispersions are often involved, the result is not always a complete separation of the solids from the liquid but is instead a separation into a more concentrated (underflow) stream and a diluted (overflow) stream. Thickeners and clarifiers are essentially identical. The only difference is that the clarifier is designed to produce a clean liquid overflow with a specified purity, whereas the thickener is designed to produce a concentrated slurry underflow product with a specified concentration (Christian, 1994; Tiller and Tarng, 1995; McCabe et al., 1993).

A schematic of a thickener/clarifier is shown in Fig. 14-5. As indicated in Fig. 14-3, several settling regions or zones can be identified, depending upon the solids concentration and interparticle interaction. For simplicity, we consider three primary zones, as indicated in Fig. 14-5 (with the understanding that there are transition zones in between). The top, or clarifying, zone contains relatively clear liquid from which most of the particles have settled. Any particles remaining in this zone will settle by free settling. The middle zone is a region of varying composition, through which the particles move by hindered settling. The size of this region and the settling rate depend upon the average solids concentration. The bottom zone is a highly concentrated settled or compressed region containing the settled particles. The particle settling rate through this zone is very slow.

In the top (clarifying) zone, the relatively clear liquid moves upward and overflows the top. In the middle zone, the particles are settling as the displaced liquid moves upward. In the bottom (compressed) zone, the solids and liquid both move downward at a rate that is determined mainly by the underflow

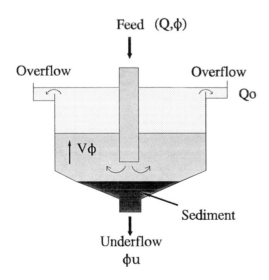

Feed (Q, ϕ)

Overflow   Overflow Q_o

$V\phi$

Sediment

Underflow ϕu

FIGURE 14-5 Thickener.

pumping rate. For a given feed rate and solids loading, the objective is to determine the area of the thickener and the optimum underflow rate to achieve a specified underflow concentration (ϕ_u) or the underflow rate and underflow concentration for optimum operating capacity.

The solids concentration can be expressed in terms of either the solids volume fraction (ϕ) or the mass ratio of solids to fluid (R). If ϕ_i is the volume fraction of solids in the feed stream (flow rate Q) and ϕ_u is the volume fraction of solids in the underflow (flow rate Q_u), then the solids ratios in the feed, R_i = [(mass of solids)/(mass of fluid)]$_{in}$, and in the underflow, R_u = [(mass of solids)/ (mass of liquid)]$_u$, are given by

$$R_i = \frac{\phi_i \rho_s}{(1 - \phi_i)\rho}, \qquad R_u = \frac{\phi_u \rho_s}{(1 - \phi_u)\rho} \qquad (14\text{-}38)$$

These relations can be rearranged to give the solids concentrations in terms of the solids ratio:

$$\phi_i = \frac{R_i}{R_i + \rho_s/\rho}, \qquad \phi_u = \frac{R_u}{R_u + \rho_s/\rho} \qquad (14\text{-}39)$$

The total (net) flux of solids and liquid moving through the thickener is given by

$$q = \frac{Q}{A} = q_s + q_L = \phi V_s + (1 - \phi)V_L \qquad (14\text{-}40)$$

where $q_s = \phi V_s$ is the solids flux, defined as the volumetric settling rate of the solids per unit cross-sectional area of the settler, and $q_L = (1 - \phi)V_L$ is the liquid flux. Now the settling characteristics of the slurry are normally determined by measuring the velocity of the interface between the top (clear) and middle (hindered settling) zones in a batch test using a closed system (e.g. a graduated cylinder, as in Fig. 14-3). A typical batch settling curve is shown in Fig. 14-6 (Foust et al., 1980). The initial (left) linear portion of this curve corresponds to free (unhindered) settling, and the slope of this region is the free settling velocity, V_o. The nonlinear region of the curve corresponds to hindered settling, and the settling velocity as a function of solids concentration can be determined from this part of the curve, as follows (Kynch, 1952). If the initial height of the suspension with a solids fraction of ϕ_o is Z_o, at some later time the height to the interface between the clear layer and the hindered settling zone will be $Z(t)$, and the average solids fraction in this zone is $\phi(t)$. Since the total amount of solids in the system is constant, assuming the amount of solids in the clear layer to be negligible it follows that

$$Z(t)\phi(t) = Z_o\phi_o \qquad (14\text{-}41)$$

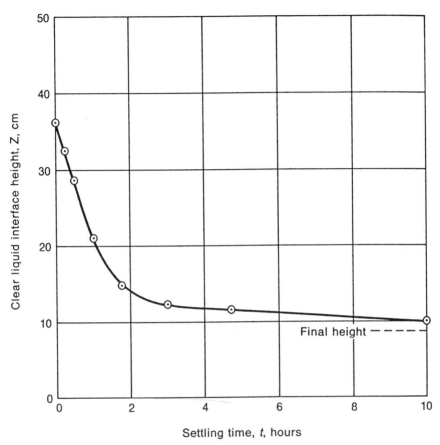

FIGURE 14-6 Typical batch settling curve for a limestone slurry (Foust et al., 1980).

Thus, if the initial height and concentration (Z_o, ϕ_o) are known, the average solids concentration $\phi(t)$ corresponding to any point on the curve $Z(t)$ can be determined. Furthermore, the hindered settling velocity of the slurry at this point is just the slope of the curve at that point, e.g. $V_s = -(dZ/dt)$. Thus, the batch settling curve can be converted to a *batch flux curve*, as shown in Fig. 14-7. The batch flux curve exhibits a maximum and a minimum, since the settling velocity is nearly constant in the free settling region (and the flux is directly proportional to the concentration), whereas the settling velocity drops rapidly with increasing solids concentration in the hindered settling region (so does the flux). This will be illustrated quantitatively later.

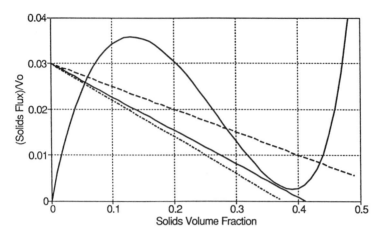

FIGURE 14-7 Typical batch flux curve with operating lines. (---) Underloaded; (———) properly loaded; (– – –) overloaded.

Since the batch flux data are obtained in a closed system with no outflow, the net solids flux (q) is zero in the batch system, and Eqn (14-40) reduces to $V_L = -\phi V_s/(1 - \phi)$. Note that V_L and V_s are of opposite sign, since the displaced liquid moves upward as the solids settle. The relative velocity between the solids and liquid is $V_r = V_s - V_L$ which, from Eqn (14-20) is $V_r = V_s/(1 - \phi)$. Thus, the relative velocity, which controls the dynamics in the thickener, may be obtained directly from settling data in a batch flux system. However, since the total flux in the thickener is different (greater) than that in the batch flux test because of the finite draw-off rate of solids in the underflow, we will refer to the settling velocity as determined from the batch flux test as V_{sb}, and the corresponding solids flux as q_{sb}. The relative velocity from the batch flux test applies directly to the thickener as well, however, so that $V_r = V_{sb}/(1 - \phi)$ is valid for both the batch and the thickener.

In the thickener, the underflow pumping rate is Q_u, and Eqn (14-40) applies. Thus, the total solids flux in the thickener (q_s) is equal to the settling flux relative to the suspension, which is identical to the batch flux q_{sb}, plus the bulk flux due to the underflow pumping rate, ϕV_u, i.e. $q_s = q_{sb} + \phi V_u$, or

$$q_{sb} = q_s - \phi V_u \qquad (14\text{-}42)$$

This represents a straight line on the batch flux curve (q_{sb} vs ϕ) with a slope of $-V_u$. The line intersects the ϕ axis at ϕ_u and the q_{sb} axis at q_s (the net solids flux in the thickener). This line is called the "operating line" for the thickener, and its tangent to the batch flux curve determines the minimum flux, and hence the required thickener area, for steady-state operation. Operating lines are shown on the batch flux plot in Fig. 14-7. The "properly loaded" line is tangent to the batch flux curve at the critical concentration, and this condition represents a steady state at which the settling rate equals the underflow solids rate. The "underloaded" line represents a condition for which the underflow pumping rate is higher than the settling rate, so that no sludge layer can build up and excess water will be taken out the underflow (the pumping rate is too high). The "overloaded" line represents the conditions at which the underflow draw-off rate is lower than the settling rate, so that solids will eventually accumulate and rise to the overflow (i.e. the pumping rate is too low).

The equations that govern the thickener operation are derived from mass balance considerations, as follows. The total underflow flux is given by

$$\frac{Q_u}{A} = q_u = V_u = q_{su} + q_{Lu} = \phi V_s + (1 - \phi)V_L \tag{14-43}$$

or, since $V_s = V_r + V_L$,

$$V_u = \phi V_r + V_L \tag{14-44}$$

Now the total flux of solids in the underflow is

$$q_s = \phi_u V_u = \phi_u(\phi V_r + V_L) \tag{14-45}$$

which is the same as the solids flux in the thickener at steady state:

$$q_s = \phi V_s = \phi(V_L + V_r) \tag{14-46}$$

If V_L is eliminated from Eqns (14-45) and (14-46), we have

$$q_s = \frac{\phi(1 - \phi)V_r}{1 - \phi/\phi_u} \tag{14-47}$$

From Eqn (14-20), $(1 - \phi)V_r = V_s$ is the solids settling rate in the batch flux test, not in the thickener. We will call this V_{sb} to denote that it is obtained from batch flux tests but does not exist as such in the thickener, i.e. $V_{sb} = (1 - \phi)V_r$. Thus, Eqn (14-47) becomes

$$q_s = \frac{Q_s}{A} = \frac{Q\phi}{A} = \frac{\phi V_{sb}}{1 - \phi/\phi_u} \tag{14-48}$$

This is the basic equation that relates the thickener area (A) and the feed rate and loading (Q, ϕ) to the solids underfow rate (Q_s), the underflow loading (ϕ_u), and the (batch flux) settling velocity (V_{sb}). Note that q_s represents the solids flux in the thickener, which is related to the solids flux in the batch flux test (q_{sb}) by Eqn (14-42). The settling velocity V_{sb}, as determined from batch settling tests on the slurry, may be expressed by e.g. one of the expressions in Section II-B.

For example, if Eqn (14-24) is used, Eqn (14-48) becomes

$$q_{sb} = q_s - \phi_u q_u = \frac{\phi V_o \exp(-k_1 \phi)}{1 - \phi/\phi_u} - \phi_u q_u \tag{14-49}$$

This equation represents the batch flux curve illustrated in Fig. 14-7.

By setting the derivative of Eqn (14-49) with respect to ϕ equal to zero, the solids loadings corresponding to the maximum and minimum solids flux can be determined as follows:

$$\phi = \frac{\phi_u}{2} \pm \sqrt{\left(\frac{\phi_u}{2}\right)^2 - \frac{\phi_u}{k_1}} \tag{14-50}$$

where the plus sign corresponds to the critical concentration at the minimum flux, and the minus sign corresponds to the concentration at the maximum flux. These concentrations are determined entirely by the underflow solids loading ϕ_u and the settling parameter k_1. Thus the total solids flux increases with the solids loading up to a point, but will then decrease as the loading is increased beyond that point, as determined by the underflow solids loading and the settling rate. Equations (14-49) and (14-50) determine the solids flux (q_s) as a function of the underflow solids loading (ϕ_u) and underflow pumping rate (q_u), as illustrated in Fig. 14-7.

The area of the thickener required for a specified underflow loading is determined from Eqn (14-48), as follows. If the underflow solids loading (ϕ_u) is specified, the operating line is drawn on the batch flux curve from ϕ_u on the ϕ axis, tangent to the batch flux curve. The intersection of the line with the flux curve is the feed flux (q) which, with a specified feed rate (Q) and concentration (ϕ), determines the thickener area A. If it is assumed that none of the solids are carried over with the overflow, the overflow rate Q_o is given by

$$Q_o = Q(1 - \phi) - Q\phi \frac{1 - \phi_u}{\phi_u} \tag{14-51}$$

or

$$\frac{Q_o}{Q} = 1 - \frac{\phi}{\phi_u} \tag{14-52}$$

Likewise, the underflow rate Q_u is given by

$$Q_u = Q - Q_o = Q - Q\left(1 - \frac{\phi}{\phi_u}\right) \tag{14-53}$$

or

$$\frac{Q_u}{Q} = \frac{\phi}{\phi_u} \qquad (14\text{-}54)$$

REFERENCES

Azbel, D. S., and N. P. Cheremisinoff, *Fluid Mechanics and Unit Operations*, Ann Arbor Science (1983)

Barnea, E. and J. Mizrahi, "A Generalized Approach to the Fluid Dynamics of Particulate Systems, Part I, *Chem. Eng. J.*, **5**, 171–189 (1973)

Christian, J. B., *Chem. Eng. Prog.*, p. 50, July (1994)

Coulson, J. M., J. F. Richardson, J. R. Blackhurst, and J. H. Harker, *Chemical Engineering*, Vol 2, 4th Ed., Pergamon Press (1991)

Darby, R., "Hydrodynamics of Slurries and Suspensions", Chapter 2 in *Encyclopedia of Fluid Mechanics*, Vol. 5, N. P. Cheremisinoff, Ed., (1986), pp. 49–92, Marcel Dekker

Davies, L., D. Dollimore, and G. B. McBride, *Powder Technol.*, **16**, 45 (1977)

Einstein, A., *Ann. Phys.*, **19**, 289 (1906)

Foust, A. S., L. A. Wenzel, C. W. Clump, L. Maus, and L. B. Anderson, *Principles of Unit Operations*, 2nd Ed., Wiley (1980)

Kynch, G. J., *Trans. Faraday Soc.*, **51**, 61 (1952)

McCabe, W. L., J. C. Smith, and P. Harriott, *Unit Operations of Chemical Engineering*, 5th Ed., McGraw-Hill (1993)

Mooney, M., *J. Colloid Sci.*, **6**, 162 (1951)

Richardson, J. F., and W. N. Zaki, *Chem. Eng. Sci.*, **3**, 65 (1954)

Steinour, H. H., *Ind. Eng. Chem.*, **36**, 618, 840, 901 (1944)

Svarovsky, L., *Hydrocyclones*, Technomic Publ. Co. (1984)

Tiller, F. M., and D. Trang, *Chem. Eng. Prog.*, p. 75, March (1995)

Vand, V., *J. Phys. Colloid Chem.*, **52**, 217 (1948)

PROBLEMS

1. Calculate the flow rate of air (ft³/min) required to fluidize a bed of sand (SG = 2.4) at 1 atm, 70°F. The sand grains have an equivalent diameter of 500 μm, and the bed is 2 ft in diameter and 1 ft deep, with a porosity of 0.35. What flow rate of air would be required to blow the sand away?

2. Calculate the flow rate of water (in gpm) required to fluidize a bed of 1/6 in. diameter lead shot (SG = 11.3). The bed is 1 ft in diameter, 1 ft deep, and has a porosity of 0.18. What water flow rate would be required to sweep the bed away?

3. Calculate the range of water velocities that will fluidize a bed of glass spheres (SG = 2.1) if the sphere diameter is: (a) 2 mm, (b) 1 mm, (c) 0.1 mm

4. A coal gasification reactor operates with particles of 500 μm diameter and density of 1.4 g/cm^3. The gas may be assumed to have properties of air at 1000°F and 30 atm. Determine the range of superficial gas velocity over which the bed is in a fluidized state.

5. A bed of coal particles, 2 ft in diameter and 6 ft deep, is to be fluidized using a hydrocarbon liquid with a viscosity of 15 cP and a density of 0.9 g/cm^3. The coal particles have a density of 1.4 g/cm^3 and an equivalent spherical diameter of 1/8 in. If the bed porosity is 0.4:
   (a) Determine the range of liquid superficial velocities over which the bed is fluidized.
   (b) Repeat the problem using the "particle swarm" (Barnea and Mizrah) hindered settling approach, assuming: (1) $\phi = 1 - \varepsilon$; (2) $\phi = 1 - \varepsilon_{mf}$.

6. A catalyst having spherical particles of d_p = 50 μm and ρ_s = 1.65 g/cm^3 is to be used to contact a hydrocarbon vapor in a fluidized reactor at 900°F, 1 atm. At operating conditions, the fluid viscosity is 0.02 cP and its density is 0.21 lb$_m$/ft^3. Determine the range of fluidized bed operation, i.e. calculate:
   (a) Minimum fluidization velocity for ε_{mf} = 0.42.
   (b) The particle terminal velocity.

7. A fluid bed reactor contains catalyst particles with a mean diameter of 500 μm and a density of 2.5 g/cm^3. The reactor feed has properties equivalent to 35° API distillate at 400°F. Determine the range of superficial velocities over which the bed will be in a fluidized state.

8. Water is pumped upward through a bed of 1 mm diameter iron oxide particles (SG = 5.3). If the bed porosity is 0.45, over what range of superficial water velocity will the bed be fluidized?

9. A fluidized bed combustor is 2 m in diameter and is fed with air at 250°F, 10 psig, at a rate of 2000 scfm. The coal has a density of 1.6 g/cm^3, and a shape factor of 0.85. The flue gas from the combustor has an average MW of 35 and leaves the combustor at a rate of 2100

scfm at 2500°F, and 1 atm. What is the size range of the coal particles that can be fluidized in this system?

10. A fluid bed incinerator, 3 m in diameter and 0.56 m high, operates at 850°C using a sand bed. The sand density is 2.5 g/cm³, and the average sand grain has a mass of 0.16 mg and a sphericity of 0.85. In the stationary (packed) state, the bed porosity is 35%. Find:
    (a) The range of air velocities which will fluidize the bed.
    (b) The compressor power required, if the bed is operated at 10 times the minimum fluidizing velocity and the compressor efficiency is 70%. The compressor takes air in from the atmosphere at 20°C, and the gases leave the bed at 1 atm.

11. Determine the range of flow rates (in gpm) that will fluidize a bed of 1 mm cubic silica particles (SG = 2.5) with water. The bed is 10 in. in diameter, 15 in. deep.

12. Calculate the velocity of water that would be required to fluidize spherical particles with SG = 1.6 and diameter of 1.5 mm, in a tube with a diameter of 10 mm. Also, determine the water velocity that would sweep the particles out of the tube. Use each of the two following methods, and compare the results:
    (a) The bed starts as a packed bed and is fluidized when the pressure drop due to friction through the bed balances the weight of the bed.
    (b) The bed is considered to be a "swarm" of particles, falling at the terminal velocity of the "swarm".
    Comment on any uncertainties or limitations in your results.

13. Solid particles with a density of 1.4 g/cm³ and a diameter of 0.01 cm are fed from a hopper into a line where they are mixed with water, which is draining by gravity from an open tank, to form a slurry having 0.4 lb_m of solids per lb_m of water. The slurry is transported by a centrifugal pump, through a 6 in. sch 40 pipeline that is 0.5 miles long, at a rate of 1000 gpm. The slurry can be described as a Bingham plastic, with a yield stress of 120 dyn/cm² and a limiting viscosity of 50 cP.
    (a) If the pipeline is at 60°F, and the pump is 60% efficient with a required NPSH of 15 ft, what horsepower motor would be required to drive the pump?
    (b) If the pump is 6 ft below the bottom of the water storage tank, and the water in the line upstream of the pump is at 90°C (P_V =

526 mm Hg), what depth of water in the tank would be required
to prevent the pump from cavitating?

(c) A venturi meter is installed in the line to measure the slurry flow
rate. If the maximum pressure drop reading for the venturi is to
be 29 in. H_2O, what diameter should the venturi throat be?

(d) The slurry is discharged from the pipeline to a thickener, where
it is desired to concentrate the slurry to 1 lb_m of solids per lb_m
of water (in the underflow). Determine the required diameter of
the thickening tank and the volumetric flow rates of the overflow
(Q_o) and underflow (Q_u), in gpm.

(e) If the slurry were to be sent to a rotary drum filter instead to
remove all of the solids, determine the required size of the drum
(assuming the drum length and diameter are equal). The drum
rotates at 3 rpm, with 25% of its surface submerged in the slurry,
and operates at a vacuum of 20 in. of mercury. Lab test data
taken on the slurry with 0.5 ft^2 of the filter medium, at a constant
flow rate of 3 gpm, indicated a pressure drop of 1.5 psi after 1
min of filtration and 2.3 psi after 2 min of operation.

14. A sludge is to be clarified in a thickener, which is 50 ft in diameter.
The sludge contains 35% solids by volume (SG = 1.8) in water, with
an average particle size of 25 μm. The sludge is pumped into the
center of the tank, where the solids are allowed to settle and the
clarified liquid overflows the top. What is the maximum flow rate of
the sludge (in gpm) that this thickener can handle? Assume that the
solids are uniformly distributed across the tank, and that all particle
motion is vertical.

15. In a batch thickener, a sludge containing 35% by volume of solids
(SG = 1.6) with an average particle size of 50 μm, is allowed to
settle. The sludge is fed to the settler at a rate of 1000 gpm, and the
clear liquid overflows the top. What is the minimum tank diameter
required for this separation?

16. Ground coal is slurried with water in a pit, and the slurry is pumped
out of the pit at a rate of 500 gpm with a centrifugal pump and into
a classifier. The classifier inlet is 50 ft above the slurry level in the
pit. The piping system consists of an equivalent length of 350 ft of
5 in. sch 40 pipe, and discharges into the classifier at 2 psig. The
slurry may be assumed to be a Newtonian fluid, with a viscosity of
30 cP, a density of 75 lb_m/ft^3, and a vapor pressure of 30 mm Hg.
The solid coal has a specific gravity of 1.5.

(a) How much power would be required to pump the slurry?
(b) Using the pump characteristic charts in Appendix H, select the best one of these for this job. Specify the pump size, motor speed (rpm), and impeller diameter that you would use. Also determine the pump efficiency and NPSH requirement.
(c) What is the maximum height above the level of the slurry in the pit that the pump could be located without cavitating?
(d) A venturi meter is located in a vertical section of the line to monitor the slurry flow rate. The meter has a 4 in. diameter throat, and the pressure taps are 1 ft apart. If a DP cell (transducer) is used to measure the pressure difference between the taps, what would it read (in inches of water)?
(e) A 90° flanged elbow is located in the line at a point where the pressure (upstream of the elbow) is 10 psig. What are the forces transmitted to the pipe by the elbow from the fluid inside the elbow (neglect the weight of the fluid)?
(f) The classifier consists of three collection tanks in series which are full of water. The slurry enters at the top on the side of the first tank and leaves at the top on the opposite side, which is 5 ft from the entrance. The solids settle into the tank as the slurry flows into it and then overflows into the next tank. The space through which the slurry flows above the tank is 2 ft wide and 3 ft high. All particles for which the settling time in the space above the collection tank is less than the residence time of the fluid flowing in the space over the collection tank will be trapped in that tank. Determine the diameter of the largest particle that will not settle into each of the three collection tanks. Assume that the particles are equivalent spheres and that they fall at their terminal velocity.
(g) The suspension leaving the classifier is transferred to a rotary drum filter to remove the remaining solids. The drum operates at a constant pressure difference of 5 psi and rotates at a rate of 2 rpm with 20% of the surface submerged. Lab tests on a sample of the suspension through the same filter medium were conducted at a constant flow rate of 1 gpm through 0.25 ft² of the medium. It was found that the pressure drop increased to 2.5 psi after 10 min, and the resistance of the medium was negligible. How much filter area would be required to filter the liquid at a rate of 40 gpm?

Appendix A

Viscosities and Other Properties of Gases and Liquids

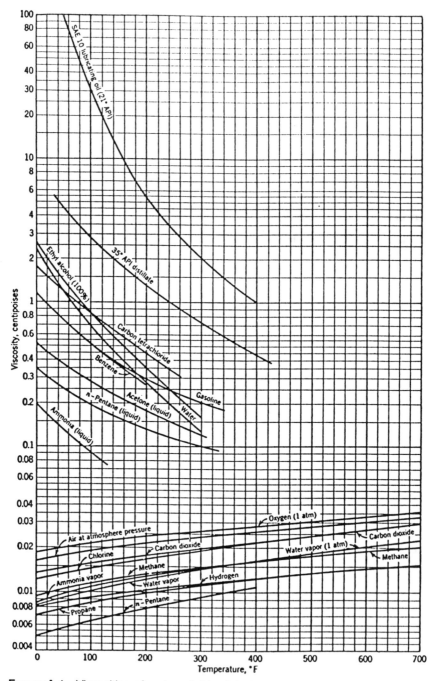

FIGURE A-1 Viscosities of various fluids at 1 atm pressure. 1 cp = 0.01 gm/cm · sec = 6.72 × 10⁻⁴ lbm/ft · sec = 2.42 lbm/ft · hr = 2.09 × 10⁻⁵ lbf · sec/ft². (From G. G. Brown et al., *Unit Operations*, Wiley, New York, 1951, p. 586. Reproduced by permission of the publisher.)

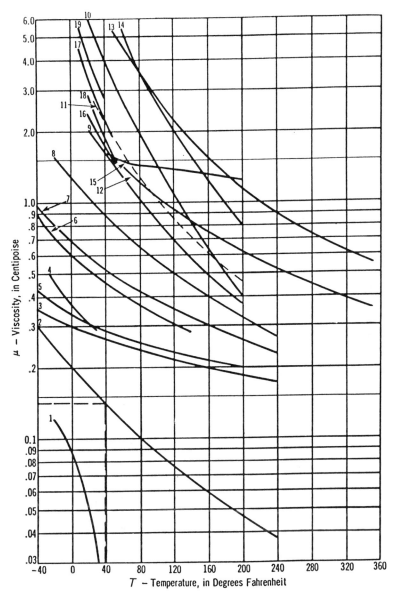

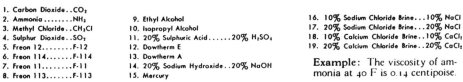

1. Carbon Dioxide..CO_2
2. Ammonia........NH_3
3. Methyl Chloride..CH_3Cl
4. Sulphur Dioxide..SO_2
5. Freon 12........F-12
6. Freon 114.......F-114
7. Freon 11........F-11
8. Freon 113.......F-113

9. Ethyl Alcohol
10. Isopropyl Alcohol
11. 20% Sulphuric Acid......20% H_2SO_4
12. Dowtherm E
13. Dowtherm A
14. 20% Sodium Hydroxide..20% NaOH
15. Mercury

16. 10% Sodium Chloride Brine...10% NaCl
17. 20% Sodium Chloride Brine...20% NaCl
18. 10% Calcium Chloride Brine..10% $CaCl_2$
19. 20% Calcium Chloride Brine..20% $CaCl_2$

Example: The viscosity of ammonia at 40 F is 0.14 centipoise.

FIGURE A-2 Viscosity of various liquids.

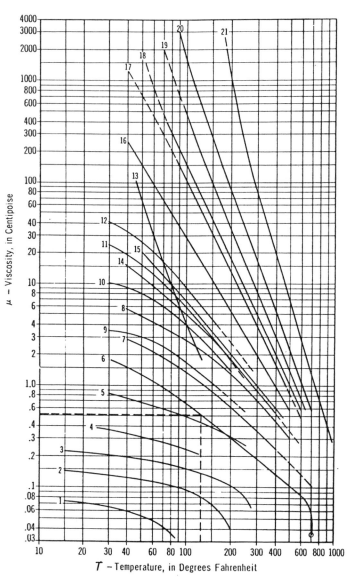

1. Ethane (C₂H₆)
2. Propane (C₃H₈)
3. Butane (C₄H₁₀)
4. Natural Gasoline
5. Gasoline
6. Water
7. Kerosene
8. Distillate
9. 48 Deg. API Crude
10. 40 Deg. API Crude
11. 35.6 Deg. API Crude
12. 32.6 Deg. API Crude
13. Salt Creek Crude
14. Fuel 3 (Max.)
15. Fuel 5 (Min.)
16. SAE 10 Lube (100 V.I.)
17. SAE 30 Lube (100 V.I.)
18. Fuel 5 (Max.) or
    Fuel 6 (Min.)
19. SAE 70 Lube (100 V.I.)
20. Bunker C Fuel (Max.) and
    M.C. Residuum
21. Asphalt

FIGURE A-3   Viscosity of water and liquid petroleum products.

TABLE A-1   Viscosities and Densities of Liquids (Coordinates Apply to Fig. A-4)

No.	Liquid	X	Y	Density at 293 K (kg/m³)
1	Acetaldehyde	15.2	4.8	783 (291 K)
2	Acetic acid, 100 per cent	12.1	14.2	1049
3	Acetic acid, 70 per cent	9.5	17.0	1069
4	Acetic anhydride	12.7	12.8	1083
5	Acetone, 100 per cent	14.5	7.2	792
6	Acetone, 35 per cent	7.9	15.0	948
7	Allyl alcohol	10.2	14.3	854
8	Ammonia, 100 per cent	12.6	2.0	817 (194 K)
9	Ammonia, 26 per cent	10.1	13.9	904
10	Amyl acetate	11.8	12.5	879
11	Amyl alcohol	7.5	18.4	817
12	Aniline	8.1	18.7	1022
13	Anisole	12.3	13.5	990
14	Arsenic trichloride	13.9	14.5	2163
15	Benzene	12.5	10.9	880
16	Brine, $CaCl_2$, 25 per cent	6.6	15.9	1228
17	Brine, NaCl, 25 per cent	10.2	16.6	1186 (298 K)
18	Bromine	14.2	13.2	3119
19	Bromotoluene	20.0	15.9	1410
20	Butyl acetate	12.3	11.0	882
21	Butyl alcohol	8.6	17.2	810
22	Butyric acid	12.1	15.3	964
23	Carbon dioxide	11.6	0.3	1101 (236 K)
24	Carbon disulphide	16.1	7.5	1263
25	Carbon tetrachloride	12.7	13.1	1595
26	Chlorobenzene	12.3	12.4	1107
27	Chloroform	14.4	10.2	1489
28	Chlorosulphonic acid	11.2	18.1	1787 (298 K)
29	Chlorotoluene, *ortho*	13.0	13.3	1082
30	Chlorotoluene, *meta*	13.3	12.5	1072
31	Chlorotoluene, *para*	13.3	12.5	1070
32	Cresol, *meta*	2.5	20.8	1034
33	Cyclohexanol	2.9	24.3	962
34	Dibromoethane	12.7	15.8	2495
35	Dichloroethane	13.2	12.2	1256
36	Dichloromethane	14.6	8.9	1336
37	Diethyl oxalate	11.0	16.4	1079
38	Dimethyl oxalate	12.3	15.8	1148 (327 K)
39	Diphenyl	12.0	18.3	992 (346 K)
40	Dipropyl oxalate	10.3	17.7	1038 (273 K)
41	Ethyl acetate	13.7	9.1	901

TABLE **A-1**   Continued

No.	Liquid	X	Y	Density at 293 K (kg/m³)
42	Ethyl alcohol, 100 per cent	10.5	13.8	789
43	Ethyl alcohol, 95 per cent	9.8	14.3	804
44	Ethyl alcohol, 40 per cent	6.5	16.6	935
45	Ethyl benzene	13.2	11.5	867
46	Ethyl bromide	14.5	8.1	1431
47	Ethyl chloride	14.8	6.0	917 (279 K)
48	Ethyl ether	14.5	5.3	708 (298 K)
49	Ethyl formate	14.2	8.4	923
50	Ethyl iodide	14.7	10.3	1933
51	Ethylene glycol	6.0	23.6	1113
52	Formic acid	10.7	15.8	1220
53	Freon-11 (CCl_3F)	14.4	9.0	1494 (290 K)
54	Freon-12 (CCl_2F_2)	16.8	5.6	1486 (293 K)
55	Freon-21 ($CHCl_2F$)	15.7	7.5	1426 (273 K)
56	Freon-22 ($CHClF_2$)	17.2	4.7	3870 (273 K)
57	Freon-113 ($CCl_2F\text{-}CClF_2$)	12.5	11.4	1576
58	Glycerol, 100 per cent	2.0	30.0	1261
59	Glycerol, 50 per cent	6.9	19.6	1126
60	Heptane	14.1	8.4	684
61	Hexane	14.7	7.0	659
62	Hydrochloric acid, 31.5 per cent	13.0	16.6	1157
63	Isobutyl alcohol	7.1	18.0	779 (299 K)
64	Isobutyric acid	12.2	14.4	949
65	Isopropyl alcohol	8.2	16.0	789
66	Kerosene	10.2	16.9	780–820
67	Linseed oil, raw	7.5	27.2	930–938 (288 K)
68	Mercury	18.4	16.4	13546
69	Methanol, 100 per cent	12.4	10.5	792
70	Methanol, 90 per cent	12.3	11.8	820
71	Methanol, 40 per cent	7.8	15.5	935
72	Methyl acetate	14.2	8.2	924
73	Methyl chloride	15.0	3.8	952 (273 K)
74	Methyl ethyl ketone	13.9	8.6	805
75	Naphthalene	7.9	18.1	1145
76	Nitric acid, 95 per cent	12.8	13.8	1493
77	Nitric acid, 60 per cent	10.8	17.0	1367
78	Nitrobenzene	10.6	16.2	1205 (291 K)
79	Nitrotoluene	11.0	17.0	1160
80	Octane	13.7	10.0	703
81	Octyl alcohol	6.6	21.1	827
82	Pentachloroethane	10.9	17.3	1671 (298 K)

TABLE **A-1** Continued

No.	Liquid	X	Y	Density at 293 K (kg/m³)
83	Pentane	14.9	5.2	630 (291 K)
84	Phenol	6.9	20.8	1071 (298 K)
85	Phosphorus tribromide	13.8	16.7	2852 (288 K)
86	Phosphorus trichloride	16.2	10.9	1574
87	Propionic acid	12.8	13.8	992
88	Propyl alcohol	9.1	16.5	804
89	Propyl bromide	14.5	9.6	1353
90	Propyl chloride	14.4	7.5	890
91	Propyl iodide	14.1	11.6	1749
92	Sodium	16.4	13.9	970
93	Sodium hydroxide, 50 per cent	3.2	25.8	1525
94	Stannic chloride	13.5	12.8	2226
95	Sulphur dioxide	15.2	7.1	1434 (273 K)
96	Sulphuric acid, 110 per cent	7.2	27.4	1980
97	Sulphuric acid, 98 per cent	7.0	24.8	1836
98	Sulphuric acid, 60 per cent	10.2	21.3	1498
99	Sulphuryl chloride	15.2	12.4	1667
100	Tetrachloroethane	11.9	15.7	1600
101	Tetrachloroethylene	14.2	12.7	1624 (288 K)
102	Titanium tetrachloride	14.4	12.3	1726
103	Toluene	13.7	10.4	866
104	Trichloroethylene	14.8	10.5	1466
105	Turpentine	11.5	14.9	861–867
106	Vinyl acetate	14.0	8.8	932
107	Water	10.2	13.0	998
108	Xylene, *ortho*	13.5	12.1	881
109	Xylene, *meta*	13.9	10.6	867
110	Xylene, *para*	13.9	10.9	861

Source: By permission from *Chemical Engineers' Handbook*, by J. H. Perry, copyright 1975, McGraw-Hill.

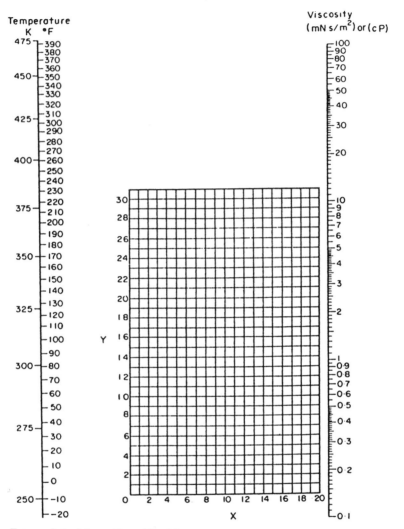

FIGURE A-4   Viscosities of liquids.

TABLE A-2 The Viscosity of Water 0°C to 100°C

°C	η(cp)	°C	η(cp)	°C	η(cp)	°C	η(cp)
0	1.787	26	0.8705	52	0.5290	78	0.3638
1	1.728	27	.8513	53	.5204	79	.3592
2	1.671	28	.8327	54	.5121	80	.3547
3	1.618	29	.8148	55	.5040	81	.3503
4	1.567	30	.7975	56	.4961	82	.3460
5	1.519	31	.7808	57	.4884	83	.3418
6	1.472	32	.7647	58	.4809	84	.3377
7	1.428	33	.7491	59	.4736	85	.3337
8	1.386	34	.7340	60	.4665	86	.3297
9	1.346	35	.7194	61	.4596	87	.3259
10	1.307	36	.7052	62	.4528	88	.3221
11	1.271	37	.6915	63	.4462	89	.3184
12	1.235	38	.6783	64	.4398	90	.3147
13	1.202	39	.6654	65	.4335	91	.3111
14	1.169	40	.6529	66	.4273	92	.3076
15	1.139	41	.6408	67	.4213	93	.3042
16	1.109	42	.6291	68	.4155	94	.3008
17	1.081	43	.6178	69	.4098	95	.2975
18	1.053	44	.6067	70	.4042	96	.2942
19	1.027	45	.5960	71	.3987	97	.2911
20	1.002	46	.5856	72	.3934	98	.2879
21	0.9779	47	.5755	73	.3882	99	.2848
22	.9548	48	.5656	74	.3831	100	.2818
23	.9325	49	.5561	75	.3781		
24	.9111	50	.5468	76	.3732		
25	.8904	51	.5378	77	.3684		

The above table was calculated from the following empirical relationships derived from measurements in viscometers calibrated with water at 20°C (and one atmosphere), modified to agree with the currently accepted value for the viscosity at 20° of 1.002 cp:

$$0° \text{ to } 20°C: \log_{10} \eta_T = \frac{1301}{998.333 + 8.1855(T - 20) + 0.00585(T - 20)^2} - 3.30233$$

(R. C. Hardy and R. L. Cottingham, J. Res. NBS *42*, 573 (1949).)

$$20° \text{ to } 100°C: \log_{10} \frac{\eta_T}{\eta_{20}} = \frac{1.3272(20 - T) - 0.001053(T - 20)^2}{T + 105}$$

(J. F. Swindells, NBS, unpublished results.)

TABLE A-3 Physical Properties of Ordinary Water and Common Liquids (SI units)

Liquid	Temperature T (°C)	Density ρ (kg/m³)	Specific gravity S	Absolute viscosity μ (N·s/m²)	Kinematic viscosity ν (m²/s)	Surface tension σ (N/m)	Isothermal bulk modulus of elasticity E_p (N/m²)	Coefficient of thermal expansion α_T (K⁻¹)
Water	0	1000	1.000	1.79 E-3	1.79 E-6	7.56 E-2	1.99 E9	6.80 E-5
	3.98	1000	1.000	1.57	1.57	—	—	—
	10	1000	1.000	1.31	1.31	7.42	2.12	8.80
	20	998	0.998	1.00	1.00	7.28	2.21	2.07 E-4
	30	996	0.996	7.98 E-4	8.01 E-7	7.12	2.26	2.94
	40	992	0.992	6.53	6.58	6.96	2.29	3.85
	50	988	0.988	5.47	5.48	6.79	2.29	4.58
	60	983	0.983	4.67	4.75	6.62	2.28	5.23
	70	978	0.978	4.04	4.13	6.64	2.24	5.84
	80	972	0.972	3.55	3.65	6.26	2.20	6.41
	90	965	0.965	3.15	3.26	—	2.14	6.96
	100	958	0.958	2.82	2.94	5.89	2.07	7.50
Mercury	0	13600	13.60	1.68 E-3	1.24 E-7	—	2.50 E10	—
	4	13590	13.59	—	—	—	—	—
	20	13550	13.55	1.55	1.14	37.5	2.50 E10	1.82 E-4
	40	13500	13.50	1.45	1.07	—	—	1.82
	60	13450	13.45	1.37	1.02	—	—	1.82
	80	13400	13.40	1.30	9.70 E-8	—	—	1.82
	100	13350	13.35	1.24	9.29	—	—	—
Ethylene glycol	0	—	—	5.70 E-2	—	—	—	—
	20	1110	1.11	1.99	1.79 E-5	—	—	—
	40	1110	1.10	9.13 E-3	8.30 E-6	—	—	—
	60	1090	1.09	4.95	4.54	—	—	—

Substance	T							
	80	1070	1.07	3.02	2.82	—	—	—
	100	1060	1.06	1.99	1.88	—	—	—
Methyl alcohol (methanol)	0	810	0.810	8.17 E-4	1.01 E-6	2.45 E-2	9.35 E8	—
	10	801	0.801	5.84	7.37 E-7	2.26	8.78	—
	20	792	0.792	5.10	6.51	—	8.23	—
	30	783	0.783	4.50	5.81	—	7.72	—
	40	774	0.774	3.96	5.18	—	7.23	—
	50	765	0.765	—	—	—	6.78	—
Ethyl alcohol (ethanol)	0	806	0.806	1.77 E-3	2.20 E-6	2.41 E-2	1.02 E9	—
	20	789	0.789	1.20	1.52	—	9.02 E8	—
	40	772	0.772	8.34 E-4	1.08	—	7.89	—
	60	754	0.754	5.92	7.85 E-7	—	6.78	—
Normal octane	0	718	0.718	7.06 E-4	9.83 E-7	—	1.00 E9	—
	16	—	—	5.74	—	—	—	—
	20	702	0.702	5.42	7.72	—	8.35 E8	—
	25	—	—	—	—	—	7.48	—
	40	686	0.686	4.33	6.31	—	—	—
Benzene	0	900	0.900	9.12 E-4	1.01 E-6	3.02 E-2	1.23 E9	—
	20	879	0.879	6.52	7.42 E-7	2.76	1.06	—
	40	858	0.857	5.03	5.86	—	9.10 E8	—
	60	836	0.836	3.92	4.69	—	7.78	—
	80	815	0.815	3.29	4.04	—	6.48	—
Kerosene	-18	841	0.841	7.06 E-3	8.40 E-6	2.9 E-2	—	—
	20	814	0.814	1.9	2.37	—	—	—
Lubricating oil	20	871	0.871	1.31 E-6	1.50 E-9	—	—	—
	40	858	0.858	6.81 E-5	7.94 E-8	—	—	—
	60	845	0.845	4.18	4.95	—	—	—
	80	832	0.832	2.83	3.40	—	—	—
	100	820	0.820	2.00	2.44	—	—	—
	120	809	0.809	1.54	1.90	—	—	—

TABLE A-4 Physical Properties of Ordinary Water and Common Liquids (EE units)

Liquid	Temperature T (°F)	Density ρ (lbm/ft³)	Specific gravity S	Absolute viscosity μ (lb·s/ft²)	Kinematic viscosity ν (ft²/s)	Surface tension σ (lb/ft)	Isothermal bulk modulus of elasticity E_p (lb/in²)	Coefficient of thermal expansion α_T (°R⁻¹)
Water	32	62.4	1.00	3.75 E-5	1.93 E-5	5.18 E-3	2.93 E5	2.03 E-3
	40	62.4	1.00	3.23	1.66	5.14	2.94	—
	60	62.4	0.999	2.36	1.22	5.04	3.11	—
	80	62.2	0.997	1.80	9.30 E-6	4.92	3.22	1.7
	100	62.0	0.993	1.42	7.39	4.80	3.27	—
	120	61.7	0.988	1.17	6.09	4.65	3.33	—
	140	61.4	0.983	9.81 E-6	5.14	4.54	3.30	—
	160	61.0	0.977	8.38	4.42	4.41	3.26	—
	180	60.6	0.970	7.26	3.85	4.26	3.13	1.52
	200	60.1	0.963	6.37	3.41	4.12	3.08	—
	212	59.8	0.958	5.93	3.19	4.04	3.00	—
Mercury	50	847	13.6	1.07 E-3	1.2 E-6	—	—	1.0 E-4
	200	834	13.4	8.4 E-4	1.0	—	—	1.0 E-4
	300	826	13.2	7.4	9.0 E-7	—	—	—
	400	817	13.1	6.7	8.0	—	—	—
	600	802	12.8	5.8	7.0	—	—	—
Ethylene glycol	68	69.3	1.11	4.16 E-4	1.93 E-4	—	—	—
	104	68.7	1.10	1.91	8.93 E-5	—	—	—
	140	68.0	1.09	1.03	4.89	—	—	—
	176	66.8	1.07	6.31 E-5	3.04	—	—	—
	212	66.2	1.06	4.12	2.02	—	—	—

Methyl alcohol (methanol)	32	50.6	0.810	1.71 E-5	1.09 E-5	1.68 E-3	1.36 E5	—
	68	50.0	0.801	—	—	1.55	1.19	—
	104	49.4	0.792	1.22	7.93 E-6	—	1.05	—
	140	48.9	0.783	1.07	7.01	—	—	—
	176	48.3	0.774	9.40 E-6	6.25	—	—	—
	212	47.8	0.765	8.27	5.58	—	—	—
Ethyl alcohol (ethanol)	32	50.3	0.806	3.70 E-5	2.37 E-5	1.65 E-3	1.48 E5	—
	68	49.8	0.798	3.03	1.96	—	1.31	—
	104	49.3	0.789	2.51	1.64	—	1.14	—
	140	48.2	0.772	1.74	1.16	—	9.83 E4	—
	176	47.7	0.754	1.24	8.45 E-6	—	—	—
	212	47.1	0.745	—	—	—	—	—
Normal octane	32	44.8	0.718	1.47 E-5	1.06 E-5	—	1.45 E5	—
	68	43.8	0.702	1.13	8.31 E-6	—	—	—
	104	42.8	0.686	9.04 E-6	6.79	—	1.08	—
Benzene	32	56.2	0.900	1.90 E-5	1.09 E-5	2.07 E-3	1.78 E5	—
	68	54.9	0.879	1.36	7.99 E-6	1.89	1.53	—
	104	53.6	0.858	1.05	6.31	—	1.32	—
	140	52.2	0.836	8.19 E-6	5.05	—	1.13	—
	176	50.9	0.815	6.87	4.35	—	9.40 E4	—
Kerosene	0	52.5	0.841	1.48 E-4	9.05 E-5	—	—	—
	77	50.8	0.814	3.97 E-5	2.55 E-5	—	—	—
Lubricating oil	68	54.4	0.871	2.74 E-8	1.61 E-8	—	—	—
	104	53.6	0.858	1.42 E-7	8.55 E-7	—	—	—
	140	52.6	0.845	8.73	5.33	—	—	—
	176	51.9	0.832	5.91	3.66	—	—	—
	212	51.2	0.820	4.18	2.63	—	—	—
	248	50.5	0.809	3.22	2.05	—	—	—

TABLE A-5 Physical Properties of SAE Oils and Lubricants

Fluid	SI Units				EE Units			
	Temp. (°C)	Specific gravity	Kinematic viscosity ν (m²/s) Minimum	Maximum	Temp. (°F)	Specific gravity	Kinematic viscosity ν (ft²/sec) Minimum	Maximum
Oil								
SAE 50	99	—	1.68 E-5	2.27 E-5	210	—	1.81 E-4	2.44 E-4
	99	—	1.29	1.68	210	—	1.08	1.81
	99	—	9.6 E-4	1.29	210	—	1.03 E-2	1.08
	99	—	—	5.7 E-4	210	—	—	6.14 E-3
	−18	0.92	2.60 E-3	1.05 E-2	0	0.92	2.80 E-2	1.13 E-1
	−18	0.92	1.30	2.60 E-2	0	0.92	1.40	2.80 E-2
	−18	0.92	—	1.30	0	0.92	—	1.40
Lubricants								
SAE 250	99	—	4.3 E-5	—	210	—	4.6 E-4	—
140	99	—	2.5	4.3 E-5	210	—	2.7	4.6 E-4
90	99	—	1.4	2.5	210	—	1.5	2.7
85W	99	—	1.1	—	210	—	1.2	—
80W	99	—	7.0 E-6	—	210	—	7.5 E-5	—
75W	99	—	4.2	—	210	—	4.5 E-5	—

TABLE A-6 Viscosity of Steam and Water

Temp. °F	Viscosity of steam and water—in centipoise (μ)														
	1 psia	2 psia	5 psia	10 psia	20 psia	50 psia	100 psia	200 psia	500 psia	1000 psia	2000 psia	5000 psia	7500 psia	10000 psia	12000 psia
Sat. water	.667	.524	.388	.313	.255	.197	.164	.138	.111	.094	.078	—	—	—	—
Sat. steam	.010	.010	.011	.012	.012	.013	.014	.015	.017	.019	.023	—	—	—	—
1500	.041	.041	.041	.041	.041	.041	.041	.041	.042	.042	.042	.044	.046	.048	.050
1450	.040	.040	.040	.040	.040	.040	.040	.040	.040	.041	.041	.043	.045	.047	.049
1400	.039	.039	.039	.039	.039	.039	.039	.039	.039	.040	.040	.042	.044	.047	.049
1350	.038	.038	.038	.038	.038	.038	.038	.038	.038	.038	.039	.041	.044	.046	.049
1300	.037	.037	.037	.037	.037	.037	.037	.037	.037	.037	.038	.040	.043	.045	.048
1250	.035	.035	.035	.035	.035	.035	.035	.036	.036	.036	.037	.039	.042	.045	.048
1200	.034	.034	.034	.034	.034	.034	.034	.034	.035	.035	.036	.038	.041	.045	.048
1150	.034	.034	.034	.034	.034	.034	.034	.034	.034	.034	.034	.037	.041	.045	.049
1100	.032	.032	.032	.032	.032	.032	.032	.032	.033	.033	.034	.037	.040	.045	.050
1050	.031	.031	.031	.031	.031	.031	.031	.031	.032	.032	.033	.036	.040	.047	.052
1000	.030	.030	.030	.030	.030	.030	.030	.030	.030	.031	.032	.035	.041	.049	.055
950	.029	.029	.029	.029	.029	.029	.029	.029	.029	.030	.031	.035	.042	.052	.059
900	.028	.028	.028	.028	.028	.028	.028	.028	.028	.028	.029	.035	.045	.057	.064
850	.026	.026	.026	.026	.026	.026	.027	.027	.027	.027	.028	.035	.052	.064	.070
800	.025	.025	.025	.025	.025	.025	.025	.025	.026	.026	.027	.040	.062	.071	.075
750	.024	.024	.024	.024	.024	.024	.024	.024	.025	.025	.026	.057	.071	.078	.081
700	.023	.023	.023	.023	.023	.023	.023	.023	.025	.024	.026	.071	.079	.085	.086
650	.022	.022	.022	.022	.022	.022	.022	.022	.023	.023	.023	.082	.088	.092	.096
600	.021	.021	.021	.021	.021	.021	.021	.021	.021	.021	.087	.091	.096	.101	.104
550	.020	.020	.020	.020	.020	.020	.020	.020	.020	.019	.095	.101	.105	.109	.113
500	.019	.019	.019	.019	.019	.019	.019	.018	.018	.103	.105	.111	.114	.119	.122
450	.018	.018	.018	.018	.017	.017	.017	.017	.115	.116	.118	.123	.127	.131	.135

TABLE A-6 Continued

Temp. °F	Viscosity of steam and water—in centipoise (μ)														
	1 psia	2 psia	5 psia	10 psia	20 psia	50 psia	100 psia	200 psia	500 psia	1000 psia	2000 psia	5000 psia	7500 psia	10000 psia	12000 psia
400	.016	.016	.016	.016	.016	.016	.016	.016	.131	.132	.134	.138	.143	.147	.150
350	.015	.015	.015	.015	.015	.015	.015	.152	.153	.154	.155	.160	.164	.168	.171
300	.014	.014	.014	.014	.014	.014	.182	.183	.183	.184	.185	.190	.194	.198	.201
250	.013	.013	.013	.013	.013	.228	.228	.228	.228	.229	.231	.235	.238	.242	.245
200	.012	.012	.012	.012	.300	.300	.300	.300	.301	.301	.303	.306	.310	.313	.316
150	.011	.011	.427	.427	.427	.427	.427	.427	.427	.428	.429	.431	.434	.437	.439
100	.680	.680	.680	.680	.680	.680	.680	.680	.680	.680	.680	.681	.682	.683	.683
50	1.299	1.299	1.299	1.299	1.299	1.299	1.299	1.299	1.299	1.298	1.296	1.289	1.284	1.279	1.275
32	1.753	1.753	1.753	1.753	1.753	1.753	1.753	1.752	1.751	1.749	1.745	1.733	1.723	1.713	1.705

Values directly below underscored viscosities are for water.
ᵃCritical point.

TABLE A-7   Viscosities of Gases (Coordinates for use with Fig. A-5)

No.	Gas	X	Y
1	Acetic acid	7.7	14.3
2	Acetone	8.9	13.0
3	Acetylene	9.8	14.9
4	Air	11.0	20.0
5	Ammonia	8.4	16.0
6	Argon	10.5	22.4
7	Benzene	8.5	13.2
8	Bromine	8.9	19.2
9	Butene	9.2	13.7
10	Butylene	8.9	13.0
11	Carbon dioxide	9.5	18.7
12	Carbon disulphide	8.0	16.0
13	Carbon monoxide	11.0	20.0
14	Chlorine	9.0	18.4
15	Chloroform	8.9	15.7
16	Cyanogen	9.2	15.2
17	Cyclohexane	9.2	12.0
18	Ethane	9.1	14.5
19	Ethyl acetate	8.5	13.2
20	Ethyl alcohol	9.2	14.2
21	Ethyl chloride	8.5	15.6
22	Ethyl ether	8.9	13.0
23	Ethylene	9.5	15.1
24	Fluorine	7.3	23.8
25	Freon-11 (CCl_3F)	10.6	15.1
26	Freon-12 (CCl_2F_2)	11.1	16.0
27	Freon-21 ($CHCl_2F$)	10.8	15.3
28	Freon-22 ($CHClF_2$)	10.1	17.0
29	Freon-113 ($CCl_2F-CClF_2$)	11.3	14.0
30	Helium	10.9	20.5
31	Hexane	8.6	11.8
32	Hydrogen	11.2	12.4
33	$3H_2 + 1N_2$	11.2	17.2
34	Hydrogen bromide	8.8	20.9
35	Hydrogen chloride	8.8	18.7
36	Hydrogen cyanide	9.8	14.9
37	Hydrogen iodide	9.0	21.3
38	Hydrogen sulphide	8.6	18.0
39	Iodine	9.0	18.4
40	Mercury	5.3	22.9
41	Methane	9.9	15.5
42	Methyl alcohol	8.5	15.6

TABLE A-7   Continued

No.	Gas	X	Y
43	Nitric oxide	10.9	20.5
44	Nitrogen	10.6	20.0
45	Nitrosyl chloride	8.0	17.6
46	Nitrous oxide	8.8	19.0
47	Oxygen	11.0	21.3
48	Pentane	7.0	12.8
49	Propane	9.7	12.9
50	Propyl alcohol	8.4	13.4
51	Propylene	9.0	13.8
52	Sulphur dioxide	9.6	17.0
53	Toluene	8.6	12.4
54	2, 3, 3-trimethylbutane	9.5	10.5
55	Water	8.0	16.0
56	Xenon	9.3	23.0

To convert to lb/ft-hr multiply by 2.42.

Source: By permission from *Chemical Engineers' Handbook*, by J. H. Perry, copyright 1975, McGraw-Hill Book Company Inc.

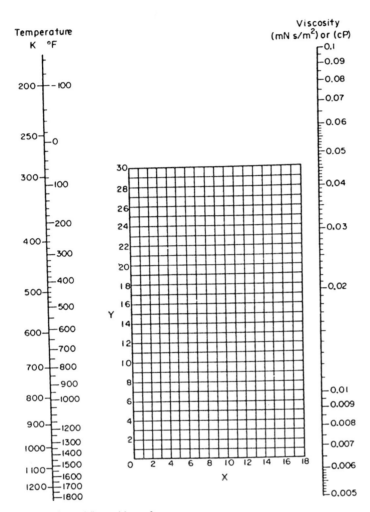

FIGURE A-5　Viscosities of gases.

The curves for hydrocarbon vapors and natural gases in the chart at the upper right are taken from Maxwell[15]; the curves for all other gases (except helium [7]) in the chart are based upon Sutherland's formula, as follows:

$$\mu = \mu_0 \left(\frac{0.555\ T_0 + C}{0.555\ T + C}\right)\left(\frac{T}{T_0}\right)^{3/2}$$

where:

μ = viscosity, in centipoise at temperature T.

μ_0 = viscosity, in centipoise at temperature T_0.

T = absolute temperature, in degrees Rankine (460 + deg. F) for which viscosity is desired.

T_0 = absolute temperature, in degrees Rankine, for which viscosity is known.

C = Sutherland's constant.

Note: The variation of viscosity with pressure is small for most gases. For gases given on this page, the correction of viscosity for pressure is less than 10 per cent for pressures up to 500 pounds per square inch.

Fluid	Approximate Values of "C"
O_2	127
Air	120
N_2	111
CO_2	240
CO	118
SO_2	416
NH_3	370
H_2	72

Upper chart example: The viscosity of sulphur dioxide gas (SO_2) at 200 F is 0.016 centipoise.

Lower chart example: The viscosity of carbon dioxide gas (CO_2) at about 80 F is 0.015 centipoise.

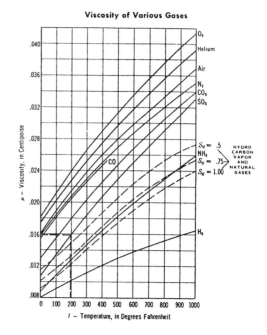

Viscosity of Various Gases

t – Temperature, in Degrees Fahrenheit

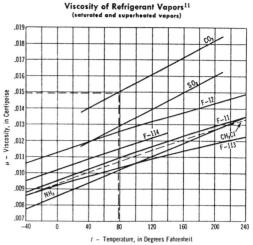

Viscosity of Refrigerant Vapors[11]
(saturated and superheated vapors)

t – Temperature, in Degrees Fahrenheit

FIGURE A-6 Viscosity of gases and vapors.

Appendix B

Generalized Viscosity Plot

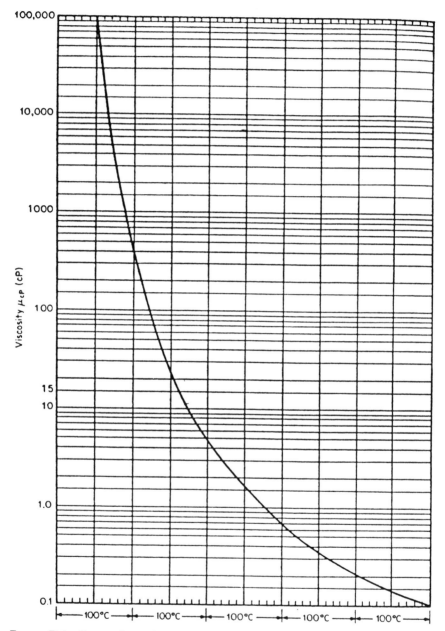

FIGURE B-1   Curves for estimating viscosity from a single measurement value.
(From Gambill, 1959.)

Appendix C

Properties of Gases

TABLE C-1 Physical Properties of Gases (Approximate Values at 68°F and 14.7 psia)

Name of gas	Chemical formula or symbol	Approx. molecular weight (M)	Weight density, pounds per cubic foot (ρ)	Specific gravity relative to air (S_g)	Individual gas constant (R)	Specific heat at room temperature (Btu/Lb °F)		Heat capacity per cubic foot		k equal to c_p/c_v
						c_p^a	c_v^b	c_p	c_v	
Acetylene (ethyne)	C_2H_2	26.0	.0682	0.907	59.4	0.350	0.269	.0239	.0184	1.30
Air	—	29.0	.0752	1.000	53.3	0.241	0.172	.0181	.0129	1.40
Ammonia	NH_3	17.0	.0448	0.596	91.0	0.523	0.396	.0234	.0178	1.32
Argon	A	39.9	.1037	1.379	38.7	0.124	0.074	.0129	.0077	1.67
Butane	C_4H_{10}	58.1	.1554	2.067	26.5	0.395	0.356	.0614	.0553	1.11
Carbon dioxide	CO_2	44.0	.1150	1.529	35.1	0.205	0.158	.0236	.0181	1.30
Carbon monoxide	CO	28.0	.0727	0.967	55.2	0.243	0.173	.0177	.0126	1.40
Chlorine	Cl_2	70.9	.1869	2.486	21.8	0.115	0.086	.0215	.0162	1.33
Ethane	C_2H_6	30.0	.0789	1.049	51.5	0.386	0.316	.0305	.0250	1.22
Ethylene	C_2H_4	28.0	.0733	0.975	55.1	0.400	0.329	.0293	.0240	1.22
Helium	He	4.0	.01039	0.1381	386.3	1.250	0.754	0.130	.0078	1.66
Hydrogen chloride	HCl	36.5	.0954	1.268	42.4	0.191	0.135	.0182	.0129	1.41
Hydrogen	H_2	2.0	.00523	0.0695	766.8	3.420	2.426	.0179	.0127	1.41

Gas	Formula								
Hydrogen sulphide	H_2S	34.1	.0895	1.190	45.2	0.243	0.187	.0217	.0167	1.30
Methane	CH_4	16.0	.0417	0.554	96.4	0.593	0.449	.0247	.0187	1.32
Methyl chloride	CH_3Cl	50.5	.1342	1.785	30.6	0.240	0.200	.0322	.0268	1.20
Natural gas	—	19.5	.0502	0.667	79.1	0.560	0.441	.0281	.0221	1.27
Nitric oxide	NO	30.0	.0780	1.037	51.5	0.231	0.165	.0180	.0129	1.40
Nitrogen	N_2	28.0	.0727	0.967	55.2	0.247	0.176	.0180	.0127	1.41
Nitrous oxide	N_2O	44.0	.1151	1.530	35.1	0.221	0.169	.0254	.0194	1.31
Oxygen	O_2	32.0	.0831	1.105	48.3	0.217	0.155	.0180	.0129	1.40
Propane	C_3H_8	44.1	.1175	1.562	35.0	0.393	0.342	.0462	.0402	1.15
Propene (propylene)	C_3H_6	42.1	.1091	1.451	36.8	0.358	0.314	.0391	.0343	1.14
Sulphur dioxide	SO_2	64.1	.1703	2.264	24.0	0.154	0.122	.0262	.0208	1.26

[a] c_p = Specific heat at constant pressure.

[b] c_v = Specific heat at constant volume.

Weight Density values were obtained by multiplying density of air by specific gravity of gas. For values at 60°F, multiply by 1.0154.

Natural Gas values are representative only. Exact characteristics require knowledge of specific constituents.

Source: Molecular Weight, Specific Gravity, Individual Gas Constant, and Specific Heat values were abstracted from, or based on, data in Table 24 of Mark's "Standard Handbook for Mechanical Engineers" (seventh edition).

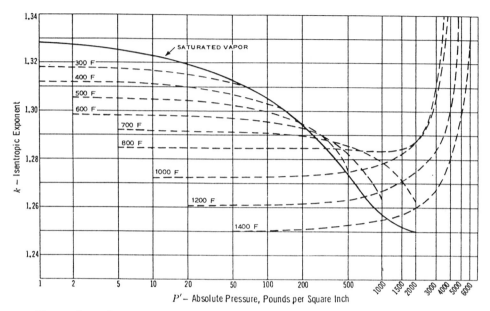

FIGURE C-1   Steam values of isentropic exponent, k (for small changes in pressure (or volume) along an isentropic, pv^k = constant).

Appendix D

Pressure-Enthalpy Diagrams for Various Compounds

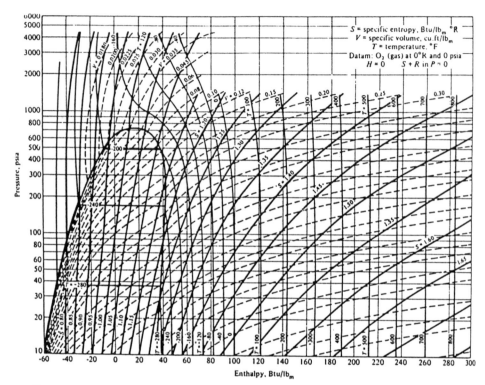

FIGURE D-1 Oxygen pressure-enthalpy diagram. (From L. N. Canjar and F. S. Manning, *Thermodynamic Properties and Reduced Correlations for Gases*, copyright Gulf Publishing Co., Houston, 1967. Reproduced by permission.)

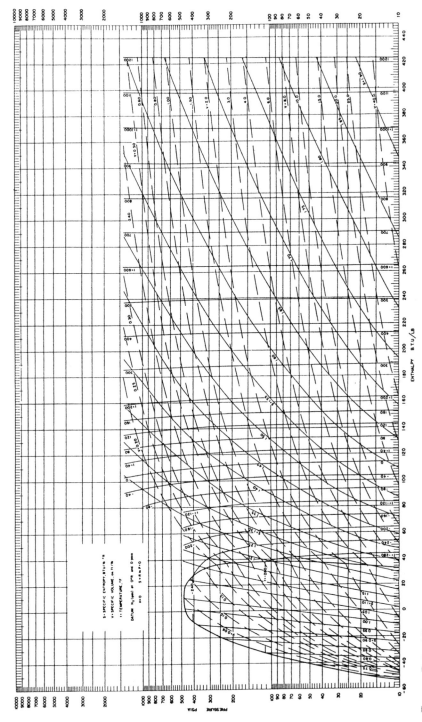

FIGURE D-2 Nitrogen pressure-enthalpy diagram. (From "Thermo Properties of Non-Hydrocarbons", by V. M. Tejada et al. *Hydrocarbon Processing & Petroleum Refiner*, March, 1966. Copyright 1966, The Gulf Publishing Co., Houston, Texas. Reprinted by permission.)

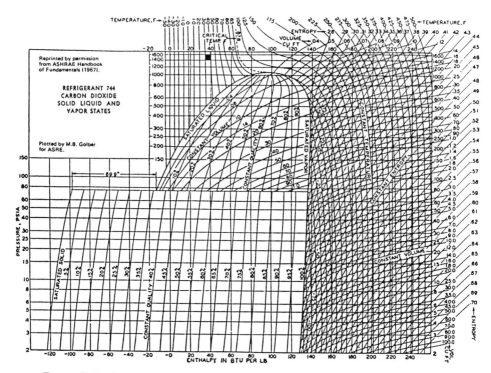

FIGURE D-3   Pressure-enthalpy chart, carbon dioxide. (From *ASHRAE Handbook of Fundamentals*, 1967.)

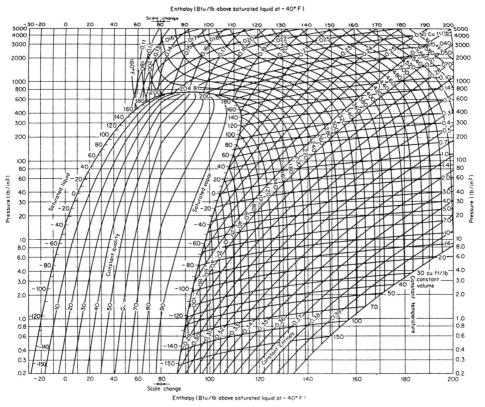

FIGURE D-4   Enthalpy-log pressure diagram for Refrigerant 22. Temperature in °F, volume in cu. ft./lb., entropy in B.t.u./(lb.)(°R.), quality in weight per cent. (Reprinted by permission of E. I. du Pont de Nemours & Company, 1967.)

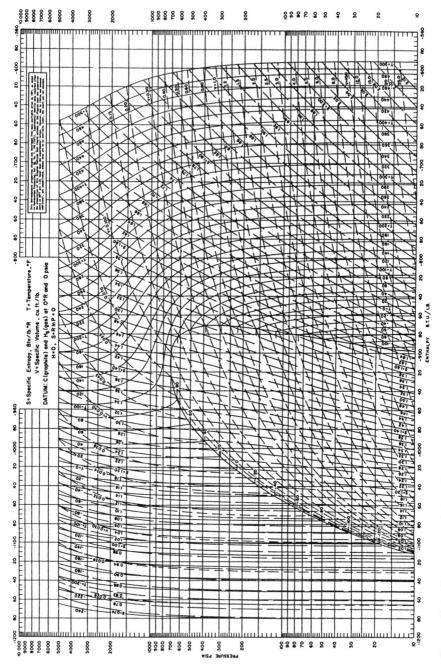

FIGURE D-5 Pressure-enthalpy diagram of ethane. [From *Hydrocarbon Processing, 50*(4):140 (1971).]

444

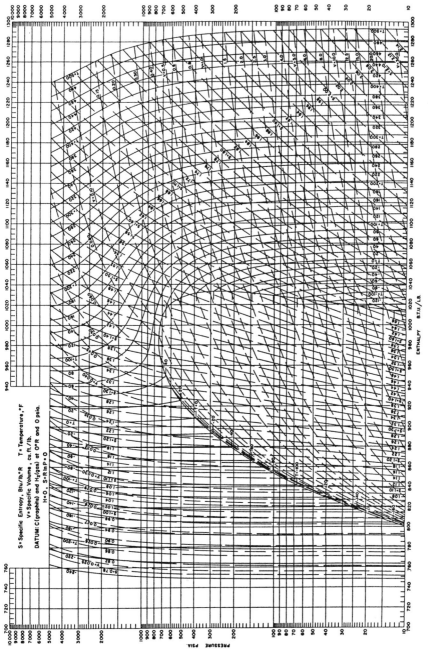

FIGURE D-6 Pressure-enthalpy diagram for ethylene. (From *Fluid Thermal Properties for Petroleum Systems*, R. E. Sterling, Copyright Gulf Publishing Co.)

445

Appendix E

Microscopic Conservation Equations in Rectangular, Cylindrical, and Spherical Coordinates

EQUATIONS OF CONTINUITY, MOMENTUM, AND ENERGY IN RECTANGULAR, CYLINDRICAL, AND SPHERICAL COORDINATES

CONTINUITY EQUATION

Rectangular coordinates (x, y, z):

$$\frac{\partial \rho}{\partial t} + \frac{\partial}{\partial x} (\rho v_x) + \frac{\partial}{\partial y} (\rho v_y) + \frac{\partial}{\partial z} (\rho v_z) = 0$$

Cylindrical coordinates (r, θ, z):

$$\frac{\partial \rho}{\partial t} + \frac{1}{r} \frac{\partial}{\partial r} (\rho r v_r) + \frac{1}{r} \frac{\partial}{\partial \theta} (\rho v_\theta) + \frac{\partial}{\partial z} (\rho v_z) = 0$$

Reprinted from Bird, Stewart, and Lightfoot, *Transport Phenomena*.

Spherical coordinates $(r,\ \theta,\ \phi)$:

$$\frac{\partial \rho}{\partial t} + \frac{1}{r^2}\frac{\partial}{\partial r}(\rho r^2 v_r) + \frac{1}{r \sin \theta}\frac{\partial}{\partial \theta}(\rho v_\theta \sin \theta) + \frac{1}{r \sin \theta}\frac{\partial}{\partial \phi}(\rho v_\phi) = 0$$

MOMENTUM EQUATION IN RECTANGULAR COORDINATES

x-component: $\displaystyle \rho\left(\frac{\partial v_x}{\partial t} + v_x\frac{\partial v_x}{\partial x} + v_y\frac{\partial v_x}{\partial y} + v_z\frac{\partial v_x}{\partial z}\right)$

$$= -\frac{\partial P}{\partial x} + \left(\frac{\partial \tau_{xx}}{\partial x} + \frac{\partial \tau_{yx}}{\partial y} + \frac{\partial \tau_{zx}}{\partial z}\right) + \rho g_x$$

y-component: $\displaystyle \rho\left(\frac{\partial v_y}{\partial t} + v_x\frac{\partial v_y}{\partial x} + v_y\frac{\partial v_y}{\partial y} + v_z\frac{\partial v_y}{\partial z}\right)$

$$= -\frac{\partial P}{\partial y} + \left(\frac{\partial \tau_{xy}}{\partial x} + \frac{\partial \tau_{yy}}{\partial y} + \frac{\partial \tau_{zy}}{\partial z}\right) + \rho g_y$$

z-component: $\displaystyle \rho\left(\frac{\partial v_z}{\partial t} + v_x\frac{\partial v_z}{\partial x} + v_y\frac{\partial v_z}{\partial y} + v_z\frac{\partial v_z}{\partial z}\right)$

$$= -\frac{\partial P}{\partial z} + \left(\frac{\partial \tau_{xz}}{\partial x} + \frac{\partial \tau_{yz}}{\partial y} + \frac{\partial \tau_{zz}}{\partial z}\right) + \rho g_z$$

MOMENTUM EQUATION IN CYLINDRICAL COORDINATES

r-component: $\displaystyle \rho\left(\frac{\partial v_r}{\partial t} + v_r\frac{\partial v_r}{\partial r} + \frac{v_\theta}{r}\frac{\partial v_r}{\partial \theta} - \frac{v_\theta^2}{r} + v_z\frac{\partial v_r}{\partial z}\right)$

$$= -\frac{\partial P}{\partial r} + \left(\frac{1}{r}\frac{\partial}{\partial r}(r\tau_{rr}) + \frac{1}{r}\frac{\partial \tau_{r\theta}}{\partial \theta} - \frac{\tau_{\theta\theta}}{r} + \frac{\partial \tau_{rz}}{\partial z}\right) + \rho g_r$$

θ-component: $\displaystyle \rho\left(\frac{\partial v_\theta}{\partial t} + v_r\frac{\partial v_\theta}{\partial r} + \frac{v_\theta}{r}\frac{\partial v_\theta}{\partial \theta} + \frac{v_r v_\theta}{r} + v_z\frac{\partial v_\theta}{\partial z}\right)$

$$= -\frac{1}{r}\frac{\partial P}{\partial \theta} + \left(\frac{1}{r^2}\frac{\partial}{\partial r}(r^2\tau_{r\theta}) + \frac{1}{r}\frac{\partial \tau_{\theta\theta}}{\partial \theta} + \frac{\partial \tau_{\theta z}}{\partial z}\right) + \rho g_\theta$$

z-component: $\displaystyle \rho\left(\frac{\partial v_z}{\partial t} + v_r\frac{\partial v_z}{\partial r} + \frac{v_\theta}{r}\frac{\partial v_z}{\partial \theta} + v_z\frac{\partial v_z}{\partial z}\right)$

$$= -\frac{\partial P}{\partial z} + \left(\frac{1}{r}\frac{\partial}{\partial r}(r\tau_{rz}) + \frac{1}{r}\frac{\partial \tau_{\theta z}}{\partial \theta} + \frac{\partial \tau_{zz}}{\partial z}\right) + \rho g_z$$

MOMENTUM EQUATION IN SPHERICAL COORDINATES

r-component: $\rho\left(\dfrac{\partial v_r}{\partial t} + v_r \dfrac{\partial v_r}{\partial r} + \dfrac{v_\theta}{r}\dfrac{\partial v_r}{\partial \theta} + \dfrac{v_\phi}{r \sin \theta}\dfrac{\partial v_r}{\partial \phi} - \dfrac{v_\theta^2 + v_\phi^2}{r}\right)$

$$= -\frac{\partial P}{\partial r} + \left(\frac{1}{r^2}\frac{\partial}{\partial r}(r^2 \tau_{rr}) + \frac{1}{r \sin \theta}\frac{\partial}{\partial \theta}(\tau_{r\theta}\sin \theta)\right.$$

$$\left. + \frac{1}{r \sin \theta}\frac{\partial \tau_{r\phi}}{\partial \phi} - \frac{\tau_{\theta\theta} + \tau_{\phi\phi}}{r}\right) + \rho g_r$$

θ-component: $\rho\left(\dfrac{\partial v_\theta}{\partial t} + v_r \dfrac{\partial v_\theta}{\partial r} + \dfrac{v_\theta}{r}\dfrac{\partial v_\theta}{\partial \theta} + \dfrac{v_\phi}{r \sin \theta}\dfrac{\partial v_\theta}{\partial \phi} + \dfrac{v_r v_\theta}{r} - \dfrac{v_\phi^2 \cot \theta}{r}\right)$

$$= -\frac{1}{r}\frac{\partial P}{\partial \theta} + \left(\frac{1}{r^2}\frac{\partial}{\partial r}(r^2 \tau_{r\theta}) + \frac{1}{r \sin \theta}\frac{\partial}{\partial \theta}(\tau_{\theta\theta}\sin \theta) + \frac{1}{r \sin \theta}\frac{\partial \tau_{\theta\phi}}{\partial \phi}\right.$$

$$\left. + \frac{\tau_{r\theta}}{r} - \frac{\cot \theta}{r}\tau_{\phi\phi}\right) + \rho g_\theta$$

φ-component: $\rho\left(\dfrac{\partial v_\phi}{\partial t} + v_r \dfrac{\partial v_\phi}{\partial r} + \dfrac{v_\theta}{r}\dfrac{\partial v_\phi}{\partial \theta} + \dfrac{v_\phi}{r \sin \theta}\dfrac{\partial v_\phi}{\partial \phi} + \dfrac{v_\phi v_r}{r} + \dfrac{v_\theta v_\phi}{r}\cot \theta\right)$

$$= -\frac{1}{r \sin \theta}\frac{\partial P}{\partial \phi} + \left(\frac{1}{r^2}\frac{\partial}{\partial r}(r^2 \tau_{r\phi}) + \frac{1}{r}\frac{\partial \tau_{\theta\phi}}{\partial \theta} + \frac{1}{r \sin \theta}\frac{\partial \tau_{\phi\phi}}{\partial \phi}\right.$$

$$\left. + \frac{\tau_{r\phi}}{r} + \frac{2 \cot \theta}{r}\tau_{\theta\phi}\right) + \rho g_\phi$$

COMPONENTS OF THE STRESS TENSOR τ

Rectangular coordinates:

$$\tau_{xx} = +\mu\left[2\frac{\partial v_x}{\partial x} - \frac{2}{3}(\nabla \cdot v)\right] + \kappa(\nabla \cdot v)$$

$$\tau_{yy} = +\mu\left[2\frac{\partial v_y}{\partial y} - \frac{2}{3}(\nabla \cdot v)\right] + \kappa(\nabla \cdot v)$$

$$\tau_{zz} = +\mu\left[2\frac{\partial v_z}{\partial z} - \frac{2}{3}(\nabla \cdot v)\right] + \kappa(\nabla \cdot v)$$

$$\tau_{xy} = \tau_{yx} = +\mu\left[\frac{\partial v_x}{\partial y} + \frac{\partial v_y}{\partial x}\right]$$

$$\tau_{yz} = \tau_{zy} = +\mu\left[\frac{\partial v_y}{\partial z} + \frac{\partial v_z}{\partial y}\right]$$

$$\tau_{zx} = \tau_{xz} = +\mu\left[\frac{\partial v_z}{\partial x} + \frac{\partial v_x}{\partial z}\right]$$

$$(\nabla \cdot v) = \frac{\partial v_x}{\partial x} + \frac{\partial v_y}{\partial y} + \frac{\partial v_z}{\partial z}$$

Cylindrical coordinates:

$$\tau_{rr} = +\mu\left[2\,\frac{\partial v_r}{\partial r} - \frac{2}{3}(\nabla \cdot v)\right] + \kappa(\nabla \cdot v)$$

$$\tau_{\theta\theta} = +\mu\left[2\left(\frac{1}{r}\frac{\partial v_\theta}{\partial\theta} + \frac{v_r}{r}\right) - \frac{2}{3}(\nabla \cdot v)\right] + \kappa(\nabla \cdot v)$$

$$\tau_{zz} = +\mu\left[2\,\frac{\partial v_z}{\partial z} - \frac{2}{3}(\nabla \cdot v)\right] + \kappa(\nabla \cdot v)$$

$$\tau_{r\theta} = \tau_{\theta r} = +\mu\left[r\,\frac{\partial}{\partial r}\left(\frac{v_\theta}{r}\right) + \frac{1}{r}\frac{\partial v_r}{\partial\theta}\right]$$

$$\tau_{\theta z} = \tau_{z\theta} = +\mu\left[\frac{\partial v_\theta}{\partial z} + \frac{1}{r}\frac{\partial v_z}{\partial\theta}\right]$$

$$\tau_{zr} = \tau_{rz} = +\mu\left[\frac{\partial v_z}{\partial r} + \frac{\partial v_r}{\partial z}\right]$$

$$(\nabla \cdot v) = \frac{1}{r}\frac{\partial}{\partial r}(rv_r) + \frac{1}{r}\frac{\partial v_\theta}{\partial\theta} + \frac{\partial v_z}{\partial z}$$

Spherical coordinates:

$$\tau_{rr} = +\mu\left[2\,\frac{\partial v_r}{\partial r} - \frac{2}{3}(\nabla \cdot v)\right] + \kappa(\nabla \cdot v)$$

$$\tau_{\theta\theta} = +\mu\left[2\left(\frac{1}{r}\frac{\partial v_\theta}{\partial\theta} + \frac{v_r}{r}\right) - \frac{2}{3}(\nabla \cdot v)\right] + \kappa(\nabla \cdot v)$$

$$\tau_{\phi\phi} = +\mu\left[2\left(\frac{1}{r\sin\theta}\frac{\partial v_\phi}{\partial\phi} + \frac{v_r}{r} + \frac{v_\theta\cot\theta}{r}\right) - \frac{2}{3}(\nabla \cdot v)\right] + \kappa(\nabla \cdot v)$$

$$\tau_{r\theta} = \tau_{\theta r} = +\mu\left[r\,\frac{\partial}{\partial r}\left(\frac{v_\theta}{r}\right) + \frac{1}{r}\frac{\partial v_r}{\partial\theta}\right]$$

$$\tau_{\theta\phi} = \tau_{\phi\theta} = +\mu\left[\frac{\sin\theta}{r}\frac{\partial}{\partial\theta}\left(\frac{v_\phi}{\sin\theta}\right) + \frac{1}{r\sin\theta}\frac{\partial v_\phi}{\partial\phi}\right]$$

$$\tau_{\phi r} = \tau_{r\phi} = +\mu \left[\frac{1}{r \sin \theta} \frac{\partial v_r}{\partial \phi} + r \frac{\partial}{\partial r} \left(\frac{v_\phi}{r} \right) \right]$$

$$(\nabla \cdot v) = \frac{1}{r^2} \frac{\partial}{\partial r} (r^2 v_r) + \frac{1}{r \sin \theta} \frac{\partial}{\partial \theta} (v_\theta \sin \theta) + \frac{1}{r \sin \theta} \frac{\partial v_\phi}{\partial \phi}$$

Appendix F

Standard Steel Pipe Dimensions and Capacities

TABLE F-1 Steel-Pipe Dimensions: Capacities and Weights

Nominal pipe size, in.	Outside diameter, in.	Schedule no.	Wall thickness, in.	Inside diameter, in.	Cross-sectional area metal, sq. in.	Inside sectional area, sq. ft	Circumference, ft, or surface, sq. ft per ft, of length		Capacity at 1 ft per sec velocity		Weight of pipe per ft, lb
							Outside	Inside	U.S. gal. per min	Lb per hr water	
$\frac{1}{8}$	0.405	40	0.068	0.269	0.072	0.00040	0.106	0.0705	0.179	89.5	0.25
		80	0.095	0.215	0.093	0.00025	0.106	0.0563	0.112	56.0	0.32
$\frac{1}{4}$	0.540	40	0.088	0.364	0.125	0.00072	0.141	0.0954	0.323	161.5	0.43
		80	0.119	0.302	0.157	0.00050	0.141	0.0792	0.224	112.0	0.54
$\frac{3}{8}$	0.675	40	0.091	0.493	0.167	0.00133	0.177	0.1293	0.596	298.0	0.57
		80	0.126	0.423	0.217	0.00098	0.177	0.1110	0.440	220.0	0.74
$\frac{1}{2}$	0.840	40	0.109	0.622	0.250	0.00211	0.220	0.1630	0.945	472.5	0.85
		80	0.147	0.546	0.320	0.00163	0.220	0.1430	0.730	365.0	1.09
		160	0.187	0.466	0.384	0.00118	0.220	0.1220	0.529	264.5	1.31
$\frac{3}{4}$	1.050	40	0.113	0.824	0.333	0.00371	0.275	0.2158	1.665	832.5	1.13
		80	0.154	0.742	0.433	0.00300	0.275	0.1942	1.345	672.5	1.48
		160	0.218	0.614	0.570	0.00206	0.275	0.1610	0.924	462.0	1.94
1	1.315	40	0.133	1.049	0.494	0.00600	0.344	0.2745	2.690	1,345.0	1.68
		80	0.179	0.957	0.639	0.00499	0.344	0.2505	2.240	1,120.0	2.17
		160	0.250	0.815	0.837	0.00362	0.344	0.2135	1.625	812.5	2.85
$1\frac{1}{4}$	1.660	40	0.140	1.380	0.669	0.01040	0.435	0.362	4.57	2,285.0	2.28
		80	0.191	1.278	0.881	0.00891	0.435	0.335	3.99	1,995.0	3.00
		160	0.250	1.160	1.107	0.00734	0.435	0.304	3.29	1,645.0	3.77
$1\frac{1}{2}$	1.990	40	0.145	1.610	0.799	0.01414	0.498	0.422	6.34	3,170.0	2.72
		80	0.200	1.500	1.068	0.01225	0.498	0.393	5.49	2,745.0	3.64
		160	0.281	1.338	1.429	0.00976	0.498	0.350	4.38	2,190.0	4.86

Nominal	OD	Schedule									
2	2.375	40	0.154	2.067	1.075	0.02330	0.622	0.542	10.45	5,225.0	3.66
		80	0.218	1.939	1.477	0.02050	0.622	0.508	9.20	4,600.0	5.03
		160	0.343	1.689	2.190	0.01556	0.622	0.442	6.97	3,485.0	7.45
2½	2.875	40	0.203	2.469	1.704	0.3322	0.753	0.647	14.92	7,460.0	5.80
		80	0.276	2.323	2.254	0.02942	0.753	0.609	13.20	6,600.0	7.67
		160	0.375	2.125	2.945	0.02463	0.753	0.557	11.07	5,535.0	10.0
3	3.500	40	0.216	3.068	2.228	0.05130	0.917	0.804	23.00	11,500.0	7.58
		80	0.300	2.900	3.016	0.04587	0.917	0.760	20.55	10,275.0	10.3
		160	0.437	2.626	4.205	0.03761	0.917	0.688	16.90	8,450.0	14.3
3½	4.000	40	0.226	3.548	2.680	0.06870	1.047	0.930	30.80	15,400.0	9.11
		80	0.318	3.364	3.678	0.06170	1.047	0.882	27.70	13,850.0	12.5
4	4.500	40	0.237	4.026	3.173	0.08840	1.178	1.055	39.6	19,800.0	10.8
		80	0.337	3.826	4.407	0.07986	1.178	1.002	35.8	17,900.0	15.0
		120	0.437	3.626	5.578	0.07170	1.178	0.950	32.2	16,100.0	19.0
		160	0.531	3.438	6.621	0.06447	1.178	0.901	28.9	14,450.0	22.6
5	5.563	40	0.258	5.047	4.304	0.1390	1.456	1.322	62.3	31,150.0	14.7
		80	0.375	4.813	6.112	0.1263	1.456	1.263	57.7	28,850.0	20.8
		120	0.500	4.563	7.953	0.1136	1.456	1.197	51.0	25,500.0	27.1
		160	0.625	4.313	9.696	0.1015	1.456	1.132	45.5	22,750.0	33.0
6	6.625	40	0.280	6.065	5.584	0.2006	1.734	1.590	90.0	45,000.0	19.0
		80	0.432	5.761	8.405	0.1810	1.734	1.510	81.1	40,500.0	28.6
		120	0.562	5.501	10.71	0.1650	1.734	1.445	73.9	36,950.0	36.4
		160	0.718	5.189	13.32	0.1469	1.734	1.360	65.8	32,900.0	45.3
8	8.625	20	0.250	8.125	6.570	0.3601	2.258	2.130	161.5	80,750.0	22.4
		30	0.277	8.071	7.260	0.3553	2.258	2.115	159.4	79,700.0	24.7
		40	0.322	7.981	8.396	0.3474	2.258	2.090	155.7	77,850.0	28.6
		60	0.406	7.813	10.48	0.3329	2.258	2.050	149.4	74,700.0	35.7
		80	0.500	7.625	12.76	0.3171	2.258	2.000	142.3	71,150.0	43.4

TABLE F-1 Continued

Nominal pipe size, in.	Outside diameter, in.	Schedule no.	Wall thickness, in.	Inside diameter, in.	Cross-sectional area metal, sq. in.	Inside sectional area, sq. ft	Circumference, ft, or surface, sq. ft per ft, of length Outside	Inside	Capacity at 1 ft per sec velocity U.S. gal per min	Lb per hr water	Weight of pipe per ft, lb
		100	0.593	7.439	14.96	0.3018	2.258	1.947	135.3	67,650.0	50.9
		120	0.718	7.189	17.84	0.2819	2.258	1.883	126.5	63,250.0	60.7
		140	0.812	7.001	19.93	0.2673	2.258	1.835	120.0	60,000.0	67.8
		160	0.906	6.813	21.97	0.2532	2.258	1.787	113.5	56,750.0	74.7
10	10.75	20	0.250	10.250	8.24	0.5731	2.814	2.685	257.0	128,500.0	28.1
		30	0.307	10.136	10.07	0.5603	2.814	2.665	252.0	126,000.0	34.3
		40	0.365	10.020	11.90	0.5475	2.814	2.620	246.0	123,000.0	40.5
		60	0.500	9.750	16.10	0.5185	2.814	2.550	233.0	116,500.0	54.8
		80	0.593	9.564	18.92	0.4989	2.814	2.503	224.0	112,000.0	64.4
		100	0.718	9.314	22.63	0.4732	2.814	2.440	212.0	106,000.0	77.0
		120	0.843	9.064	26.24	0.4481	2.814	2.373	201.0	100,500.0	89.2
		140	1.000	8.750	30.63	0.4176	2.814	2.290	188.0	93,750.0	105.0
		160	1.125	8.500	34.02	0.3941	2.814	2.230	177.0	88,500.0	116.0
12	12.75	20	0.250	12.250	9.82	0.8185	3.338	3.31	367.0	183,500.0	33.4
		30	0.330	12.090	12.87	0.7972	3.338	3.17	358.0	179,000.0	43.8
		40	0.406	11.938	15.77	0.7773	3.338	3.13	349.0	174,500.0	53.6
		60	0.562	11.626	21.52	0.7372	3.338	3.05	331.0	165,500.0	73.2
		80	0.687	11.376	26.03	0.7058	3.338	2.98	317.0	158,500.0	88.6
		100	0.843	11.064	31.53	0.6677	3.338	2.90	299.0	149,500.0	108.0
		120	1.000	10.750	36.91	0.6303	3.338	2.82	283.0	141,500.0	126.0
		140	1.125	10.500	41.08	0.6013	3.338	2.75	270.0	135,000.0	140.0
		160	1.312	10.126	47.14	0.5592	3.338	2.66	251.0	125,500.0	161.0

14	14.0	10	0.250	13.500	10.80	0.9940	3.665	3.54	446.0	223,000.0	36.8
		20	0.312	13.376	13.42	0.9750	3.665	3.51	438.0	219,000.0	45.7
		30	0.375	13.250	16.05	0.9575	3.665	3.47	430.0	215,000.0	54.6
		40	0.437	13.126	18.61	0.9397	3.665	3.44	422.0	211,000.0	63.3
		60	0.593	12.814	24.98	0.8956	3.665	3.36	402.0	201,000.0	85.0
		80	0.750	12.500	31.22	0.8522	3.665	3.28	382.0	191,000.0	107.0
		100	0.937	12.126	38.45	0.8020	3.665	3.18	360.0	180,000.0	131.0
		120	1.062	11.876	43.17	0.7693	3.665	3.11	345.0	172,500.0	147.0
		140	1.250	11.500	50.07	0.7213	3.665	3.01	324.0	162,000.0	171.0
		160	1.406	11.188	55.63	0.6827	3.665	2.93	306.0	153,000.0	190.0
16	16.0	10	0.250	15.500	12.37	1.3104	4.189	4.06	587.0	293,500.0	42.1
		20	0.312	15.376	15.38	1.2895	4.189	4.03	578.0	289,000.0	52.3
		30	0.375	15.250	18.41	1.2680	4.189	4.00	568.0	284,000.0	62.6
		40	0.500	15.000	24.35	1.2272	4.189	3.93	550.0	275,000.0	82.8
		60	0.656	14.688	31.62	1.1766	4.189	3.85	528.0	264,000.0	108.0
		80	0.843	14.314	40.14	1.1175	4.189	3.76	500.0	250,000.0	137.0
		100	1.031	13.938	48.48	1.0596	4.189	3.65	474.0	237,000.0	165.0
		120	1.218	13.564	56.56	1.0035	4.189	3.56	450.0	225,000.0	193.0
		140	1.437	13.126	65.74	0.9397	4.189	3.44	422.0	211,000.0	224.0
		160	1.562	12.876	70.85	0.9043	4.189	3.37	405.0	202,500.0	241.0
18	18.0	10	0.250	17.50	13.94	1.6703	4.712	4.59	748.0	374,000.0	47.4
		20	0.312	17.376	17.34	1.6468	4.712	4.55	738.0	369,000.0	59.0
		30	0.437	17.126	24.11	1.5993	4.712	4.49	717.0	358,500.0	82.0
		40	0.562	16.876	30.79	1.5533	4.712	4.42	697.0	348,500.0	105.0
		60	0.718	15.564	38.98	1.4964	4.712	4.34	670.0	335,000.0	133.0
		80	0.937	16.126	50.23	1.4183	4.712	4.23	635.0	317,500.0	171.0
		100	1.156	15.688	61.17	1.3423	4.712	4.11	602.0	301,000.0	208.0
		120	1.343	15.314	70.28	1.2791	4.712	4.02	573.0	286,500.0	239.0
		140	1.562	14.876	80.66	1.2070	4.712	3.90	540.0	270,000.0	275.0
		160	1.750	14.500	89.34	1.1467	4.712	3.80	514.0	257,000.0	304.0

TABLE F-1 Continued

Nominal pipe size, in.	Outside diameter, in.	Schedule no.	Wall thickness, in.	Inside diameter, in.	Cross-sectional area metal, sq. in.	Inside sectional area, sq. ft	Circumference, ft, or surface, sq. ft per ft, of length Outside	Inside	Capacity at 1 ft per sec velocity U.S. gal per min	Lb per hr water	Weight of pipe per ft, lb
20	20.0	10	0.250	19.500	15.51	2.0740	5.236	5.11	930.0	465,000.0	52.8
		20	0.375	19.250	23.12	2.0211	5.236	5.05	902.0	451,000.0	78.6
		30	0.500	19.000	30.63	1.9689	5.236	4.98	883.0	441,500.0	105.0
		40	0.593	18.814	36.15	1.9305	5.236	4.94	866.0	433,000.0	123.0
		60	0.812	18.376	48.95	1.8317	5.236	4.81	826.0	413,000.0	167.0
		80	1.031	17.938	61.44	1.7550	5.236	4.70	787.0	393,500.0	209.0
		100	1.250	17.500	73.63	1.6703	5.236	4.59	750.0	375,000.0	251.0
		120	1.500	17.000	87.18	1.5762	5.236	4.46	707.0	353,500.0	297.0
		140	1.750	16.500	100.3	1.4849	5.236	4.32	665.0	332,500.0	342.0
		160	1.937	16.126	109.9	1.4183	5.236	4.22	635.0	317,500.0	374.0

The schedule number corresponds roughly to 10^3 allowable pressure per allowable stress. Thus, for a material with an allowable stress of 10,000 lb/in^2 a schedule 40 pipe would have an allowable pressure of 400 lb/in^2. [From *Chemical Engineers' Handbook*, by Perry, Chilton, and Kirkpatrick. Copyright © 1963, McGraw-Hill, Inc. Used by permission of the publisher.]

Appendix G

Flow of Water/Air Through Schedule 40 Pipe

TABLE G-1 Flow of Water Through Schedule 40 Steel Pipe

Pressure Drop per 100 feet and Velocity in Schedule 40 Pipe for Water at 60 F.

Velocity values are in Feet per Second; Pressure Drop values are in Lbs per Sq. In. In each column pair the upper (smaller) pipe size applies to the lighter flows and the lower (larger) pipe size applies to the heavier flows.

Gallons per Minute	Cubic Ft per Second	1/8″ · 2″ Veloc.	1/8″ · 2″ Press.	1/4″ · 2½″ Veloc.	1/4″ · 2½″ Press.	3/8″ · 3″ Veloc.	3/8″ · 3″ Press.	1/2″ · 3½″ Veloc.	1/2″ · 3½″ Press.	3/4″ · 4″ Veloc.	3/4″ · 4″ Press.	1″ · 5″ Veloc.	1″ · 5″ Press.	1¼″ · 6″ Veloc.	1¼″ · 6″ Press.	1½″ · 8″ Veloc.	1½″ · 8″ Press.
.2	0.000446	1.13	1.86	0.616	0.359												
.3	0.000668	1.69	4.22	0.924	0.903	0.504	0.159	0.317	0.061								
.4	0.000891	2.26	6.98	1.23	1.61	0.672	0.345	0.422	0.086								
.5	0.00111	2.82	10.5	1.54		0.840	0.539	0.528	0.167	0.301	0.033						
.6	0.00134	3.39	14.7	1.85	3.29	1.01	0.751	0.633	0.240	0.361	0.041						
.8	0.00178	4.52	25.0	2.46	5.44	1.34	1.25	0.844	0.408	0.481	0.102						
1	0.00223	5.65	37.2	3.08	8.28	1.68	1.85	1.06	0.600	0.602	0.155	0.371	0.048				
2	0.00446	11.29	134.4	6.16	30.1	3.36	6.58	2.11	2.10	1.20	0.526	0.743	0.164				
3	0.00668			9.25	64.1	5.04	13.9	3.17	4.33	1.81	1.09	1.114	0.336				
4	0.00891			12.33	111.2	6.72	23.9	4.22	7.42	2.41	1.83	1.49	0.565				
5	0.01114					8.40	36.7	5.28	11.2	3.01	2.75	1.86	0.835	1.073	0.223	0.788	0.104
6	0.01337	0.574	0.044			10.08	51.9	6.33	15.8	3.61	3.84	2.23	1.17	1.29	0.309	0.946	0.145
8	0.01782	0.765	0.073			13.44	91.1	8.45	27.7	4.81	6.60	2.97	1.99	1.72	0.518	1.26	0.241
10	0.02228	0.956	0.108	0.670	0.046			10.56	42.4	6.02	9.99	3.71	2.99	2.15	0.774	1.58	0.361
15	0.03342	1.43	0.224	1.01	0.094					9.03	21.6	5.57	6.36	3.22	1.63	2.37	0.755
20	0.04456	1.91	0.375	1.34	0.158	0.868	0.056			12.03	37.8	7.43	10.9	4.29	2.78	3.16	1.28
25	0.05570	2.39	0.561	1.68	0.234	1.09	0.083	0.812	0.041			9.28	16.7	5.37	4.22	3.94	1.93
30	0.06684	2.87	0.786	2.01	0.327	1.30	0.114	0.974	0.056			11.14	23.8	6.44	5.92	4.73	2.72
35	0.07798	3.35	1.05	2.35	0.436	1.52	0.151	1.14	0.071	0.882	0.041	13.0	32.2	7.51	7.90	5.52	3.64
40	0.08912	3.83	1.35	2.68	0.556	1.74	0.192	1.30	0.095	1.01	0.052	14.85	41.5	8.59	10.24	6.30	4.65
45	0.1003	4.30	1.67	3.02	0.668	1.95	0.239	1.46	0.117	1.13	0.064			9.67	12.80	7.09	5.85
50	0.1114	4.78	2.03	3.35	0.839	2.17	0.288	1.62	0.142	1.26	0.076			10.74	15.66	7.88	7.15
60	0.1337	5.74	2.87	4.02	1.18	2.60	0.406	1.95	0.204	1.51	0.107			12.89	22.2	9.47	10.21
70	0.1560	6.70	3.84	4.69	1.59	3.04	0.540	2.27	0.261	1.76	0.143	1.12	0.047			11.05	13.71
80	0.1782	7.65	4.97	5.36	2.03	3.47	0.687	2.60	0.334	2.02	0.180	1.28	0.060			12.62	17.59
90	0.2005	8.60	6.20	6.03	2.53	3.91	0.861	2.92	0.416	2.27	0.224	1.44	0.074			14.20	22.0
100	0.2228	9.56	7.59	6.70	3.09	4.34	1.05	3.25	0.509	2.52	0.272	1.60	0.090	1.11	0.036		
125	0.2785	11.97	11.76	8.38	4.71	5.43	1.61	4.06	0.769	3.15	0.415	2.01	0.135	1.39	0.055		
150	0.3342	14.36	16.70	10.05	6.69	6.51	2.24	4.87	1.08	3.78	0.580	2.41	0.190	1.67	0.077		
175	0.3899	16.75	22.3	11.73	8.97	7.60	3.08	5.68	1.44	4.41	0.774	2.81	0.253	1.94	0.102		
200	0.4456	19.14	28.8	13.42	11.68	8.68	3.87	6.49	1.85	5.04	0.985	3.21	0.323	2.22	0.130		
225	0.5013			15.09	14.63	9.77	4.83	7.30	2.32	5.67	1.23	3.61	0.401	2.50	0.162	1.44	0.043
250	0.557					10.85	5.93	8.12	2.84	6.30	1.46	4.01	0.495	2.78	0.195	1.60	0.051
275	0.6127					11.94	7.14	8.93	3.40	6.93	1.79	4.41	0.583	3.05	0.234	1.76	0.061
300	0.6684					13.00	8.36	9.74	4.02	7.56	2.11	4.81	0.683	3.33	0.275	1.92	0.072
325	0.7241					14.12	9.89	10.53	4.09	8.19	2.47	5.21	0.797	3.61	0.320	2.08	0.083

		10"		12"		14"		16"		18"		20"		24"		
350	0.7798	1.93														2.24	0.095
375	0.8355	2.03														2.40	0.108
400	0.8912	2.24	0.054												2.56	0.121
425	0.9469	2.44	0.059												2.73	0.136
450	1.003	2.64	0.071												2.89	0.151
475	1.059	2.85	0.112	2.01	0.047										3.04	0.166
500	1.114	3.05	0.127	2.15	0.054										3.21	0.182
550	1.225	3.25	0.143	2.29	0.061										3.53	0.219
600	1.337	3.46	0.160	2.44	0.068										3.85	0.258
650	1.448	3.66	0.179	2.58	0.075										4.17	0.301
700	1.560	3.86	0.198	2.72	0.083	2.25	0.052								4.49	0.343
750	1.671	4.07	0.218	2.87	0.091	2.37	0.057								4.81	0.392
800	1.782	4.48	0.260	3.15	0.110	2.61	0.068								5.13	0.443
850	1.894	4.88	0.306	3.44	0.128	2.85	0.080								5.45	0.497
900	2.005	5.29	0.355	3.73	0.150	3.08	0.093								5.77	0.554
950	2.117	5.70	0.409	4.01	0.171	3.32	0.107	2.54	0.055						6.09	0.613
1000	2.228	6.10	0.466	4.30	0.195	3.56	0.122	2.72	0.063						6.41	0.675
1100	2.451	6.51	0.527	4.59	0.219	3.79	0.138	2.90	0.071						7.05	0.807
1200	2.674	7.32	0.663	5.16	0.276	4.27	0.172	3.27	0.088						7.70	0.948
1300	2.896	8.14	0.808	5.73	0.339	4.74	0.209	3.63	0.107						8.33	1.11
1400	3.119	10.17	1.24	7.17	0.515	5.93	0.321	4.54	0.163						8.98	1.28
1500	3.342	12.20	1.76	8.60	0.731	7.11	0.451	5.45	0.232						9.62	1.46
1600	3.565	14.24	2.38	10.03	0.982	8.30	0.607	5.02	0.312						10.26	1.65
1800	4.010	16.27	3.08	11.47	1.27	9.48	0.787	5.74	0.401						11.54	2.08
2000	4.456	18.31	3.87	12.90	1.60	10.67	0.990	6.46	0.503						12.82	2.55
2500	5.570	20.35	4.71	14.33	1.95	11.85	1.21	8.08	0.617	3.46	0.091		3.19	0.052	16.03	3.94
3000	6.684	24.41	6.74	17.20	2.77	14.21	1.71	9.69	0.877	4.04	0.129		3.59	0.065	19.24	5.59
3500	7.798	28.49	9.11	20.07	3.74	16.60	2.31	11.31	1.18	4.62	0.173				22.44	7.56
4000	8.912			22.93	4.84	18.96	2.99	12.71	1.51	5.20	0.222				25.65	9.80
4500	10.03			25.79	6.09	21.34	3.76	16.34	1.90		0.280				28.87	12.2
5000	11.14			28.66	7.46	23.71	4.61	18.15	2.34	5.77	0.340		3.99	0.079		
6000	13.37			34.40	10.7	28.45	6.59	21.79	3.33	6.93	0.483		4.79	0.111		
7000	15.60					33.19	8.89	25.42	4.49	8.08	0.652		5.59	0.150		
8000	17.82							29.05	5.83	9.23	0.839		6.38	0.192		
9000	20.05							32.68	7.31	10.39	1.05		7.18	0.242		
10000	22.28							36.31	9.03	11.54	1.28		7.98	0.294		
12000	26.74									13.85	1.83		9.58	0.416		
14000	31.19									16.16	2.45		11.17	0.562		
16000	35.65									18.47	3.18		12.77	0.723		
18000	40.10									20.77	4.03		14.36	0.907		
20000	44.56									23.08	4.93		15.96	1.12		

For pipe lengths other than 100 feet, the pressure drop is proportional to the length. Thus, for 50 feet of pipe, the pressure drop is approximately one-half the value given in the table ... for 300 feet, three times the given value, etc.

Velocity is a function of the cross sectional flow area; thus, it is constant for a given flow rate and is independent of pipe length.

For calculations for pipe other than Schedule 40, see explanation on next page.

TABLE G-2 Flow of Air Through Schedule 40 Steel Pipe

For lengths of pipe other than 100 feet, the pressure drop is proportional to the length. Thus, for 50 feet of pipe, the pressure drop is approximately one-half the value given in the table . . . for 300 feet, thre times the given value, etc.

The pressure drop is also inversely proportional to the absolute pressure and directly proportional to the absolute temperature.

Therefore, to determine the pressure drop for inlet or average pressures other than 100 psi and at temperatures other than 60 F, multiply the values given in the table by the ratio:

$$\left(\frac{100+14.7}{P+14.7}\right)\left(\frac{460+t}{520}\right)$$

where:

"P" is the inlet or average gauge pressure in pounds per square inch, and,

"t" is the temperature in degrees Fahrenheit under consideration.

The cubic feet per minute of compressed air at any pressure is inversely proportional to the absolute pressure and directly proportional to the absolute temperature.

To determine the cubic feet per minute of compressed air at any temperature and pressure other than standard conditions, multiply the value of cubic feet per minute of free air by the ratio:

$$\left(\frac{14.7}{14.7+P}\right)\left(\frac{460+t}{520}\right)$$

Calculations for Pipe Other than Schedule 40

To determine the velocity of water, or the pressure drop of water or air, through pipe other than Schedule 40, use the following formulas:

$$v_a = v_{40}\left(\frac{d_{40}}{d_a}\right)^2$$

$$\Delta P_a = \Delta P_{40}\left(\frac{d_{40}}{d_a}\right)^5$$

Subscript "a" refers to the Schedule of pipe through which velocity or pressure drop is desired.

Subscript "40" refers to the velocity or pressure drop through Schedule 40 pipe, as given in the tables on these facing pages.

Free Air q'_m Cubic Feet Per Minute at 60 F and 14.7 psia	Compressed Air Cubic Feet Per Minute at 60 F and 100 psig	Pressure Drop of Air In Pounds per Square Inch Per 100 Feet of Schedule 40 Pipe For Air at 100 Pounds per Square Inch Gauge Pressure and 60 F Temperature								
		1/8"	1/4"	3/8"	1/2"					
1	0.128	0.361	0.083	0.018						
2	0.256	1.31	0.285	0.064	0.020					
3	0.384	3.06	0.605	0.133	0.042	3/4"				
4	0.513	4.83	1.04	0.226	0.071					
5	0.641	7.45	1.58	0.343	0.106	0.027				
6	0.769	10.6	2.23	0.408	0.148	0.037	1"			
8	1.025	18.6	3.89	0.848	0.255	0.062	0.019			
10	1.282	28.7	5.96	1.26	0.356	0.094	0.029	1 1/4"	1 1/2"	
15	1.922	...	13.0	2.73	0.834	0.201	0.062			
20	2.563	...	22.8	4.76	1.43	0.345	0.102	0.026		
25	3.204	...	35.6	7.34	2.21	0.526	0.156	0.039	0.019	
30	3.845	...	...	10.5	3.15	0.748	0.219	0.055	0.026	
35	4.486	...	...	14.2	4.24	1.00	0.293	0.073	0.035	
40	5.126	...	...	18.4	5.49	1.30	0.379	0.095	0.044	
45	5.767	...	...	23.1	6.90	1.62	0.474	0.116	0.055	2"
50	6.408			28.5	8.49	1.99	0.578	0.149	0.067	0.019
60	7.690	2 1/2"		40.7	12.2	2.85	0.819	0.200	0.094	0.027
70	8.971			...	16.5	3.83	1.10	0.270	0.126	0.036
80	10.25	0.019		...	21.4	4.96	1.43	0.350	0.162	0.046
90	11.53	0.023		...	27.0	6.25	1.80	0.437	0.203	0.058
100	12.82	0.029	3"		33.2	7.69	2.21	0.534	0.247	0.070
125	16.02	0.044			...	11.9	3.39	0.825	0.380	0.107
150	19.22	0.062	0.021		...	17.0	4.87	1.17	0.537	0.151
175	22.43	0.083	0.028		...	23.1	6.60	1.58	0.727	0.205
200	25.63	0.107	0.036	3 1/2"	...	30.0	8.54	2.05	0.937	0.264
225	28.84	0.134	0.045	0.022		37.9	10.8	2.59	1.19	0.331
250	32.04	0.164	0.055	0.027		...	13.3	3.18	1.45	0.404
275	35.24	0.191	0.066	0.032		...	16.0	3.83	1.75	0.484
300	38.45	0.232	0.078	0.037		...	19.0	4.56	2.07	0.573
325	41.65	0.270	0.090	0.043		...	22.3	5.32	2.42	0.673
350	44.87	0.313	0.104	0.050	4"	...	25.8	6.17	2.80	0.776
375	48.06	0.356	0.119	0.057	0.030	...	29.6	7.05	3.20	0.887
400	51.26	0.402	0.134	0.064	0.034	...	33.6	8.02	3.64	1.00
425	54.47	0.452	0.151	0.072	0.038	...	37.9	9.01	4.09	1.13
450	57.67	0.507	0.168	0.081	0.042	...	...	10.2	4.59	1.26
475	60.88	0.562	0.187	0.089	0.047		...	11.3	5.09	1.40
500	64.08	0.623	0.206	0.099	0.052		...	12.5	5.61	1.55
550	70.49	0.749	0.248	0.118	0.062		...	15.1	6.79	1.87
600	76.90	0.887	0.293	0.139	0.073		...	18.0	8.04	2.21
650	83.30	1.04	0.342	0.163	0.086	5"	...	21.1	9.43	2.60
700	89.71	1.19	0.395	0.188	0.099	0.032		24.3	10.9	3.00
750	96.12	1.36	0.451	0.214	0.113	0.036		27.9	12.6	3.44
800	102.5	1.55	0.513	0.244	0.127	0.041		31.8	14.2	3.90
850	108.9	1.74	0.576	0.274	0.144	0.046		35.9	16.0	4.40
900	115.3	1.95	0.642	0.305	0.160	0.051	6"	40.2	18.0	4.91
950	121.8	2.18	0.715	0.340	0.178	0.057	0.023	...	20.0	5.47
1 000	128.2	2.40	0.788	0.375	0.197	0.063	0.025	...	22.1	6.06
1 100	141.0	2.89	0.948	0.451	0.236	0.075	0.030	...	26.7	7.29
1 200	153.8	3.44	1.13	0.533	0.279	0.089	0.035	...	31.8	8.63
1 300	166.6	4.01	1.32	0.626	0.327	0.103	0.041	...	37.3	10.1
1 400	179.4	4.65	1.52	0.718	0.377	0.119	0.047			11.8
1 500	192.2	5.31	1.74	0.824	0.431	0.136	0.054			13.5
1 600	205.1	6.04	1.97	0.932	0.490	0.154	0.061	8"		15.3
1 800	230.7	7.65	2.50	1.18	0.616	0.193	0.075			19.3
2 000	256.3	9.44	3.06	1.45	0.757	0.237	0.094	0.023		23.9
2 500	320.4	14.7	4.76	2.25	1.17	0.366	0.143	0.035	10"	37.3
3 000	384.5	21.1	6.82	3.20	1.67	0.524	0.204	0.051	0.016	
3 500	448.6	28.8	9.23	4.33	2.26	0.709	0.276	0.068	0.022	
4 000	512.6	37.6	12.1	5.66	2.94	0.919	0.358	0.088	0.028	
4 500	576.7	47.6	15.3	7.16	3.69	1.16	0.450	0.111	0.035	12"
5 000	640.8	...	18.8	8.85	4.56	1.42	0.552	0.136	0.043	0.018
6 000	769.0	...	27.1	12.7	6.57	2.03	0.794	0.195	0.061	0.025
7 000	897.1	...	36.9	17.2	8.94	2.76	1.07	0.262	0.082	0.034
8 000	1025	...	...	22.5	11.7	3.59	1.39	0.339	0.107	0.044
9 000	1153	...	...	28.5	14.9	4.54	1.76	0.427	0.134	0.055
10 000	1282	...	...	35.2	18.4	5.60	2.16	0.526	0.164	0.067
11 000	1410	...	...	...	22.2	6.78	2.62	0.633	0.197	0.081
12 000	1538	...	...	...	26.4	8.07	3.09	0.753	0.234	0.096
13 000	1666	...	...	...	31.0	9.47	3.63	0.884	0.273	0.112
14 000	1794	...	...	...	36.0	11.0	4.21	1.02	0.316	0.129
15 000	1922	...	...	...	...	12.6	4.84	1.17	0.364	0.148
16 000	2051	...	...	...	...	14.3	5.50	1.33	0.411	0.167
18 000	2307	...	...	...	...	18.2	6.96	1.68	0.520	0.213
20 000	2563	...	...	...	...	22.4	8.60	2.01	0.642	0.260
22 000	2820	...	...	...	...	27.1	10.4	2.50	0.771	0.314
24 000	3076	...	...	...	...	32.3	12.4	2.97	0.918	0.371
26 000	3332	...	...	...	...	37.9	14.5	3.49	1.12	0.435
28 000	3588	...	...	...	...	...	16.9	4.04	1.25	0.505
30 000	3845	...	...	...	...	...	19.3	4.64	1.42	0.520

Appendix H

Typical Pump Head Capacity Range Charts

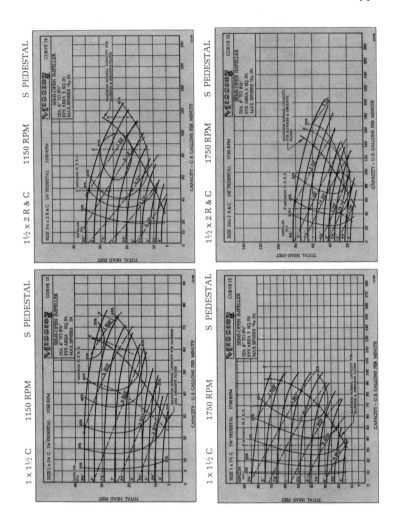

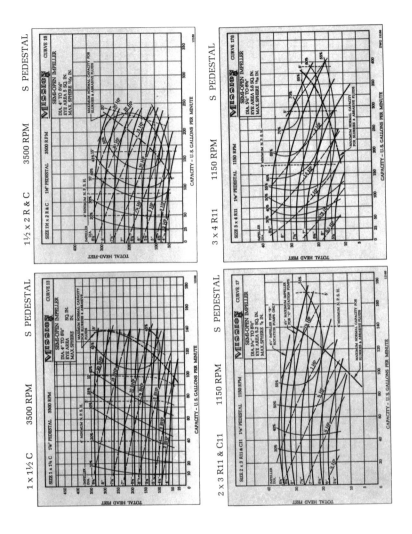

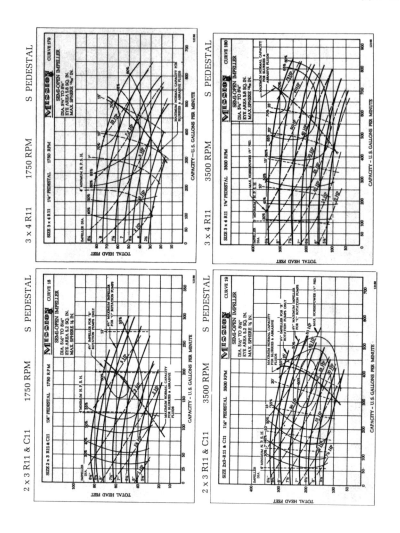

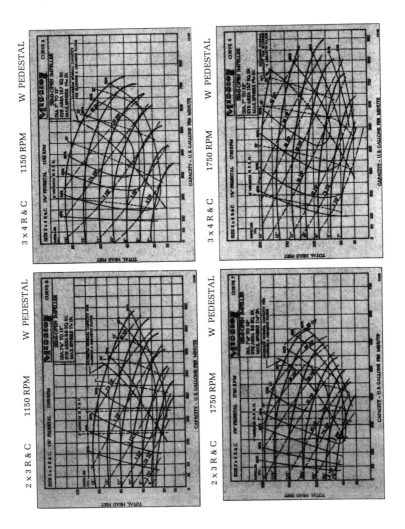

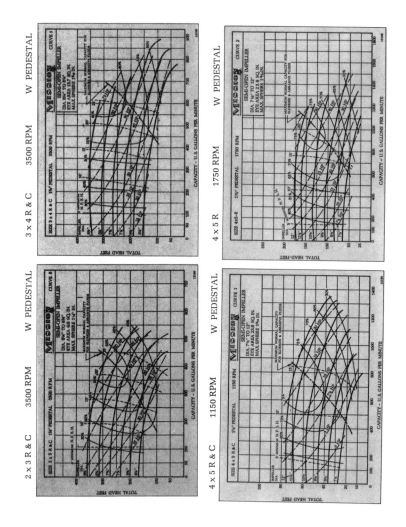

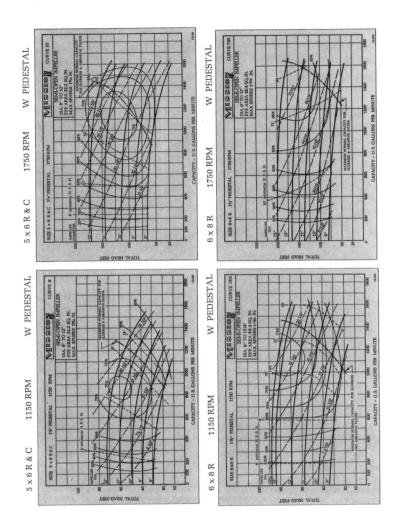

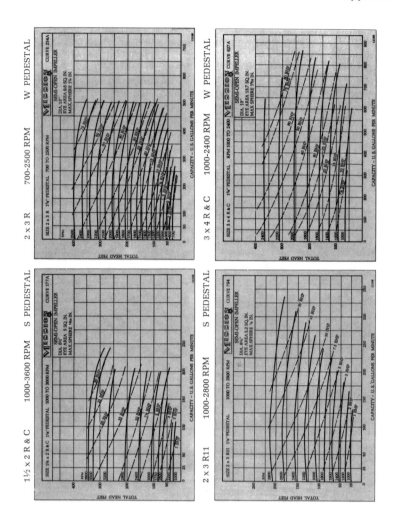

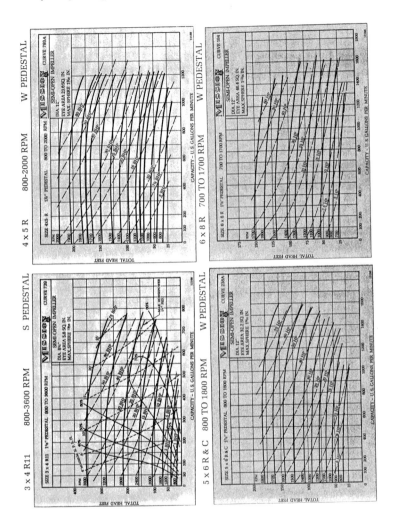

Appendix I

Fanno Line Tables for Adiabatic Flow of Air in a Constant Area Duct

TABLE I-1 Fanno Line—Adiabatic, Constant-Area Flow (k = 1.400)

N_{Ma}	$\dfrac{T}{T^\star}$	$\dfrac{P}{P^\star}$	$\dfrac{P^0}{P^{0\star}}$	$\dfrac{V}{V^\star}$	$\dfrac{F}{F^\star}$	$\dfrac{4fL}{D}$
0	1.2000	∞	∞	0	∞	∞
.01	1.2000	10,9.544	5,7.874	.01095	4,5.650	7,134.40
.02	1.1999	5,4.770	2,8.942	.02191	22.,834	1,778.45
.03	1.1998	3,6.511	1,9.300	.03286	15.,232	7,87.08
.04	1.1996	27.,382	14.,482	.04381	11.,435	4,40.35
.05	1.1994	21.,903	11.,5914	.05476	9.,1584	2,80.02
.06	1.1991	18.,251	9.,6659	.06570	7.6,428	19,3.03
.07	1.1988	15.6,42	8.2,915	.07664	6.5,620	14,0.66
.08	1.1985	13.6,84	7.2,616	.08758	5.7,529	10,6.72
.09	1.1981	12.1,62	6.4,614	.09851	5.1,249	8,3.496
.10	1.1976	10.9,435	5.8,218	.10943	4.6,236	66.,922
.11	1.1971	9.9,465	5.2,992	.12035	4.2,146	54.,688
.12	1.1966	9.1,156	4.8,643	.13126	3.8,747	45.,408
.13	1.1960	8.4,123	4.4,968	.14216	3.58,80	38.,207
.14	1.1953	7.8,093	4.1,824	.15306	3.34,32	32.,511
.15	1.1946	7.2,866	3.91,03	.16395	3.13,17	27.,932
.16	1.1939	6.82,91	3.67,27	.17482	2.94,74	24.,198
.17	1.1931	6.42,52	3.46,35	.18568	2.78,55	21.,115
.18	1.1923	6.06,62	3.27,79	.19654	2.64,22	18.5,43
.19	1.1914	5.74,48	3.11,23	.20739	2.51,46	16.3,75
.20	1.1905	5.45,55	2.96,35	.21822	2.40,04	14.5,33
.21	1.1895	5.19,36	2.82,93	.22904	2.29,76	12.9,56
.22	1.1885	4.95,54	2.70,76	.23984	2.20,46	11.5,96
.23	1.1874	4.73,78	2.59,68	.25063	2.12,03	10.4,16
.24	1.1863	4.53,83	2.49,56	.26141	2.04,34	9.3,865
.25	1.1852	4.35,46	2.40,27	.27217	1.97,32	8.4,834
.26	1.1840	4.18,50	2.31,73	.28291	1.90,88	7.6,876
.27	1.1828	4.02,80	2.23,85	.29364	1.84,96	6.9,832
.28	1.1815	3.88,20	2.16,56	.30435	1.795,0	6.3,572
.29	1.1802	3.74,60	2.09,79	.31504	1.744,6	5.7,989
.30	1.1788	3.61,90	2.035,1	.32572	1.697,9	5.2,992
.31	1.1774	3.50,02	1.976,5	.33637	1.654,6	4.8,507
.32	1.1759	3.38,88	1.921,9	.34700	1.614,4	4.44,68
.33	1.1744	3.28,40	1.870,8	.35762	1.576,9	4.08,21
.34	1.1729	3.18,53	1.822,9	.36822	1.542,0	3.75,20
.35	1.1713	3.09,22	1.778,0	.37880	1.509,4	3.45,25
.36	1.1697	3.004,2	1.735,8	.38935	1.478,9	3.18,01
.37	1.1680	2.920,9	1.696,1	.39988	1.450,3	2.93,20
.38	1.1663	2.842,0	1.658,7	.41039	1.423,6	2.70,55
.39	1.1646	2.767,1	1.623,4	.42087	1.398,5	2.49,83

TABLE I-1   Continued

N_{Ma}	$\dfrac{T}{T*}$	$\dfrac{P}{P*}$	$\dfrac{P^0}{P^{0*}}$	$\dfrac{V}{V*}$	$\dfrac{F}{F*}$	$\dfrac{4fL}{D}$
.40	1.1628	2.695,8	1.590,1	.43133	1.374,9	2.30,85
.41	1.1610	2.628,0	1.558,7	.44177	1.352,7	2.13,44
.42	1.1591	2.563,4	1.528,9	.45218	1.331,8	1.97,44
.43	1.1572	2.501,7	1.500,7	.46257	1.312,2	1.82,72
.44	1.1553	2.442,8	1.473,9	.47293	1.293,7	1.69,15
.45	1.1533	2.386,5	1.448,6	.48326	1.276,3	1.56,64
.46	1.1513	2.332,6	1.424,6	.49357	1.259,8	1.45,09
.47	1.1492	2.280,9	1.401,8	.50385	1.244,3	1.34,42
.48	1.1471	2.231,4	1.380,1	.51410	1.229,6	1.24,53
.49	1.1450	2.183,8	1.359,5	.52433	1.215,8	1.15,39
.50	1.1429	2.138,1	1.339,9	.53453	1.202,7	1.06,908
.51	1.1407	2.094,2	1.321,2	.54469	1.190,3	.99,042
.52	1.1384	2.051,9	1.303,4	.55482	1.178,6	.91,741
.53	1.1362	2.011,2	1.286,4	.56493	1.167,5	.84,963
.54	1.1339	1.971,9	1.270,2	.57501	1,157,1	.786,62
.55	1.1315	1.934,1	1.254,9	.58506	1.147,2	.728,05
.56	1.1292	1.897,6	1.240,3	.59507	1.137,8	.673,57
.57	1.1268	1.862,3	1.226,3	.60505	1.128,9	.622,86
.58	1.1244	1.828,2	1.213,0	.61500	1.120,5	.575,68
.59	1.1219	1.795,2	1.200,3	.62492	1.112,6	.531,74
.60	1.1194	1.763,4	1.188,2	.63481	1.1050,4	.490,81
.61	1.1169	1.732,5	1.176,6	.64467	1.0979,3	.452,70
.62	1.1144	1.702,6	1.165,6	.65449	1.0912,0	.417,20
.63	1.1118	1.673,7	1.155,1	.66427	1.0848,5	.384,11
.64	1.1091	1.645,6	1.145,1	.67402	1.0788,3	.353,30
.65	1.10650	1.618,3	1.135,6	.68374	1.0731,4	.324,60
.66	1.10383	1.591,9	1.126,5	.69342	1.0677,7	.297,85
.67	1.10114	1.566,2	1.117,9	.70306	1.0627,1	.272,95
.68	1.09842	1.541,3	1.109,7	.71267	1.0579,2	.249,78
.69	1.09567	1.517,0	1.101,8	.72225	1.0534,0	.228,21
.70	1.09290	1.493,4	1.0943,6	.73179	1.0491,5	.208,14
.71	1.09010	1.470,5	1.0872,9	.74129	1.0451,4	.189,49
.72	1.08727	1.448,2	1.0805,7	.75076	1.0413,7	.172,15
.73	1.08442	1.426,5	1.0741,9	.76019	1.0378,3	.156,06
.74	1.08155	1.405,4	1.0681,5	.76958	1.0345,0	.141,13
.75	1.07856	1.384,8	1.0624,2	.77893	1.0313,7	.127,28
.76	1.07573	1.364,7	1.0570,0	.78825	1.0284,4	.114,46
.77	1.07279	1.345,1	1.0518,8	.79753	1.0257,0	.102,62
.78	1.06982	1.326,0	1.0470,5	.80677	1.0231,4	.091,67
.79	1.06684	1.3074	1.0425,0	.81598	1.0207,5	.081,59

TABLE I-1   Continued

N_{Ma}	$\dfrac{T}{T^*}$	$\dfrac{P}{P^*}$	$\dfrac{P^0}{P^{0*}}$	$\dfrac{V}{V^*}$	$\dfrac{F}{F^*}$	$\dfrac{4fL}{D}$
.80	1.06383	1.2892	1.0382,3	.82514	1.0185,3	.072,29
.81	1.06080	1.2715	1.0342,2	.83426	1.0164,6	.063,75
.82	1.05775	1.2542	1.0304,7	.84334	1.0145,5	.055,93
.83	1.05468	1.2373	1.0269,6	.85239	1.0127,8	.048,78
.84	1.05160	1.2208	1.0237,0	.86140	1.0111,5	.042,26
.85	1.04849	1.2047	1.0206,7	.87037	1.0096,6	.036,32
.86	1.04537	1.1889	1.0178,7	.87929	1.0082,9	.030,97
.87	1.04223	1.1735	1.0152,9	.88818	1.0070,4	.026,13
.88	1.03907	1.1584	1.0129,4	.89703	1.0059,1	.021,80
.89	1.03589	1.1436	1.0108,0	.90583	1.0049,0	.017,93
.90	1.03270	1.1291,3	1.0088,7	.91459	1.0039,9	.0145,13
.91	1.02950	1.1150,0	1.0071,4	.92332	1.0031,8	.0115,19
.92	1.02627	1.1011,4	1.0056,0	.93201	1.0024,8	.0089,16
.93	1.02304	1.0875,8	1.0042,6	.94065	1.0018,8	.0066,94
.94	1.01978	1.0743,0	1.0031,1	.94925	1.0013,6	.0048,15
.95	1.01652	1.0612,9	1.0021,5	.95782	1.0009,3	.0032,80
.96	1.01324	1.0485,4	1.0013,7	.96634	1.0005,9	.0020,56
.97	1.00995	1.0360,5	1.0007,6	.97481	1.0003,3	.0011,35
.98	1.00664	1.0237,9	1.0003,3	.98324	1.0001,4	.0004,93
.99	1.00333	1.0117,8	1.0000,8	.99164	1.0000,3	.0001,20
1.00	1.00000	1.0000,0	1.0000,0	1.00000	1,0000,0	0
1.01	.99666	.9884,4	1.0000,8	1.00831	1.0000,3	.0001,14
1.02	.99331	.9771,1	1.0003,3	1.01658	1.0001,3	.0004,58
1.03	.98995	.9659,8	1.0007,3	1.02481	1.0003,0	.0010,13
1.04	.98658	.9550,6	1.0013,0	1.03300	1.0005,3	.0017,71
1.05	.98320	.9443,5	1.0020,3	1.04115	1.0008,2	.0027,12
1.06	.97982	.9338,3	1.0029,1	1.04925	1.0011,6	.0038,37
1.07	.97642	.9235,0	1.0039,4	1.05731	1.0015,5	.0051,29
1.08	.97302	.9133,5	1.0051,2	1.06533	1.0020,0	.0065,82
1.09	.96960	.9033,8	1.0064,5	1.07331	1.0025,0	.0081,85
1.10	.96618	.8935,9	1.0079,3	1.08124	1.00305	.0099,33
1.11	.96276	.8839,7	1.0095,5	1.08913	1.00365	.0118,13
1.12	.95933	.8745,1	1.0113,1	1.09698	1.00429	.0138,24
1.13	.95589	.8652,2	1.0132,2	1.10479	1.00497	.0159,49
1.14	.95244	.8560,8	1.0152,7	1.11256	1.00569	.0181,87
1.15	.94899	.8471,0	1.0174,6	1.1203	1.00646	.0205,3
1.16	.94554	.8382,7	1.0197,8	1.1280	1.00726	.0229,8
1.17	.94208	.8295,8	1.0222,4	1.1356	1.00810	.0255,2
1.18	.93862	.8210,4	1.0248,4	1.1432	1.00897	.0281,4
1.19	.93515	.8126,3	1.0275,7	1.1508	1.00988	.0308,5

TABLE I-1   Continued

N_{Ma}	$\dfrac{T}{T^*}$	$\dfrac{P}{P^*}$	$\dfrac{P^0}{P^{0*}}$	$\dfrac{V}{V^*}$	$\dfrac{F}{F^*}$	$\dfrac{4fL}{D}$
1.20	.93168	.8043,6	1.0304,4	1.1583	1.01082	.0336,4
1.21	.92820	.7962,3	1.0334,4	1.1658	1.01178	.0365,0
1.22	.92473	.7882,2	1.0365,7	1.1732	1.01278	.0394,2
1.23	.92125	.7803,4	1.0398,3	1.1806	1.01381	.0424,1
1.24	.91777	.7725,8	1.0432,3	1.1879	1.01486	.0454,7
1.25	.91429	.7649,5	1.0467,6	1.1952	1.01594	.04858
1.26	.91080	.7574,3	1.0504,1	1.2025	1.01705	.05174
1.27	.90732	.7500,3	1.0541,9	1.2097	1.01818	.05494
1.28	.90383	.7427,4	1.0580,9	1.2169	1.01933	.05820
1.29	.90035	.7355,6	1.0621,3	1.2240	1.02050	.06150
1.30	.89686	.7284,8	1.0663,0	1.2311	1.02169	.06483
1.31	.89338	.7215,2	1.0706,0	1.2382	1.02291	.06820
1.32	.88989	.7146,5	1.0750,2	1.2452	1.02415	.07161
1.33	.88641	.7078,9	1.0795,7	1.2522	1.02540	.07504
1.34	.88292	.7012,3	1.0842,4	1.2591	1.02666	.07850
1.35	.87944	.6946,6	1.0890,4	1.2660	1.02794	.08199
1.36	.87596	.6881,8	1.0939,7	1.2729	1.02924	.08550
1.37	.87249	.6818,0	1.0990,2	1.2797	1.03056	.08904
1.38	.86901	.6755,1	1.1041,9	1.2864	1.03189	.09259
1.39	.06554	.6693,1	1.1094,8	1.2932	1.03323	.09616
1.40	.86207	.6632,0	1.1149	1.2999	1.03458	.09974
1.41	.85860	.6571,7	1.1205	1.3065	1.03595	.10333
1.42	.85514	.6512,2	1.1262	1.3131	1.03733	.10694
1.43	.85168	.6453,6	1.1320	1.3197	1.03872	.11056
1.44	.84822	.6395,8	1.1379	1.3262	1.04012	.11419
1.45	.84477	.6338,7	1.1440	1.3327	1.04153	.11782
1.46	.84133	.6282,4	1.1502	1.3392	1.04295	.12146
1.47	.83788	.6226,9	1.1565	1.3456	1.04438	.12510
1.48	.83445	.6172,2	1.1629	1.3520	1.04581	.12875
1.49	.83101	.6118,1	1.1695	1.3583	1.04725	.13240
1.50	.82759	.6064,8	1.1762	1.3646	1.04870	.13605
1.51	.82416	.6012,2	1.1830	1.3708	1.05016	.13970
1.52	.82075	.5960,2	1.1899	1.3770	1.05162	.14335
1.53	.81734	.5908,9	1.1970	1.3832	1.05309	.14699
1.54	.81394	.5858,3	1.2043	1.3894	1.05456	.15063
1.55	.81054	.5808,4	1.2116	1.3955	1.05604	.15427
1.56	.80715	.5759,1	1.2190	1.4015	1.05752	.15790
1.57	.80376	.5710,4	1.2266	1.4075	1.05900	.16152
1.58	.80038	.5662,3	1.2343	1.4135	1.06049	.16514
1.59	.79701	.5614,8	1.2422	1.4195	1.06198	.16876

TABLE I-1   Continued

N_{Ma}	$\dfrac{T}{T^\star}$	$\dfrac{P}{P^\star}$	$\dfrac{P^0}{P^{0\star}}$	$\dfrac{V}{V^\star}$	$\dfrac{F}{F^\star}$	$\dfrac{4fL}{D}$
1.60	.79365	.5567,9	1.2502	1.4254	1.06348	.17236
1.61	.79030	.5521,6	1.2583	1.4313	1.06498	.17595
1.62	.78695	.5475,9	1.2666	1.4371	1.06648	.17953
1.63	.78361	.5430,8	1.2750	1.4429	1.06798	.18311
1.64	.78028	.5386,2	1.2835	1.4487	1.06948	.18667
1.65	.77695	.5342,1	1.2922	1.4544	1.07098	.19022
1.66	.77363	.5298,6	1.3010	1.4601	1.07249	.19376
1.67	.77033	.5255,6	1.3099	1.4657	1.07399	.19729
1.68	.76703	.5213,1	1.3190	1.4713	1.07550	.20081
1.69	.76374	.5171,1	1.3282	1.4769	1.07701	.20431
1.70	.76046	.5129,7	1.3376	1.4825	1.07851	.20780
1.71	.75718	.5088,7	1.3471	1.4880	1.08002	.21128
1.72	.75392	.5048,2	1.3567	1.4935	1.08152	.21474
1.73	.75067	.5008,2	1.3665	1.4989	1.08302	.21819
1.74	.74742	.4968,6	1.3764	1.5043	1.08453	.22162
1.75	.74419	.4929,5	1.3865	1.5097	1.08603	.22504
1.76	.74096	.4890,9	1.3967	1.5150	1.08753	.22844
1.77	.73774	.4852,7	1.4070	1.5203	1.08903	.23183
1.78	.73453	.4814,9	1.4175	1.5256	1.09053	.23520
1.79	.73134	.4777,6	1.4282	1.5308	1.09202	.23855
1.80	.72816	.47407	1.4390	1.5360	1.09352	.24189
1.81	.72498	.47042	1.4499	1.5412	1.09500	.24521
1.82	.72181	.46681	1.4610	1.5463	1.09649	.24851
1.83	.71865	.46324	1.4723	1.5514	1.09798	.25180
1.84	.71551	.45972	1.4837	1.5564	1.00946	.25507
1.85	.71238	.45623	1.4952	1.5614	1.1009	.25832
1.86	.70925	.45278	1.5069	1.5664	1.1024	.26156
1.87	.70614	.49937	1.5188	1.5714	1.1039	.26478
1.88	.70304	.44600	1.5308	1.5763	1.1054	.26798
1.89	.69995	.44266	1.5429	1.5812	1.1068	.27116
1.90	.69686	.43936	1.5552	1.5861	1.1083	.27433
1.91	.69379	.43610	1.5677	1.5909	1.1097	.27748
1.92	.69074	.43287	1.5804	1.5957	1.1112	.28061
1.93	.68769	.42967	1.5932	1.6005	1.1126	.28372
1.94	.68465	.42651	1.6062	1.6052	1.1141	.28681
1.95	.68162	.42339	1.6193	1.6099	1.1155	.28989
1.96	.67861	.42030	1.6326	1.6146	1.1170	.29295
1.97	.67561	.41724	1.6461	1.6193	1.1184	.29599
1.98	.67262	.41421	1.6597	1.6239	1.1198	.29901
1.99	.66964	.41121	1.6735	1.6824	1.1213	.30201

TABLE I-1   Continued

N_{Ma}	$\dfrac{T}{T^*}$	$\dfrac{P}{P^*}$	$\dfrac{P^0}{P^{0*}}$	$\dfrac{V}{V^*}$	$\dfrac{F}{F^*}$	$\dfrac{4fL}{D}$
2.00	.66667	.40825	1.6875	1.6330	1.1227	.30499
2.01	.66371	.40532	1.7017	1.6375	1.1241	.30796
2.02	.66076	.40241	1.7160	1.6420	1.1255	.31091
2.03	.65783	.39954	1.7305	1.6465	1.1269	.31384
2.04	.65491	.39670	1.7452	1.6509	1.1283	.31675
2.05	.65200	.39389	1.7600	1.6553	1.1297	.31965
2.06	.64910	.39110	1.7750	1.6597	1.1311	.32253
2.07	.64621	.38834	1.7902	1.6640	1.1325	.32538
2.08	.64333	.38562	1.8056	1.6683	1.1339	.32822
2.09	.64047	.38292	1.8212	1.6726	1.1352	.33104
2.10	.63762	.38024	1.8369	1.6769	1.1366	.33385
2.11	.63478	.37760	1.8528	1.6811	1.1380	.33664
2.12	.63195	.37498	1.8690	1.6853	1.1393	.33940
2.13	.62914	.37239	1.8853	1.6895	1.1407	.34215
2.14	.62633	.36982	1.9018	1.6936	1.1420	.34488
2.15	.62354	.36728	1.9185	1.6977	1.1434	.34760
2.16	.62076	.36476	1.9354	1.7018	1.1447	.35030
2.17	.61799	.36227	1.9525	1.7059	1.1460	.35298
2.18	.61523	.35980	1.9698	1.7099	1.1474	.35564
2.19	.61249	.35736	1.9873	1.7139	1.1487	.35828
2.20	.60976	.35494	2.0050	1.7179	1.1500	.36091
2.21	.60704	.35254	2.0228	1.7219	1.1513	.36352
2.22	.60433	.35017	2.0409	1.7258	1.1526	.36611
2.23	.60163	.34782	2.0592	1.7297	1.1539	.36868
2.24	.59895	.34550	2.0777	1.7336	1.1552	.37124
2.25	.59627	.34319	2.0964	1.7374	1.1565	.37378
2.26	.59361	.34091	2.1154	1.7412	1.1578	.37630
2.27	.59096	.33865	2.1345	1.7450	1.1590	.37881
2.28	.58833	.33641	2.1538	1.7488	1.1603	.38130
2.29	.58570	.33420	2.1733	1.7526	1.1616	.38377
2.30	.58309	.33200	2.1931	1.7563	1.1629	.38623
2.31	.58049	.32983	2.2131	1.7600	1.1641	.38867
2.32	.57790	.32767	2.2333	1.7637	1.1653	.39109
2.33	.57532	.32554	2.2537	1.7673	1.1666	.39350
2.34	.57276	.32342	2.2744	1.7709	1.1678	.39589
2.35	.57021	.32133	2.2953	1.7745	1.1690	.39826
2.36	.56767	.31925	2.3164	1.7781	1.1703	.40062
2.37	.56514	.31720	2.3377	1.7817	1.1715	.40296
2.38	.56262	.31516	2.3593	1.7852	1.1727	.40528
2.39	.56011	.31314	2.3811	1.7887	1.1739	.40760

TABLE I-1   Continued

N_{Ma}	$\dfrac{T}{T^\star}$	$\dfrac{P}{P^\star}$	$\dfrac{P^0}{P^{0\star}}$	$\dfrac{V}{V^\star}$	$\dfrac{F}{F^\star}$	$\dfrac{4fL}{D}$
2.40	.55762	.31114	2.4031	1.7922	1.1751	.40989
2.41	.55514	.30916	2.4254	1.7956	1.1763	.41216
2.42	.55267	.30720	2.4479	1.7991	1.1775	.41442
2.43	.55021	.30525	2.4706	1.8025	1.1786	.41667
2.44	.54776	.30332	2.4936	1.8059	1.1798	.41891
2.45	.54533	.30141	2.5168	1.8092	1.1810	.42113
2.46	.54291	.29952	2.5403	1.8126	1.1821	.42333
2.47	.54050	.29765	2.5640	1.8159	1.1833	.42551
2.48	.53810	.29579	2.5880	1.8192	1.1844	.42768
2.49	.53571	.29395	2.6122	1.8225	1.1856	.42983
2.50	.53333	.29212	2.6367	1.8257	1.1867	.43197
2.51	.53097	.29031	2.6615	1.8290	1.1879	.43410
2.52	.52862	.28852	2.6865	1.8322	1.1890	.43621
2.53	.52627	.28674	2.7117	1.8354	1.1901	.43831
2.54	.52394	.28498	2.7372	1.8386	1.1912	.44040
2.55	.52163	.28323	2.7630	1.8417	1.1923	.44247
2.56	.51932	.28150	2.7891	1.8448	1.1934	.44452
2.57	.51702	.27978	2.8154	1.8479	1.1945	.44655
2.58	.51474	.27808	2.8420	1.8510	1.1956	.44857
2.59	.51247	.27640	2.8689	1.8541	1.1967	.45059
2.60	.51020	.27473	2.8960	1.8571	1.1978	.45259
2.61	.50795	.27307	2.9234	1.8602	1.1989	.45457
2.62	.50571	.27143	2.9511	1.8632	1.2000	.45654
2.63	.50349	.26980	2.9791	1.8662	1.2011	.45850
2.64	.50127	.26818	3.0074	1.8691	1.2021	.46044
2.65	.49906	.26658	3.0359	1.8721	1.2031	.46237
2.66	.49687	.26499	3.0647	1.8750	1.2042	.46429
2.67	.49469	.26342	3.0938	1.8779	1.2052	.46619
2.68	.49251	.26186	3.1234	1.8808	1.2062	.46807
2.69	.49035	.26032	3.1530	1.8837	1.2073	.46996
2.70	.48820	.25878	3.1830	1.8865	1.2083	.47182
2.71	.48606	.25726	3.2133	1.8894	1.2093	.47367
2.72	.48393	.25575	3.2440	1.8922	1.2103	.47551
2.73	.48182	.25426	3.2749	1.8950	1.2113	.47734
2.74	.47971	.25278	3.3061	1.8978	1.2123	.47915
2.75	.47761	.25131	3.3376	1.9005	1.2133	.48095
2.76	.47553	.24985	3.3695	1.9032	1.2143	.48274
2.77	.47346	.24840	3.4017	1.9060	1.2153	.48452
2.78	.47139	.24697	3.4342	1.9087	1.2163	.48628
2.79	.46933	.24555	3.4670	1.9114	1.2173	.48803

TABLE I-1 Continued

N_{Ma}	$\dfrac{T}{T^*}$	$\dfrac{P}{P^*}$	$\dfrac{P^0}{P^{0*}}$	$\dfrac{V}{V^*}$	$\dfrac{F}{F^*}$	$\dfrac{4fL}{D}$
2.80	.46729	.24414	3.5001	1.9140	1.2182	.48976
2.81	.46526	.24274	3.5336	1.9167	1.2192	.49148
2.82	.46324	.24135	3.5674	1.9193	1.2202	.49321
2.83	.46122	.23997	3.6015	1.9220	1.2211	.49491
2.84	.45922	.23861	3.6359	1.9246	1.2221	.49660
2.85	.45723	.23726	3.6707	1.9271	1.2230	.49828
2.86	.45525	.23592	3.7058	1.9297	1.2240	.49995
2.87	.45328	.23458	3.7413	1.9322	1.2249	.50161
2.88	.45132	.23326	3.7771	1.9348	1.2258	.50326
2.89	.44937	.23196	3.8133	1.9373	1.2268	.50489
2.90	.44743	.23066	3.8498	1.9398	1.2277	.50651
2.91	.44550	.22937	3.8866	1.9423	1.2286	.50812
2.92	.44358	.22809	3.9238	1.9448	1.2295	.50973
2.93	.44167	.22682	3.9614	1.9472	1.2304	.51133
2.94	.43977	.22556	3.9993	1.9497	1.2313	.51291
2.95	.43788	.22431	4.0376	1.9521	1.2322	.51447
2.96	.43600	.22307	4.0763	1.9545	1.2331	.51603
2.97	.43413	.22185	4.1153	1.9569	1.2340	.51758
2.98	.43226	.22063	4.1547	1.9592	1.2348	.51912
2.99	.43041	.21942	4.1944	1.9616	1.2357	.52064
3.00	.42857	.21822	4.2346	1.9640	1.2366	.52216
3.50	.34783	.16850	6.7896	2.0642	1.2743	.58643
4.00	.28571	.13363	10.719	2.1381	1.3029	.63306
4.50	.23762	.10833	16.562	2.1936	1.3247	.66764
5.00	.20000	.08944	25.000	2.2361	1.3416	.69381
6.00	.14634	.06376	53.180	2.2953	1.3655	.72987
7.00	.11111	.04762	104.14	2.3333	1.3810	.75281
8.00	.08696	.03686	190.11	2.3591	1.3915	.76820
9.00	.06977	.02935	327.19	2.3772	1.3989	.77898
10.00	.05714	.02390	535.94	2.3905	1.4044	.78683
∞	0	0	∞	2.4495	1.4289	.82153

Index